In Honor of Professor Serge Galam for His 70th Birthday and Forty Years of Sociophysics

Guest Editors

André Martins
Taksu Cheon
Xijin Tang
Bastien Chopard
Soumyajyoti Biswas

MDPI

Basel • Beijing • Wuhan • Barcelona • Belgrade • Novi Sad • Cluj • Manchester

In Honor of Professor Serge Galam for His 70th Birthday and Forty Years of Sociophysics

Guest Editors

André Martins
Escola de Artes, Ciências e Humanidades
Universidade de São Paulo
São Paulo
Brazil

Taksu Cheon
Laboratory of Physics
Kochi University of Technology
Kochi
Japan

Xijin Tang
Key Lab on Management, Decision-Making and Information Systems,
Institute of Systems Science, Academy of Mathematics and Systems Sciences
Chinese Academy of Sciences
Beijing
China

Bastien Chopard
Computer Science Department
University of Geneva
Geneva
Switzerland

Soumyajyoti Biswas
Department of Physics
SRM University
Andhra Pradesh
India

Editorial Office
MDPI AG
Grosspeteranlage 5
4052 Basel, Switzerland

This is a reprint of the Special Issue, published open access by the journal *Physics* (ISSN 2624-8174), freely accessible at: https://www.mdpi.com/journal/physics/special_issues/SergeGalam70.

For citation purposes, cite each article independently as indicated on the article page online and as indicated below:

Lastname, A.A.; Lastname, B.B. Article Title. *Journal Name* **Year**, *Volume Number*, Page Range.

ISBN 978-3-7258-2717-6 (Hbk)
ISBN 978-3-7258-2718-3 (PDF)
https://doi.org/10.3390/books978-3-7258-2718-3

Cover image courtesy of Aurore Papegay/Sciences Po

Contents

Preface

In 1982, Serge Galam and his colleagues published five papers that opened up a new field of research, which they coined sociophysics. Sociophysics involves the use of ideas, methods, and theories from physics to describe social and political systems. It does not rely on metaphors, maps, or narratives.

After a challenging beginning (see Galam's testimony paper, Physica A 336, 49-55, 2004), sociophysics is now a flourishing field of research among physicists, mathematicians, and computer scientists.

The current applications of sociophysics are diverse, including topics such as the emergence and dissemination of marginal ideas, polarization, economic and financial applications, the description of cities, models of crime spread, the modeling of terrorism, and the spread of radicalism, among others. A few unexpected predictions have been successful.

Yet, a current challenge is that social systems often lack first principles, conservation properties, and fundamental laws, such as those at the heart of exact sciences.

Beyond the academic interest sociophysics has generated, its societal impact is of great importance. Currently, policymakers are facing increasing global challenges, for which there is a critical need to understand, predict, and prepare for the evolution of our societies. Sociophysics is a vital source of novel insights that can help us tackle these global societal issues, for which new ideas and investigation methods are lacking.

In keeping with the adventurous spirit of the early days of sociophysics, we have put together this Special Issue in honor of Serge Galam for his 70th birthday in a new journal entitled *Physics*, with 18 open access papers available online.

We hope you find these papers insightful and inspiring.

André Martins, Taksu Cheon, Xijin Tang, Bastien Chopard, and Soumyajyoti Biswas
Guest Editors

physics

MDPI

Editorial

Foreword to the Special Issue "In Honor of Professor Serge Galam for His 70th Birthday and Forty Years of Sociophysics"

Serge Galam

CEVIPOF—Centre for Political Research, Sciences Po and CNRS, 1, Place Saint Thomas d'Aquin, 75007 Paris, France; serge.galam@sciencespo.fr

I am deeply moved and honored by this Special Issue of the journal *Physics* celebrating my seventieth birthday and forty years of sociophysics [1]. Retrospectively, such an issue seems amazing, especially when recalling that, over forty years ago, in the late seventies, the founding paper coining the word sociophysics [2] and co-signed with Yuval Gefen and Yonathan Shapir, was confiscated by the Chairman of the Physics Department at Tel-Aviv University. Furthermore, almost all academics at that time supported the confiscation, deeming our paper a threat to the Department's reputation [3].

We then brandished academic freedom but, not being tenured, the faculties asserted that we did not have the academic freedom to mention an affiliation to the department without the approval of a tenured Professor. Eventually, Alexander Voronel, a newly emigrated Soviet physicist who had suffered censorship there, endorsed our paper. Thanks to him, we got our paper back and submitted it to the *Journal of Mathematical Sociology*, where it underwent about two years of reviewing before being accepted and published in 1982 [2].

Throughout the subsequent decades, I kept fighting on my own to gain acceptance for sociophysics within the worldwide community of physicists, first in the US, and then in France [4–6]. That was a significant challenge, and I only managed to convince a very few of them, which was already an accomplishment. I believe that most physicists currently engaged in sociophysics are unaware of, and even unable to, figure out the scale of the rejection sociophysics faced in the 1980s.

At the same time, I kept developing sociophysics, supporting its feasibility through the publication of additional papers [7–9], always with long and tedious prior exchanges with referees of the related journals.

Over time, after a long and arduous journey, sociophysics has become an established field of physics, with the papers now published in most of the major physics research journals and physics magazines, as seen with the following non-exhaustive list of journals, given in alphabetical order: *Chaos, Solitons and Fractals* [10], *Entropy* [11], *European Physical Journal Plus* [12], *European Physics Letters* [13], *Frontiers in Physics* [14], *International Journal of Modern Physics* [15], *Journal of Statistical Mechanics* [16], *Journal of Statistical Physics* [17], *Review of Modern Physics* [18], *Scientific Reports* [19], *Physica A* [20], *Physics Letters A* [21], *Physics Reports* [22], *Physical Review E* [23], *Physical Review Letters* [24], *Physics Today* [25], *Physics World* [26], and *PLoS One* [27]. In addition, arXiv [28] has a section titled "Popular Physics, Physics and Society", which is devoted, in good part, to sociophysics preprints.

With this Special Issue dedicated to sociophysics, the newly born journal *Physics* is joining the trend, thus enriching the numerous recent Special Issues on the topic [29–33]. In this respect, I express my gratitude to the five Guest Editors, Soumyajyoti Biswas from India, Taksu Cheon from Japan, Bastien Chopard from Switzerland, André Martins from Brazil and Xijin Tang from China.

The Special Issue is now completed, featuring a total of 17 papers published and available online in open access. This is a significant achievement and I warmly thank all the authors who contributed, making this a truly special issue. I also thank the authors whose submissions were not accepted by the journal.

Citation: Galam, S. Foreword to the Special Issue "In Honor of Professor Serge Galam for His 70th Birthday and Forty Years of Sociophysics". *Physics* **2024**, *6*, 1032–1034. https://doi.org/10.3390/physics6030063

Received: 24 April 2024
Accepted: 30 April 2024
Published: 6 August 2024

In this regard, the richness and diversity of this Special Issue demonstrates that, after forty or so years, sociophysics has gone a tremendous way towards becoming solid and promising. I believe that we, as a community of researchers, will succeed in discovering the basic laws governing the social interactions among humans, thus turning sociophysics into an established hard science to tackle social and political phenomena.

That goal is also desirable given the many social and political issues that today societies are currently facing, with seemingly no solution at hand, amidst multiple growing and threatening polarizations.

Given the current level of research associated with in sociophysics, my optimism is sound. Yet, to build a tool that could intervene, as an active key in changing the fate of a society, is both exciting and scary. However, that is inherent to all fields of knowledge and related discoveries. The ethical responsibility of applying sociophysics to real social community will lie among the future users, policy makers and others. For us, the makers of sociophysics, one safety measure to prevent its misuse could be to ensure that the results of our respective works are freely accessible to all.

The next decade should tell us to what extent these hopes, expectations and fears were justified. See you then!

Conflicts of Interest: The author declares no conflicts of interest.

References

1. Martins, A.; Cheon, T.; Tang, X.; Chopard, B.; Biswas, S. (Eds.) *Special Issue: In Honor of Professor Serge Galam for His 70th Birthday and Forty Years of Sociophysics. Physics* **2023–2024**. Available online: https://www.mdpi.com/si/153410 (accessed on 25 April 2024).
2. Galam, S.; Gefen, Y.; Shapir, Y. Sociophysics: A mean behavior model for the process of strike. *J. Math. Sociol.* **1982**, *9*, 1–13. [CrossRef]
3. Galam, S. Sociophysics: A personal testimony. *Phys. A Stat. Mech. Appl.* **2004**, *336*, 49–55. [CrossRef]
4. Galam, S. Physicists as a revolutionary catalyst. *Fundam. Scientiae* **1980**, *1*, 351–353.
5. Galam, S.; Pfeuty, P. Physicists are "frustrated"? *Phys. Today* **1982**, *35*, 89–91. [CrossRef]
6. Galam, S. Physicists, non physical topics, and interdisciplinarity. *Front. Phys.* **2022**, *10*, 986782. [CrossRef]
7. Galam, S. Majority rule, hierarchical structures, and democratic totalitarianism: A statistical approach. *J. Math. Psychol.* **1986**, *30*, 426–434. [CrossRef]
8. Galam, S. Social paradoxes of majority rule voting and renormalization group. *J. Stat. Phys.* **1990**, *61*, 943–951. [CrossRef]
9. Galam, S.; Moscovici, S. Towards a theory of collective phenomena: Consensus and attitude changes in groups. *Eur. J. Soc. Psychol.* **1991**, *21*, 49–74. [CrossRef]
10. Oliveira, I.V.G.; Wang, C.; Dong, G.; Du, R.; Fiore, C.E.; Vilela, A.L.M.; Stanley, H.E. Entropy production on cooperative opinion dynamics. *Chaos Solitons Fractals* **2024**, *181*, 114694. [CrossRef]
11. Weron, T.; Nyczka, P.; Szwabiński, J. Composition of the influence group in the *q*-voter model and its impact on the dynamics of opinions. *Entropy* **2024**, *26*, 132. [CrossRef]
12. Noorazar, H. Recent advances in opinion propagation dynamics: A 2020 survey. *Eur. Phys. J. Plus* **2020**, *135*, 521. [CrossRef]
13. Shen, C.; Guo, H.; Hu, S.; Shi, L.; Wang, Z.; Tanimoto, J. How Committed individuals shape social dynamics: A survey on coordination games and social dilemma games. *Eur. Phys. Lett.* **2023**, *144*, 11002. [CrossRef]
14. Galam, S.; Cheon, T. Tipping points in opinion dynamics: A universal formula in five dimensions. *Front. Phys.* **2020**, *8*, 446. [CrossRef]
15. Crokidakis, N. Radicalization phenomena: Phase transitions, extinction processes and control of violent activities. *Inter. J. Mod. Phys. C* **2023**, *34*, 2350100. [CrossRef]
16. Javarone, M.A.; Singh, S.P. Strategy revision phase with payoff threshold in the public goods game. *J. Stat. Mech.* **2024**, 023404. [CrossRef]
17. Gärtner, B.; Zehmakan, A.N. Threshold behavior of democratic opinion dynamics. *J. Stat. Phys.* **2020**, *178*, 1442–1466. [CrossRef]
18. Castellano, C.; Fortunato, S.; Loreto, V. Statistical physics of social dynamics. *Rev. Mod. Phys.* **2009**, *81*, 591–646. [CrossRef]
19. Carro, A.; Toral, R.; Miguel, M.S. The noisy voter model on complex networks. *Sci. Rep.* **2016**, *6*, 24775. [CrossRef] [PubMed]
20. Cui, P.-B. Exploring the foundation of social diversity and coherence with a novel attraction-repulsion model framework. *Phys. A Stat. Mech. Appl.* **2023**, *618*, 128714. [CrossRef]
21. Cheon, T.; Morimoto, J. Balancer effects in opinion dynamics. *Phys. Lett. A* **2016**, *380*, 429–434. [CrossRef]
22. Jusup, M.; Holme, P.; Kanazawa, K.; Takayasu, M.; Romić, I.; Wang, Z.; Geček, S.; Lipić, T.; Podobnik, B.; Wang, L.; et al. Social physics. *Phys. Rep.* **2022**, *948*, 1–148.
23. Pal, R.; Kumar, A.; Santhanam, M.S. Depolarization of opinions on social networks through random nudges. *Phys. Rev. E* **2023**, *108*, 034307. [CrossRef] [PubMed]

24. Baumann, F.; Lorenz-Spreen, P.; Sokolov, I.M.; Starnini, M. Modeling echo chambers and polarization dynamics in social networks. *Phys. Rev. Lett.* **2020**, *124*, 048301. [CrossRef] [PubMed]
25. Schweitzer, F. Sociophysics. *Phys. Today* **2018**, *71*, 40–47. [CrossRef]
26. Brazil, R. The physics of public opinion. *Phys. World* **2020**, *33*, 24–28. [CrossRef]
27. Vilone, D.; Polizzi, E. Modeling opinion misperception and the emergence of silence in online social system. *PLoS ONE* **2024**, *19*, e0296075. [CrossRef]
28. arXiv. Physics: Physics and Society. Cornel Tech. Cornell University: New York, NY, USA. Available online: https://arxiv.org/list/physics.soc-ph/recent (accessed on 25 April 2024).
29. De Sousa Lima, F.W. (Ed.) *Special Issue: Entropy-Based Applications in Sociophysics. Entropy* **2023–2024**. Available online: https://www.mdpi.com/journal/entropy/special_issues/entropy_sociophys (accessed on 25 April 2024).
30. Vazquez, F. (Ed.) *Special Issue: Statistical Physics of Opinion Formation and Social Phenomena. Entropy* **2022–2023**. Available online: https://www.mdpi.com/journal/entropy/special_issues/Opin_Form_Soc_Phenom (accessed on 25 April 2024).
31. Diep, H.T. (Ed.) *Special Issue: Computational and Statistical Physics Approaches for Complex Systems and Social. Entropy* **2019–2020**. Available online: https://www.mdpi.com/journal/entropy/special_issues/Statis_Social (accessed on 25 April 2024).
32. Malarz, K.; Sznajd-Weron, K. (Eds.) *Special Issue: Modern Trends in Sociophysics. Entropy* **2022–2023**. Available online: https://www.mdpi.com/journal/entropy/special_issues/Trends_Sociophysics (accessed on 24 May 2024).
33. Barra, A.; Agliari, E.; Javarone, M.A. (Eds.) *Research Topic: Social Spreading: Opinions, Behaviours and Strategies. Front. Phys.* **2020**. Available online: https://www.frontiersin.org/research-topics/11146/social-spreading-opinions-behaviours-and-strategies (accessed on 25 April 2024).

physics

MDPI

Article

Agent Mental Models and Bayesian Rules as a Tool to Create Opinion Dynamics Models

André C. R. Martins

Interdisciplinary Research Group in Complex Systems Modelling (GRIFE), Escola de Artes, Ciências e Humanidades (EACH), Universidade de São Paulo, Rua Arlindo Bétio, 1000, São Paulo 03828-000, Brazil; amartins@usp.br

Abstract: Traditional models of opinion dynamics provide a simplified approach to understanding human behavior in basic social scenarios. However, when it comes to issues such as polarization and extremism, a more nuanced understanding of human biases and cognitive tendencies are required. This paper proposes an approach to modeling opinion dynamics by integrating mental models and assumptions of individuals agents using Bayesian-inspired methods. By exploring the relationship between human rationality and Bayesian theory, this paper demonstrates the usefulness of these methods in describing how opinions evolve. The analysis here builds upon the basic idea in the Continuous Opinions and Discrete Actions (CODA) model, by applying Bayesian-inspired rules to account for key human behaviors such as confirmation bias, motivated reasoning, and human reluctance to change opinions. Through this, This paper updates rules that are compatible with known human biases. The current work sheds light on the role of human biases in shaping opinion dynamics. I hope that by making the model more realistic this might lead to more accurate predictions of real-world scenarios.

Keywords: opinion dynamics; Bayesian methods; cognition; CODA; agent-based models; sociophysics

Citation: Martins, A.C.R. Agent Mental Models and Bayesian Rules as a Tool to Create Opinion Dynamics Models. *Physics* **2024**, *6*, 1013–1031. https://doi.org/10.3390/physics6030062

Received: 18 March 2024
Revised: 14 June 2024
Accepted: 17 June 2024
Published: 31 July 2024

1. Introduction: The Need for General Methods

Opinion dynamics modeling [1–9] is a fascinating area of research that seeks to understand how opinions spread through society. A plethora of models have been developed to describe this process, ranging from simple to complex, and covering topics such as the formation of consensus [10], the emergence of polarization [11–14], the different ways one can define it [15], and the spread of extreme opinions [16–24]. Extremism can be defined as the end of a range over a continuous variable [7,25], or as inflexibles who do not change their minds [26], or using mixed models [8,9,27]. To explore the problem of extremism in the real world, not only opinions matter [28], and one must also consider actions as part of what defines an extremist [29,30].

However, despite the wealth of knowledge already gathered, there are still many aspects of opinion dynamics that require greater attention. Community efforts are necessary to fill gaps in research and promote progress in the field [31]. Currently, most models are only comparable to similar implementations, with a lack of translation between different types. While attempts to propose general frameworks and universal formulas exist [9,32,33], they are, so far, isolated efforts. To achieve greater understanding, one needs to explore how different models relate to each other [34] and develop methods to incorporate new effects and assumptions.

For a deeper understanding of the spread of polarization and extremism, we must also consider actions, not just opinions. Incorporating decision making and behavioral aspects is crucial in modeling opinions [35,36], as it allows for a more accurate depiction of how individuals perceive and react to complex information [37]. One promising avenue of exploration is the use of Bayesian-inspired models [9].

Bayesian rules to model opinions have been introduced both in the opinion dynamics community, as extensions of the Continuous Opinions and Discrete Actions (CODA) model [8,9] and similar opinion models [9,21,27,28,34,38–67], by the use of Bayesian belief networks [68], as well as, independently, in models associated with economical reasoning [69–74]. Despite their popularity, there are two aspects of Bayesian-inspired models for opinion dynamics that have not been properly debated so far. They are how to turn assumptions on how the agents reason into dynamical model equations and the problem of the relationship between Bayesianism and rationality.

It is also worth mentioning that Bayesian update rules can even offer deeper insights into the meaning of extreme opinions, by making it straightforward the need to distinguish between being quite sure about something (high probability one is right) from the inability or strong difficulty to change one's position [28]. More than that, it is possible to demonstrate that if each agent includes the possibility that when the neighbor agree with its choice that might be because the agent has influenced the neighbor, one can obtain the dynamics of other discrete models as a limit case [34]. Extensions of concepts such as contrarians [40] and inflexibles [27] have been studied and Bayesian updates have been used even to obtain bounded confidence-like models [38,66].

In this paper, I explore both aspects of the problem. First, I provide a brief explanation on why the use of Bayesian-inspired rules is both supported by experimental evidence [73,75–78] and not the same as assuming rationality [79,80]. And, to illustrate how Bayesian rules can be used in a general problem, I explain how one can create mental models for the agents. And, one can see how models may include any kind of bias and bounded rationality effects [81,82], and turn those assumptions into update rules. More exactly, I show how to introduce agents who distrust opinions opposed either to a certain choice (a direct bias) or against their current preference. Update rules for both kinds of agents are calculated and the effects of those biases on how extreme the positions of the agents become are studied.

2. Bayesian Models and Rationality

Bayesian methods are one of the gold standards for rationality and inductive arguments [80]. If one starts with quite simple rules about how induction about plausibilities must be performed, one can show that plausibilities should be updated using Bayes' theorem [83,84]. The same theorem can be obtained from other axioms, such as maximization of entropy [85] or the considerably weaker basis of "dutch books". However, using those ideas to describe how people reason has been considered problematic [86]. On the other hand, Bayesian ideas can be used both in a complicated way, strictly following its rules, or as a soft version, where its basic ideas are used to represent aspects such as updating subjective opinions [87]. That poses the question of whether Bayesian rules can be used to describe human reasoning well. Certainly, one should also ask about the requirements been imposed from human rationality models. Even the definition of bounded rationality can be challenged, as it assumes there is someone to judge if any behavior is entirely rational [88]. Indeed, using Bayesian methods with perfection is impossible as they require infinite abilities [80]. One can only approximate them by considering a limited set of possibilities, and that is compatible with how human brains work.

There is quite a good evidence that this study reasons in ways that are similar to Bayesian methods [73,75–78]. But, if one wants to use them in mathematical and computational models, one needs to go further than just similarity. Indeed, it is well understood that humans are not ideally rational nor good enough at statistics. One sometimes fails at straightforward problems [89,90] and tends to be too confident about human mental skills [91]. At first, experiments about human cognitive abilities seemed to point to a remarkable amount of incompetence.

But that is not the whole story. When looking closer, some of human mistakes are not as serious as they look. While people are not good enough abstract logicians, when the same problems are presented associated with normal day-to-day circumstances, those problems are answered correctly [92]. And, there is evidence that many of human mistakes can be described as the use of reasonable heuristics [93], short-cuts that allow us to arrive at answers quite fast and with less effort [94,95]. As simplified rules, heuristics fail under some circumstances. If one goes looking for such cases, one for sure finds the answers. But they are not a sign of complete incompetence.

That does not explain human overconfidence problem, surely. But it has been also observed that human reasoning skills might not have evolved to find the best answers, even if we can use the skills for that purpose. Instead, humans show a tendency to defend their identity-defining beliefs [96,97]. More than that, human ancestors had solid reasons to be quite good at fitting inside their groups and, if possible, ascending socially inside them. Human reasoning and argumentative skills were more valuable from an evolutionary perspective if they worked towards agreeing or convincing human peers. Group belonging mattered more than being right about things that affect direct survival [98,99]. Being sure and defending ideas from people social group become more important than looking for better but unpopular answers.

And there is one more issue. In many laboratory problems where it has been observed that humans make mistakes, scientists used questions that never appear in real life [79]. Take the case of the observation of weighting functions, that one seems to alter probability values that are heard [100]. Using changed values might seem to serve no purpose at first. However, the scientists who performed those experiments assumed the values they presented were known with certainty. But there is no such certainty in real life. If someone tells you there is a 30% chance of rain tomorrow, even if based on highly calibrated models, you know that is an estimate. As with any estimates, at best, the actual value is close to 30%. A Bayesian assessment of the problem combines the previous estimate of rain with the information about the forecast to obtain a final opinion. Applying this to the many experiments that showed people use probability values wrong, one can see that human behavior is compatible with Bayesian rules. The observed changes match reasonable assumptions for everyday situations when one hears uncertain estimates [76]. Applying this is actually wrong in artificial cases, such as the laboratory experiments, where there is no (or quite little) uncertainty about the probability values. Even human tendency to change the opinions far less than one should, called conservatism (no relationship to politics implied in the technical term) [101], can straightforwardly be explained. One just needs to include in a Bayesian model a tendency to skepticism about the data heard [76]. Human brains might just have heuristics that mimic Bayesian estimates for an uncertain and unreliable world.

That is, people are not ideal, but human bounded abilities are not those of incompetents. One makes reasonable approximations. People are motivated reasoners, more interested in defending ideas than looking for better answers. Given the right preferences, these can even be described as rational, despite ethical considerations. Even when it seemed one is making mistakes, one might have been behaving closer to Bayesian rules than it was initially assumed. A more complete discussion about these themes, while of high interest, is beyond the scope of the paper and, for further reading, see Refs. [76,77,79,95,102–104].

3. Update Rules from Agent Mental Models

Humans are not well rational, but can still be described by Bayesian rules. Therefore, it is reasonable trying to apply Bayesian methods as a way to represent human opinions. Thus, the next question one must answer is how one can include human biases and cognitive characteristics in the models. For this, one must consider how the agents think, what the agents show to others, and what the agents expect to see from their neighbors. That is, one needs to describe agent mental models. However, first, let us consider how Bayesian methods work.

3.1. A Brief Introduction to Bayesian Methods

While Bayesian statistics, performed correctly, can become complicated fast enough, it is based on an elementary, almost trivial basis, Bayes' theorem. The theorem works as follows. There is an issue we want to learn about, and let us represent it by a random variable X, where each x represents one possible value. Here, x can be a quantity, but it can also be nothing more than a label. Let us start with a probability distribution, the initial guess on how likely each possible x is, represented by a probability distribution $f(x)$, called the prior opinion. Once one observes data D, one must change the opinions on x. For this, one needs to know, for each possible value x, how likely it is that one observes D. That is, one needs the likelihood, $f(D|x)$. From that, calculating the posterior estimate $f(x|D)$ is achieved just by a straightforward multiplication $f(x|D) \propto f(x)f(D|x)$. The proportionality constant is calculated by imposing that the final distribution must add (or integrate) to one. Everything in Bayesian methods is a consequence of that update rule and considerations on how to use it. The basic idea, already using an opinion dynamics problem, is represented in Figure 1.

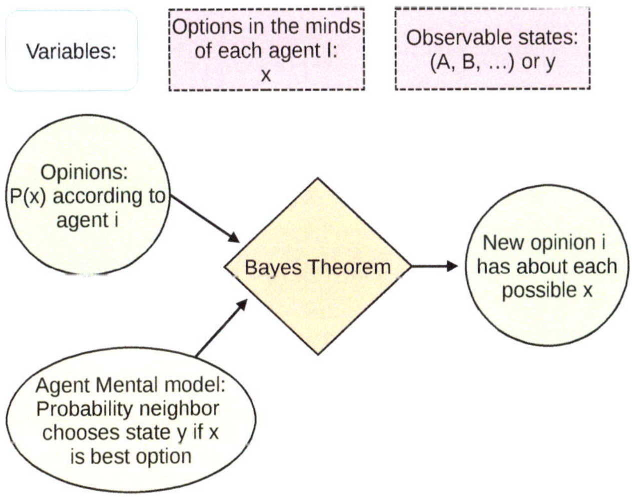

Figure 1. Schematics for the general use of Bayes' theorem as a tool for creating opinion update equations, highlighting the role of agent mental models for what others communicate to them (likelihoods).

To illustrate how it works in practice, let us look at the demonstration of the CODA model rules. In CODA, the agents try to decide between two possible choices, A or B (sometimes represented as values of spin, $+1$ or -1). Each agent i has a time t, and a probability opinion $p_i(t)$ that A is better than B (and, $1 - p_i(t)$ that B is better). But, instead of expressing their probabilistic opinion, they only show their neighbors the option they consider more likely to be better. They also assume their neighbors have a larger than 50% chance, α, to pick the best option. In general, there can be asymmetric different chances, α to choose A when A is better and β to choose B when B is better. As a first example, let us assume the $\alpha = \beta$ symmetry here. From Bayes' rule, one obtains the update model for $p_i(t+1) \propto p_i(t)\alpha$ and, similarly, $1 - p_i(t+1) \propto (1 - p_i(t))(1 - \alpha)$. As the probabilities must add to one, one divides by their sum and obtains the update rule:

$$p_i(t+1) = \frac{p_i(t)\alpha}{p_i(t)\alpha + (1 - p_i(t))(1 - \alpha)}. \quad (1)$$

At this step, one has an update rule and can be used it as it is. In this case, however, it is trivial to make a change in variables that provides us a significantly more computationally light rule. The update rule becomes essentially straightforward if let us make the transformation

$$\nu = \ln(\frac{p}{1-p}). \quad (2)$$

The denominators cancel and one obtains

$$\nu(t+1) = \nu(t) \pm C, \tag{3}$$

where $C = \ln(\frac{\alpha}{1-\alpha})$ and the sign on the sum depends on whether the neighbor prefers A (+) or B (−). One can obtain an even more simplified model by renormalizing ν and making $C = 1$.

And that is it. Let us start from the initial opinion, use the Bayes theorem, and is reached an update rule. In this case, the final rule is to add one when a neighbor prefers A and subtract one when it prefers B, flipping opinions at $\nu = 0$.

3.2. *Agent Communication Rules and Their Mental Assumptions*

There are two major assumptions in the CODA model. One is how agents communicate. While they have a probabilistic estimate of which option is better, everyone else observes only their best estimate. The second assumption is the mental model of the agents. They think their neighbors are more likely than not to pick the best choice. And, all those neighbors are assumed to have the same chance, $\alpha > 0.5$, to be right. Naturally, one can introduce some agents that think their neighbors are more likely to be wrong, that is, $\alpha < 0.5$. These is known as contrarians [40], that is, agents who tend to disagreement [105].

Making the model assumption explicit makes it straightforward to investigate what happens if agents behave or act differently. For example, one can have a case when agents look for the best choice between A and B. Still, the agents communicate their probability estimates that A is the better choice, $p_i(t)$. In that case, while one can keep the probability $p_i(t)$ as a measure of the opinion of the agents, those agents no longer state a binary choice but a continuous specific probability value.

3.2.1. Changing What Is Communicated

Mental models become crucial, including which question the agents are trying to answer. Agents may still want to determine which is better, A or B, sharing information on their uncertainty. Or they might see the probability values as an ideal mixture. They might accept, for example, that the best position is 60% of A and 40% of B. First, let us consider the case where they just want to pick the best choice.

In that case, $p_i(t)$ is still just a value associated with A, while, trivially, $1 - p_i(t)$ gives us the probability B is better. But one has to evaluate the chance other agents may have any continuous value for $p_j(t)$ if A is better (or if B is). That is, one needs a distribution probability of probabilities to represent the agent not belonging to agents; unified over the paper. mental model. In mathematical terms, one needs a model that says, assuming A is better, how likely is it that neighbor j has an opinion p_j if A is better, $f(p_j|A)$. Certainly, one also needs $f(p_j|B)$, but in many situations of interest, that can be obtained by symmetry assumptions. This model was implemented originally by assuming $f(p_j|A)$ was a beta function $Be(p_j|\alpha,\beta,A)$, that is,

$$f(p_j|A) = Be(p_j|\alpha,\beta,A) = \frac{1}{B(\alpha,\beta)} p_j^{\alpha-1}(1-p_j)^{\beta-1},$$

where $B(\alpha,\beta)$ is obtained from gamma functions by

$$B(\alpha,\beta) = \frac{\Gamma(\alpha)\Gamma(\beta)}{\Gamma(\alpha+\beta)}.$$

Here, α and β are the traditional parameters of the beta function. Interestingly, the update rule can once more be defined in terms of the log-odds variable $\nu = \ln(\frac{p}{1-p})$, and that leads once more to an additive model. However, the term to be added depends on the probability p_j communicated by j, and as the agents become more certain, the size of the additive term explodes. Consequently, extreme opinions become remarkably stronger than the already extreme opinions in the original CODA model [28,53].

3.2.2. Changing the Mental Variables

While in the example in Section 3.2.1 one needs a new likelihood, the beta function that tells us how likely agents think others provide each answer, that need was caused by a change in the communication. But it is not only what is communicated that can be changed. Inner assumptions agents make can also be changed, including the questions they want to answer. That is, one can have quite different assumptions in their mental model.

Assume that instead of having "wisher" agents looking for the best option between A and B, each "mixer" agent has an estimate about the best mixture between A and B. In this case, p_i is the percentage of A in the best blend of two options, and, as such, each value $0 \leq p_i \leq 1$ must have a probability. One needs probability densities $f(p)$ as prior and posterior opinion distributions. The more straightforward way to implement such a model is to look for conjugate distributions [106]. Conjugate distributions correspond to those cases where, for a certain likelihood, the prior and the posterior are represented by the same function. In that case, update rules can just update the parameters of that distribution function and not complete distributions. However, this is not the general case for a random application of the Bayes rule. As one builds more detailed models, finding conjugate distributions might not be possible for a given mental model.

Luckily, for the more straightforward cases of interest here, conjugate options exist. Before proceeding to a model, one needs to decide how the agents communicate. Here, once again, one can have a discrete communication, where each agent just tells others which option, A or B, should appear in a more significant amount, or communication may include the average estimate for the proportion p of A, $E[p]$.

In the first case, with discrete communication, one can find a natural conjugate family by using beta distributions for the opinions and a binomial distribution for the likelihood, that is, the chance that a neighbor chooses A or B depending on its average estimate. Surely, other choices of distributions are possible, and they correspond to a similar thought structure but with different dynamics. Under the binomial–beta option, interestingly, the dynamics of the preferred choice, given the more straightforward choice of proper functions for the prior and the likelihood, mimics the original CODA dynamics. However, while the evolution of the preferred option in the mixture is the same, the probability (or proportion) values never reach the same extreme values [53].

3.2.3. Other Mental Models Already Explored

Other variations are possible and have been explored. An initial approach to trust was implemented in a fully connected setting by adding one assumption to the agent mental model [47]. Instead of assuming that every other agent had a probability α to obtain the best answer, each agent assumed there were two types of agents. Agent i assumed there was a probability τ_{ij} that agent j was a reliable source who picks the best option with chance α. But other agents may also be untrustworthy (or useless) and pick the best option with probability μ, so that $\mu < \alpha$, possibly even 0.5 or lower. That is, if A was the best choice, instead of a chance α neighbor j prefers A, there was a chance given by $\alpha\tau_{ij} + \mu(1 - \tau_{ij})$. Applying Bayes' theorem with this new likelihood led to update rules for p_i and τ_{ij}. Each agent updated both its opinions on whether A or B is better. And it also changed its estimates about the trustworthiness of the observed agent. The update rule cannot be simplified by a transformation of variables because no exact way to uncouple the evolution of the opinion and how much agents trusted their neighbors was found.

A similar idea was used for a Bayesian version for continuous communication and the "mixer" type of agents. That model [38] led to an evolution of opinions qualitatively equivalent to what one observes in bounded confidence models [7,25]. That continuous model was later extended to study the problem of several independent issues when agents adjusted their trust based not only on the debated subject but also on their neighbor's positions on other matters [66]. Interestingly, that caused opinions to become more clustered and aligned, similar to the irrational consistency one can observe in humans [107].

Even the agent's influence on its neighbors can be used for their mental models. That was introduced as a simplified version by assuming that there were different chances a and c that a neighbor prefers A in the case a was indeed better, depending on whether the observing agent selected A or not [34]. That actually weakened the reinforcement effects of agreement, as the other agent may think A was better not because it was but because the observer also thought that. In the limit of strong influence, the dynamics of the voter model [108,109]—or other types of discrete models, such as majority [110,111] or Sznajd [6] rules, depending on the interaction rules—was recovered. That shows that Bayesian-inspired models are highly more general than the traditional discrete versions.

3.3. Introducing Other Behavioral Questions

Bayesian rules can help us explain how humans reason [76,79]. And, one just saw a few examples of introducing extra details in the agent mental models so that new assumptions can be included. In this Section, I discuss how one can move ahead and model some biases observed in human behavior by applying those concepts to create an original model for a specific human tendency.

Let us start by considering a most straightforward one, confirmation bias [112], as that does not even need new mental models. Confirmation bias is just a human tendency to look for information from sources who agree with us. As such, it can be better modeled by introducing rules that reconnect the influence network so that agents are more likely surrounded by those who agree with them. The co-evolution of the CODA model over a network that evolved based on the agreement or disagreement of the agents and their physical location plus thermal noise was studied. Depending on the noise and the strength of the agreement term in the network rewiring, the tendency to polarization and confirmation bias was apparent [63].

Motivated reasoning [96], on the other hand, is not only about who one learns from but how one interprets information depending on whether one agrees with the reasoning or not. This can be implemented in more than one way. Quite a simple version is an approach where trust was introduced in the CODA model [47]. In that model, depending on the initial conditions, as agents became more confident about their estimates, they eventually distrust those who disagreed with them, even when they met for the first time.

3.3.1. Direct Bias

Surely, there are other possibilities, including more heavy-handed approaches. For example, agents might think being untrustworthy is associated with one of the two options. Instead of a trust matrix signaling how much each agent i trusts each agent j, one can introduce trust based on the possible choices. For two options, A and B, one can assume that each agent has a prior preference. Each agent believes that untrustworthy people defend only the side the agent is biased against. That can be represented by a small addition to the CODA model. Assume agent i prefers A, and it thinks that people only go wrong to defend B. One way to describe that is to assume that there is a proportion λ of reasonable agents who behave like CODA. They pick the better alternative with chance $\alpha > 0.5$. However, the remaining $(1-\lambda)$ agents choose B more often, regardless of whether it is true or not, with probability $\beta > 0.5$. That is, for agents biased towards A, the chance a neighbor chooses A, represented by CA, if A (or B) is better is given by the equations

$$P(CA|A) = \lambda\alpha + (1-\lambda)(1-\beta)$$

and

$$P(CA|B) = \lambda(1-\alpha) + (1-\lambda)(1-\beta).$$

Also,

$$P(CB|A) = \lambda(1-\alpha) + (1-\lambda)\beta$$

and

$$P(CB|B) = \lambda\alpha + (1-\lambda)\beta.$$

Actually, one can introduce an update rule for both the probability p_i that A is better and λ. However, for this exercise, let us assume there is an initial fixed value for λ. For example, to illustrate how this bias can change the CODA model, the agents can suppose that a majority of honest people are given by $\lambda = 0.8$. Let us also assume that honest people obtain the better answer with a chance of $\alpha = 0.6$, while biased people provide their wrong estimate of B with a probability of $\beta = 0.9$. That makes $P(CA|A) = 0.5$, $P(CA|B) = 0.34$, $P(CB|A) = 0.5$, and $P(CB|B) = 0.66$. So, if agent i observes someone who prefers A, it updates its opinion by $p_i(t+1) = \frac{p_i(t)0.5}{p_i(t)0.5+(1-p_i)0.34}$. That is, for the CODA-transformed variable $\nu = \ln(\frac{p}{1-p})$: $\nu_i(t+1) = \nu_i(t) + \ln(0.5/0.34) \approx \nu_i(t) + 0.386$. On the other hand, if B is observed, the update rule provides (again for an agent biased towards A), $\nu_i(t+1) = \nu_i(t) + \ln(0.5/0.66) \approx \nu_i(t) - 0.278$. That means that steps in favor of A are larger than those in favor of B for such an agent. While that agent can be convinced by a majority, the agent moves towards its preference if there is a tie in the neighbors. Depending on the exact values of the parameters, the ratio between step size can become larger.

Let us assume a renormalization of the additive term to implement this bias, as normally performed in CODA applications. Assuming agent i is biased in favor of A (the equations for the case where i is biased in favor of B are symmetric but are not considered here), one has for the variable $\nu_i(t)$, as defined in Equation (2), when the neighbor also prefers A

$$\nu_i(t+1) = \nu_i(t) + \ln\left[\frac{\lambda\alpha + (1-\lambda)(1-\beta)}{\lambda(1-\alpha) + (1-\lambda)(1-\beta)}\right]. \tag{4}$$

When the neighbor prefers B, then:

$$\nu_i(t+1) = \nu_i(t) + \ln\left[\frac{\lambda(1-\alpha) + (1-\lambda)\beta}{\lambda\alpha + (1-\lambda)\beta}\right]. \tag{5}$$

These equations trivially revert to the standard CODA model in Equation (3) when all agents are considered honest, that is, when $\lambda = 1$. For ease of further manipulations, one can define the size of the steps in both Equations (4) and (5) as

$$SA = \ln\left[\frac{\lambda\alpha + (1-\lambda)(1-\beta)}{\lambda(1-\alpha) + (1-\lambda)(1-\beta)}\right],$$

$$SD = \ln\left[\frac{\lambda(1-\alpha) + (1-\lambda)\beta}{\lambda\alpha + (1-\lambda)\beta}\right].$$

The first question one can ask about those steps is how they relate to each other and the original step size $C = \ln(\frac{\alpha}{1-\alpha})$. Straghtforward algebraic manipulations show that $SA < C$ and that $SD > -C$, as long as $\alpha > 0$ in both cases. As $SD < 0$, both step sizes are smaller. That was to be expected. Introducing a chance the neighbor might not know what it is talking about should, indeed, decrease the information content of its opinion.

Figure 2 shows how the step sizes change as a function of the estimated proportion λ of honest agents among those each agent is biased against. In Figure 2, upper, one can see both the step sizes when the neighbor agrees with the opinion favored by the bias, SA, and when the neighbor disagrees with it, SD. Notice that when λ tends to 1.0, both steps become equal. That corresponds to the scenario where everyone is honest; one recovers the original CODA model with identical steps. The apparent equality when λ tends to zero is not real and is only a product of visualizing minimal steps. Indeed, Figure 2, lower, shows that the ratio between the steps, $s = SA/SD$, increases continuously as λ becomes close to zero.

The changes in the step size are indeed not of the same size. If one needs to normalize the steps, as in CODA, one can choose either SA or SD as the step made equal to 1.0. Let us initially make, for the implementations, the smaller $SD = 1.0$. And, as a simplification, instead of carrying dependencies on λ, α, and β, let us just assume there is a ratio s such that $SA = sSD$. That is, it is assumed that disagreement with the bias corresponds to a step size of 1.0 and agreement with the bias to a step size of $s \geq 1.0$, where $s = 1.0$ corresponds to the case when there is no bias. Finally, one has elementary update rules one can implement given by

$$\nu_i(t+1) = \nu_i(t) + \begin{cases} \text{sign}(\nu_j) & \text{if neighbor } j \text{ disagrees with } i \text{ bias,} \\ s \cdot \text{sign}(\nu_j) & \text{if neighbor } j \text{ agrees with } i \text{ bias.} \end{cases} \tag{6}$$

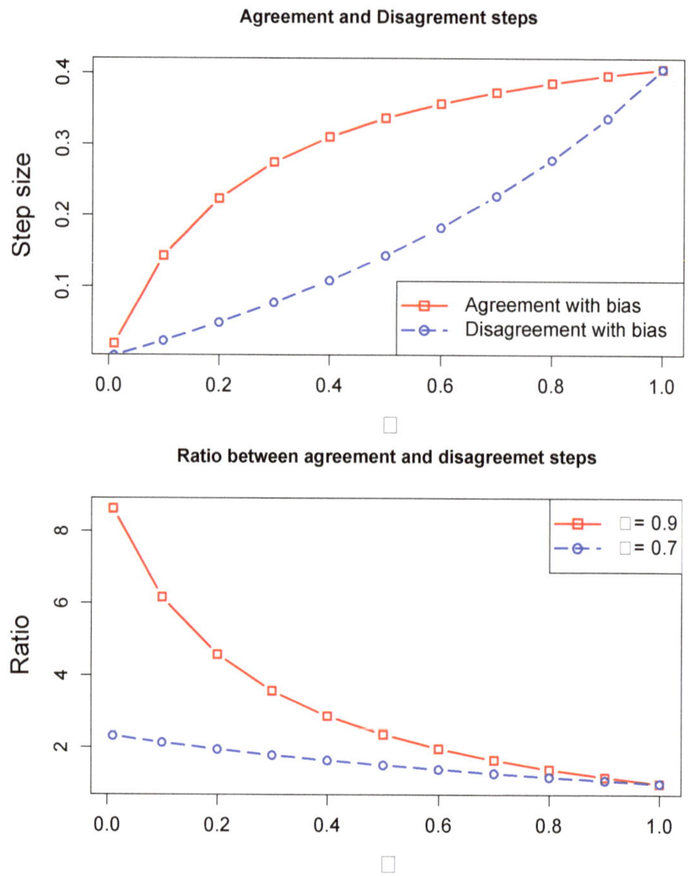

Figure 2. Upper: size of the steps for agreement, SA, and disagreement, SD, as a function of the estimated proportion, λ, of honest agents among those each agent is biased against when it is believed dishonest agents lie with probability $\beta = 0.9$. **Lower**: ratio $s = SA/SD$ for two values of β. See text for details.

3.3.2. Conservatism Bias

As a related example, let us introduce the effect called conservatism [101], where people change their opinions less than they should. That can be quickly introduced using a mental model where the agent thinks there is a chance the data are reliable and a chance that the information is only non-informative noise. When that happens, it is only natural that the update's size is considerably smaller. The larger the chance associated with noisy data, the smaller the update's size.

Here, there is the same mathematical problem as with the direct bias in Section 3.3.1, except that now, instead of believing that defenders of a specific side might be lying, the agents think that defenders of the side who disagree with them might be dishonest. Suppose an agent changes its opinion from A to B. In that case, the agent changes its assessment of where there might be dishonesty from B to A. That means one has quite similar rules as those in Equation (6). However, the bias is always the same as opinion i, so the rules depend directly on $\text{sign}(\nu_i)$. That is, then:

$$\nu_i(t+1) = \nu_i(t) + \begin{cases} \text{sign}(\nu_j) & \text{if sign}(\nu_j) \neq \text{sign}(\nu_i), \\ s \cdot \text{sign}(\nu_j) & \text{if sign}(\nu_j) = \text{sign}(\nu_i). \end{cases} \tag{7}$$

4. Results

To observe how the system might evolve under each rule, I implemented the models using the R software (Version 2.7.0) environment [113]. All cases shown here correspond to an initial neighborhood of agents defined as a square, bi-dimensional lattice with 40^2 agents, no periodic conditions, and second-level neighbors. As commented in Section 4.2 below, in some cases, the network is let to evolve into a polarized case before allowing opinions to change using this polarized and quenched network. Once that initial network, lattice or rearranged, is established, agents interact, observing the choice of one neighbor and updating their own opinion based on that observation, according to the update rules of each case. There was an average number of 50 interactions per agent. For all simulations, initial opinions were drawn from a uniform continuous distribution in the range $-2.0 \leq \nu_i \leq +2.0$. All the results for the distribution of opinions correspond to averages over 20 realizations of each case.

4.1. Simulating a Direct Bias

Results for the distribution of opinions in the direct bias case, with no initial rewiring, for three distinct values of the ratio s can be seen in Figure 3. Each curve corresponds to a different ratio s between the agreement and the disagreement step. Figure 3, upper, shows how extreme the opinion is measured in the number of disagreement steps (following the algorithm implementation). Figure 3, lower, shows the renormalized distribution if one measures opinions in terms of agreement steps.

One can observe in Figure 3, upper, with the distribution measured in disagreement steps, that as s increases, the opinions spread further away from the central position. That suggests that the opinions become more extreme. While the peaks associated with the more extreme opinions become softer (and seem to disappear for $s = 2.0$), that happens because the existing extremists become distributed over an extensive range of even stronger opinions.

A different picture emerges if one looks at a renormalized step size, using the agreement step as unit, $SA = 1$, instead of the implemented disagreement step. The distribution of opinions for such a case can be seen in Figure 3, lower. What one observes in this case is that when one measures the strength of opinions using SA as the measuring unit, the distributions become more similar. As s increases, one observes an increase in the number of agents around the weaker opinions and smaller peaks of extremists.

The apparently contradictory conclusions one can arrive at by looking at only one graphic are another example of the problem with adequately defining what an extremist is [28]. In this model, there are two different ways to define extremism. One of those definitions arises if one was to transform back the number of steps into probability values by inverting the renormalizations and transformations of variables. It is worth noticing that, as can be seen in the upper graphic in Figure 2, both step sizes, for agreement and disagreement, become smaller when bias is introduced when compared to the unbiased case $s = 1.0$. That leads to the strange conclusion that when agents think others might be biased, their final opinion becomes less extreme. That is reasonable as soon as one only cares about the agents' confidence in absolute terms. After all, other agents become less reliable. Their information should convince less.

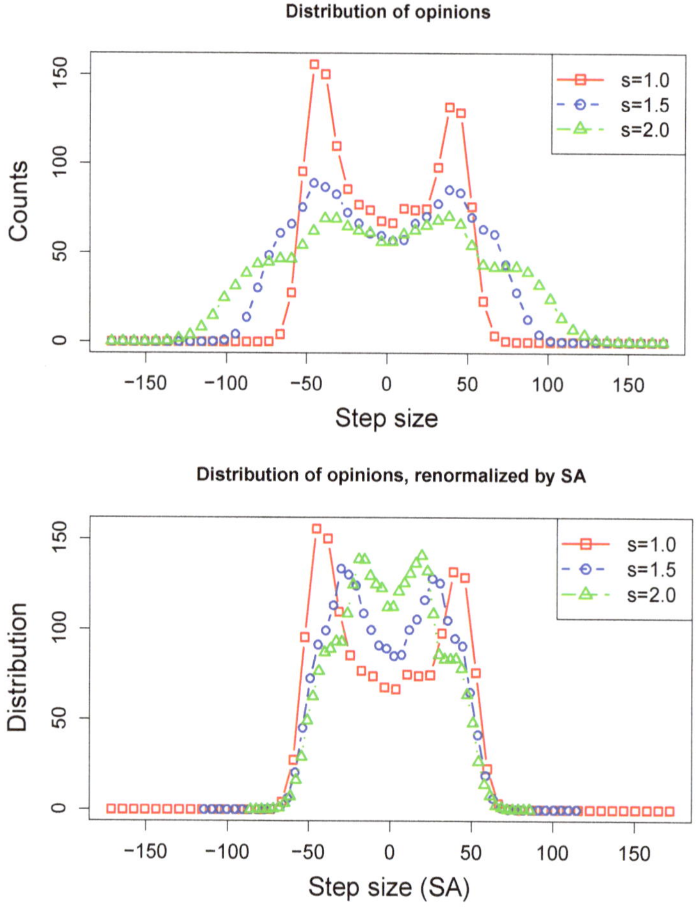

Figure 3. Distribution of opinions after these opinions are (**upper**) measured as disagreement steps and (**lower**) renormalized to agreement steps. See text for details.

But there is another definition of extremism that is also natural and reasonable. That definition comes from asking how straightforward (or complicated) it is to change the choice of a specific agent. This does correspond to the number of steps away from the central opinion. More precisely, since it is necessary to move to the opposite view to change one's choice, using the disagreement step as the unit is the best choice.

That is the case as soon as one was studying results from the conservatism bias, as presented in Section 4.2.1. Here, however, bias was fixed, corresponding to the initial choice of the agent. And that means that there is an essential proportion of agents whose biases do not conform to their opinions. While the proportion of agreement between bias and final opinion increases with s (observed averages were 57.7% for $s = 1$, 66.4% for $s = 1.5$, and 74.5% for $s = 2.0$), one a best match between bias and opinion. Even at $s = 2.0$, about one in four agents moves towards the opposite choice using agreement steps.

That might seem unrealistic. While there may be a bias toward initial opinions, people usually defend their current position. What size of step humans use when returning to a previously held belief is an interesting question, but that is beyond the scope of this paper.

On the other hand, confirmation bias is described not as an agreement with initial views but as the tendency to look for sources of information that agree with people current beliefs.

That bias was already studied before, with a network that evolved simultaneously with the opinion updates [63]. As agents stopped interacting with the agents they disagreed with, one has a case of traditional confirmation bias. That also means that the results been analyzed in this Section correspond to a direct bias independent of opinions. Real or not, it was introduced here as an exercise and as an example that one can model different modes of thinking using Bayesian tools, regardless of if those tools correspond to reality.

4.2. *Rewiring the Network*

Naturally, one wants to explore more realistic cases. One achieves this in two steps: first, by moving closer to a confirmation bias by introducing rewired networks, and second, by implementing the model of conservatism bias, as defined in Equation (7).

Here, the study follows the rewiring algorithm previously [63] used to study the simultaneous evolution of networks and opinions to generate an initial, quenched network before opinion updates start. At each step, the algorithm tries to destroy an existing link between two agents (1 and 2) and create a new one between two other agents (3 and 4). The decision of whether to accept that change depends on the Euclidean distance between agents 1 and 2, d_{12}, and agents 3 and 4, d_{34}, measured by considering a coordinate system over the square lattice where first neighbor distances correspond to 1. That way, there is a tendency to preserve the initial square lattice. And, one also uses a term that makes it more likely to accept the change when the old link was between disagreeing agents and the new one is between agreeing agents. That is, each rewiring is accepted with probability

$$P = \exp(-\Delta H) = \exp[-\beta(d_{34} - d_{12} - J(\sigma_3\sigma_4 - \sigma_1\sigma_2))], \quad (8)$$

where J is the relative importance between physical proximity and opinion agreement and σs denote the choice of each agent. Notice that if one needs a reasonable chance that distant agents connect, J should be comparable with the initial side of the square network.

Figure 4 shows a typical case, as an example, of how an initially square lattice is altered after an average of twenty rewirings per agent: $\beta = 1.0$ and $J = 20$. In Figure 4, brown and blue colors correspond to the two choices, and darker hues show a stronger final opinion, obtained after the opinion update phase. As the network did not change during the opinion update phase, its final shape after implementing the rule defined in Equation (8) is preserved.

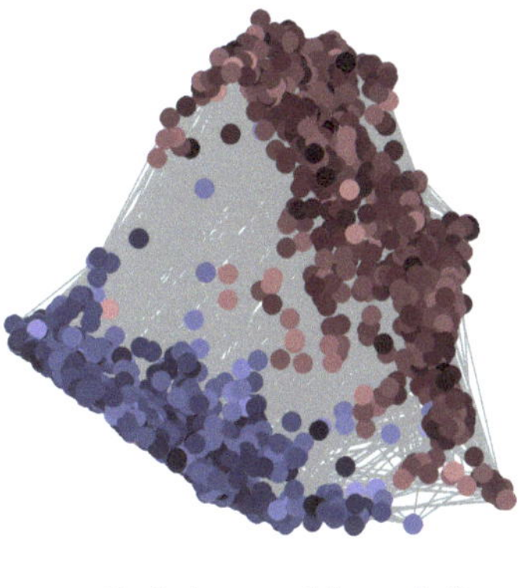

Figure 4. Typical network formed after applying the rewiring algorithm with an average of twenty rewirings per agent: $\beta = 1.0$ and $J = 20$. Brown and blue colors correspond to the two choices, and darker hues show a stronger final opinion obtained after the opinion update phase.

Figure 5 shows the distribution of opinions when one uses the direct bias rules after obtaining a quenched network using the parameters to generate networks similar to those in Figure 4. One can see significant differences between the results with no initial rewiring (Figure 3) and the new ones (Figure 5). With the initial rewiring, one finds almost no moderates in the final opinions, and the extremist peaks, when one uses the disagreement step, SD, for normalization, move to even stronger values as s increases. That dislocation also corresponds to smaller peaks distributed over a more extensive range. That, however, is an artifact of using SD. When renormalized to SA size steps, One can see that the three curves for different values of s match almost ideally. That happens because, with the implemented rewiring, most interactions involve agents who already share the same opinions.

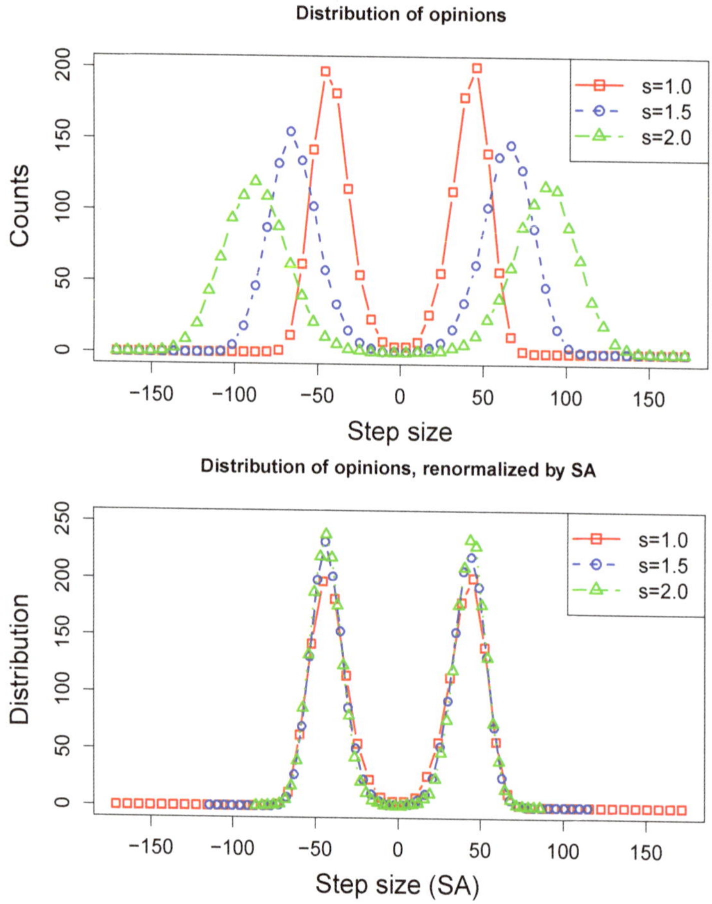

Figure 5. Distribution of opinions after these opinions are (**upper**) measured as disagreement steps and (**lower**) renormalized to agreement steps.

However, the problem of defining extremism under these circumstances becomes even more straightforward. One no longer has a case when the agent bias does not agree with its opinion. Indeed, the observed averages for the proportion of agreement between bias and opinions were 99.8% for $s = 1$, 100% for $s = 1.5$, and 100% for $s = 2.0$. That means that to change position, all agents move one disagreement step at a time. While observing how the curves match when one measures opinions in agreement steps is interesting, the disagreement step case is more informative. And here, as expected, the more bias one introduces, the more complicated it becomes for all agents to change their opinions.

4.2.1. Simulating the Conservatism Bias

The conservatism model defined in Equation (7) was also implemented using the same parameter values used in the previous cases in Section 4.1. As in this scenario there is no distinction between opinion and bias, the simulations presented here do not include an initial rewiring phase to guarantee that most agents are aligned with their own biases.

Figure 6 shows the distribution of opinions as disagreement steps (Figure 6, upper) and agreement steps (Figure 6, lower). As have been discussed in Section 4.2, in this case, the measure that tells us how complicated it is for an agent to change its choice is associated with the disagreement steps. One can see in Figure 6, upper, that, as was observed with the direct bias case with no initial rewiring (Figure 2), the peaks of extreme opinions move to more distant values and become less pronounced. That happens because, contrary to the rewired case when there were few links between disagreeing agents, there is still more extensive borders where one can find agents whose neighbors have distinct choices. Despite that, moderate agents, close to zero, become rarer as the conservatism bias increases. While not as relevant for understanding how extreme opinions are in this model, Figure 6, lower, renormalized for agreement steps of size one, is still of interest. It shows the same tendency of preserving the general shape for different values of s observed before (in Figure 5), except that one can see that when no conservatism is expected ($s = 1.0$), there are still more agents in the moderate region.

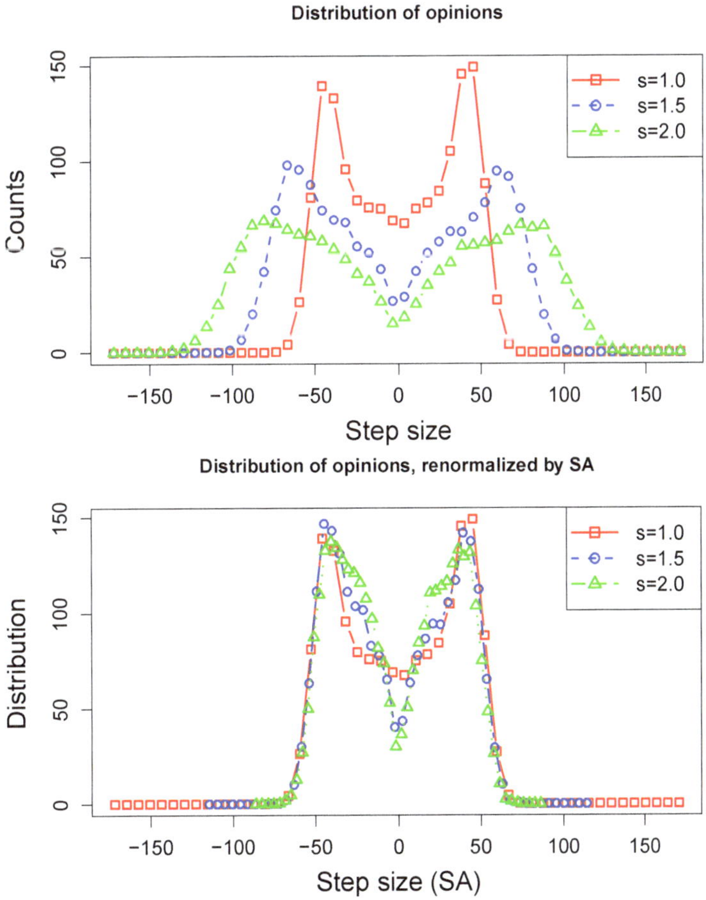

Figure 6. Distribution of opinions after these opinions opinions are (**upper**) measured as disagreement steps and (**lower**) renormalized to agreement steps.

5. Conclusions

Approximating the complete but impossible Bayesian rules provides a reasonable description of human behavior, especially when one accounts for individuals' imperfect trust in others and their tendency to reason in a motivated manner. By modeling these approximations, one can create realistic models of human behavior, as demonstrated in this paper. This paper analysis focuses on introducing variations to the CODA model where agents exhibit biases in favor of a particular opinion. Through the current investigation, it is found that conservatism implementation, where the agents distrust information that goes against their beliefs, results in more extreme opinions and a greater resistance to change.

Furthermore, this research examines how to obtain update rules from assumptions about the agents' mental models, using both previously published cases and the new examples. Another crucial aspect of using Bayesian-inspired rules is that it allows for a better understanding of the relationship between distinct models. By exploring which assumptions lead to the current opinion models, one gets insight into their inter-relatedness and identify cases when each model might be more applicable. Overall, this study sheds light on the potential of Bayesian-inspired modeling to offer a more nuanced description of agent behavior and its impact on opinion dynamics.

Funding: This work was supported by the Fundação de Amparo a Pesquisa do Estado de São Paulo (FAPESP) under grant 2019/26987-2.

Data Availability Statement: The code is available at https://www.comses.net/codebase-release/d4ab2a25-4233-4e6e-a8c5-a3b919cfd6e2/ (accessed on 13 June 2024).

Conflicts of Interest: The author declares no conflicts of interest.

References

1. Castellano, C.; Fortunato, S.; Loreto, V. Statistical physics of social dynamics. *Rev. Mod. Phys.* **2009**, *81*, 591–646. [CrossRef]
2. Galam, S. *Sociophysics: A Physicist's Modeling of Psycho-Political Phenomena*; Springer Science+Business Media, LLC: New York, NY, USA, 2012. [CrossRef]
3. Latan, B. The psychology of social impact. *Am. Psychol.* **1981**, *36*, 343–365. [CrossRef]
4. Galam, S.; Gefen, Y.; Shapir, Y. Sociophysics: A new approach of sociological collective behavior: Mean-behavior description of a strike. *J. Math. Sociol.* **1982**, *9*, 1–13. [CrossRef]
5. Galam, S.; Moscovici, S. Towards a theory of collective phenomena: Consensus and attitude changes in groups. *Eur. J. Soc. Psychol.* **1991**, *21*, 49–74. [CrossRef]
6. Sznajd-Weron, K.; Sznajd, J. Opinion evolution in a closed community. *Int. J. Mod. Phys. C* **2000**, *11*, 1157. [CrossRef]
7. Deffuant, G.; Neau, D.; Amblard, F.; Weisbuch, G. Mixing beliefs among interacting agents. *Adv. Compl. Sys.* **2000**, *3*, 87–98. [CrossRef]
8. Martins, A.C.R. Continuous opinions and discrete actions in opinion dynamics problems. *Int. J. Mod. Phys. C* **2008**, *19*, 617–624. [CrossRef]
9. Martins, A.C.R. Bayesian updating as basis for opinion dynamics models. *AIP Conf. Proc.* **2012**, *1490*, 212–221.
10. Schawe, H.; Fontaine, S.; Hernández, L. When network bridges foster consensus. Bounded confidence models in networked societies. *Phys. Rev. Res.* **2021**, *3*, 023208. [CrossRef]
11. DiMaggio, P.; Evans, J.; Bryson, B. Have American's social attitudes become more polarized? *Am. J. Sociol.* **1996**, *102*, 690–755. [CrossRef]
12. Baldassarri, D.; Gelman, A. Partisans without constraint: Political polarization and trends in american public opinion. *Am. J. Sociol.* **2008**, *114*, 408–446. [CrossRef]
13. Taber, C.S.; Cann, D.; Kucsova, S. The motivated processing of political arguments. *Polit. Behav.* **2009**, *31*, 137–155. [CrossRef]
14. Dreyer, P.; Bauer, J. Does voter polarisation induce party extremism? the moderating role of abstention. *West Eur. Politics* **2019**, *42*, 824–847. [CrossRef]
15. Bramson, A.; Grim, P.; Singer, D.J.; Berger, W.J.; Sack, G.; Fisher, S.; Flocken, C.; Holman, B. Understanding polarization: Meanings, measures, and model evaluation. *Philos. Sci.* **2017**, *84*, 115–159. [CrossRef]
16. Deffuant, G.; Amblard, F.; Weisbuch, G.; Faure, T. How can extremism prevail? A study based on the relative agreement interaction model. *J. Artif. Soc. Soc. Simul. (JASSS)* **2002**, *5*, 1. Available online: https://www.jasss.org/5/4/1.html (accessed on 15 June 2024).
17. Amblard, F.; Deffuant, G. The role of network topology on extremism propagation with the relative agreement opinion dynamics. *Phys. A Stat. Mech. Appl.* **2004**, *343*, 725–738. [CrossRef]

18. Galam, S. Heterogeneous beliefs, segregation, and extremism in the making of public opinions. *Phys. Rev. E* **2005**, *71*, 046123. [CrossRef] [PubMed]
19. Weisbuch, G.; Deffuant, G.; Amblard, F. Persuasion dynamics. *Phys. A Stat. Mech. Appl.* **2005**, *353*, 555–575. [CrossRef]
20. Franks, D.W.; Noble, J.; Kaufmann, P.; Stagl, S. Extremism propagation in social networks with hubs. *Adapt. Behav.* **2008**, *16*, 264–274. [CrossRef]
21. Martins, A.C.R. Mobility and social network effects on extremist opinions. *Phys. Rev. E* **2008**, *78*, 036104. [CrossRef]
22. Li, L.; Scaglione, A.; Swami, A.; Zhao, Q. Consensus, polarization and clustering of opinions in social networks. *IEEE J. Sel. Areas Commun.* **2013**, *31*, 1072–1083. [CrossRef]
23. Parsegov, S.E.; Proskurnikov, A.V.; Tempo, R.; Friedkin, N.E. Novel multidimensional models of opinion dynamics in social networks. *IEEE Trans. Autom. Control* **2017**, *62*, 2270–2285. [CrossRef]
24. Amelkin, V.; Bullo, F.; Singh, A.K. Polar opinion dynamics in social networks. *IEEE Trans. Autom. Control* **2017**, *62*, 5650–5665. [CrossRef]
25. Hegselmann, R.; Krause, U. Opinion dynamics and bounded confidence models, analysis and simulation. *J. Artif. Soc. Soc. Simul. (JASSS)* **2002**, *5*, 2. Available online: https://www.jasss.org/5/3/2.html (accessed on 15 June 2024).
26. Galam, S.; Jacobs, F. The role of inflexible minorities in the breaking of democratic opinion dynamics. *Phys. A Stat. Mech. Appl.* **2007**, *381*, 366–376. [CrossRef]
27. Martins, A.C.R.; Galam, S. The building up of individual inflexibility in opinion dynamics. *Phys. Rev. E* **2013**, *87*, 042807. [CrossRef] [PubMed]
28. Martins, A.C.R. Extremism definitions in opinion dynamics models. *Phys. A Stat. Mech. Appl.* **2022**, *589*, 126623. [CrossRef]
29. Tileaga, C. Representing the 'other': A discurive analysis of prejudice and moral exclusion in talk about romanies. *J. Community Appl. Soc. Psychol.* **2006**, *16*, 19–41. [CrossRef]
30. Bafumi, J.; Herron, M.C. Leapfrog representation and extremism: A study of american voters and their members in congress. *Am. Polit. Sci. Rev.* **2010**, *104*, 519–542. [CrossRef]
31. Sobkowicz, P. Whither now, opinion modelers? *Front. Phys.* **2020**, *8*, 461 [CrossRef]
32. Böttcher, L.; Nagler, J.; Herrmann, H.J. Critical behaviors in contagion dynamics. *Phys. Rev. Lett.* **2017**, *118*, 088301. [CrossRef] [PubMed]
33. Galam, S.; Cheon, T. Tipping points in opinion dynamics: A universal formula in five dimensions. *Front. Phys.* **2020**, *8*, 446. [CrossRef]
34. Martins, A.C.R. Discrete opinion models as a limit case of the coda model. *Phys. A Stat. Mech. Appl.* **2014**, *395*, 352–357. [CrossRef]
35. Kowalska-Pyzalska, A.; Maciejowska, K.; Suszczyński, K.; Sznajd-Weron, K.; Weron, R. Turning green: Agent-based modeling of the adoption of dynamic electricity tariffs. *Energy Policy* **2014**, *72*, 164–174. [CrossRef]
36. Müller-Hansen, F.; Schlüter, M.; Mäs, M.; Donges, J.F.; Kolb, J.J.; Thonicke, K.; Heitzig, J. Towards representing human behavior and decision making in earth system models—An overview of techniques and approaches. *Earth Syst. Dyn.* **2017**, *8*, 977–1007. [CrossRef]
37. Haghtalab, N.; Jackson, M.O.; Procaccia, A.D. Belief polarization in a complex world: A learning theory perspective. *Proc. Natl. Acad. Sci. USA (PNAS)* **2021**, *118*, e2010144118. [CrossRef]
38. Martins, A.C.R. Bayesian updating rules in continuous opinion dynamics models. *J. Stat. Mech. Theo. Exp.* **2009**, *2009*, P02017. [CrossRef]
39. Martins, A.C.R.; de Pereira, C.; Vicente, R. An opinion dynamics model for the diffusion of innovations. *Phys. A Stat. Mech. Appl.* **2009**, *388*, 3225–3232. [CrossRef]
40. Martins, A.C.R.; Kuba, C.D. The importance of disagreeing: Contrarians and extremism in the coda model. *Adv. Compl. Sys.* **2010**, *13*, 621–634. [CrossRef]
41. Vicente, R.; Martins, A.C.R.; Caticha, N. Opinion dynamics of learning agents: Does seeking consensus lead to disagreement? *J. Stat. Mech. Theo. Exp.* **2009**, *2009*, P03015. [CrossRef]
42. Si, X.-M.; Liu, Y.; Xiong, F.; Zhang, Y.-C.; Ding, F.; Cheng, H. Effects of selective attention on continuous opinions and discrete decisions. *Phys. A Stat. Mech. Appl.* **2010**, *389*, 3711–3719. [CrossRef]
43. Si, X.-M.; Yun; Cheng, H.; Zhang, Y.-C. An opinion dynamics model for online mass incident. In *ICASTE 2010. 2010 3rd International Conference on Advanced Computer Theory and Engineering*; Proceedings Volume 5; Desheng, W., Ruofeng, W., Yi, X., Eds.; The Institute of Electrical and Electronics Engineers, Inc.: New York, NY, USA, 2010; pp. V5-96–V5-99. [CrossRef]
44. Martins, A.C.R. A middle option for choices in the continuous opinions and discrete actions model. *Adv. Appl. Stat. Sci.* **2010**, *2*, 333–346.
45. Martins, A.C.R. Modeling scientific agents for a better science. *Adv. Compl. Sys.* **2010**, *13*, 519–533. [CrossRef]
46. Deng, L.; Liu, Y.; Xiong, F. An opinion diffusion model with clustered early adopters. *Phys. A Stat. Mech. Appl.* **2013**, *392*, 3546–3554. [CrossRef]
47. Martins, A.C.R. Trust in the coda model: Opinion dynamics and the reliability of other agents. *Phys. Lett. A* **2013**, *377*, 2333–2339. [CrossRef]

48. Diao, S.-M.; Liu, Y.; Zeng, Q.-A.; Luo, G.-X.; Xiong, F. A novel opinion dynamics model based on expanded observation ranges and individuals' social influences in social networks. *Phys. A Stat. Mech. Appl.* **2014**, *415*, 220–228. [CrossRef]
49. Luo, G.-X.; Liu, Y.; Zeng, Q.-A.; Diao, S.-M.; Xiong, F. A dynamic evolution model of human opinion as affected by advertising. *Phys. A Stat. Mech. Appl.* **2014**, *414*, 254–262. [CrossRef]
50. Caticha, N.; Cesar, J.; Vicente, R. For whom will the bayesian agents vote? *Front. Phys.* **2015**, *3*, 25. [CrossRef]
51. Martins, A.C.R. Opinion particles: Classical physics and opinion dynamics. *Phys. Lett. A* **2015**, *379*, 89–94. [CrossRef]
52. Lu, X.; Mo, H.; Deng, Y. An evidential opinion dynamics model based on heterogeneous social influential power. *Chaos Solitons Fractals* **2015**, *73*, 98–107. [CrossRef]
53. Martins, A.C.R. Thou shalt not take sides: Cognition, logic and the need for changing how we believe. *Front. Phys.* **2016**, *4*, 7. [CrossRef]
54. Chowdhury, N.R.; Morărescu, I.-C.; Martin, S.; Srikant, S. Continuous opinions and discrete actions in social networks: A multi-agent system approach. In *2016 IEEE 55th Conference on Decision and Control (CDC)*; The Institute of Electrical and Electronics Engineers, Inc.: New York, NY, USA, 2016; pp. 1739–1744 . [CrossRef]
55. Cheng, Z.; Xiong, Y.; Xu, Y. An opinion diffusion model with decision-making groups: The influence of the opinion's acceptability. *Phys. A Stat. Mech. Appl.* **2016**, *461*, 429–438. [CrossRef]
56. Huang, C.; Hu, B.; Jiang, G.; Yang, R. Modeling of agent-based complex network under cyber-violence. *Phys. A Stat. Mech. Appl.* **2016**, *458*, 399–411. [CrossRef]
57. Garcia, L.M.T.; Roux, A.V.D.; Martins, A.C.R.; Yang, Y.; Florindo, A.A. Development of a dynamic framework to explain population patterns of leisure-time physical activity through agent-based modeling. *Int. J. Behav. Nutr. Phys. Act.* **2017**, *14*, 111. [CrossRef]
58. Sun, R.; Mendez, D. An application of the continuous opinions and discrete actions (coda) model to adolescent smoking initiation. *PLoS ONE* **2017**, *12*, e0186163. [CrossRef] [PubMed]
59. Sobkowicz, P. Opinion dynamics model based on cognitive biases of complex agents. *J. Artif. Soc. Soc. Simul. (JASSS)* **2018**, *21*, 8. [CrossRef]
60. Lee, H.K.; Kim, Y.W. Public opinion by a poll process: Model study and bayesian view. *J. Stat. Mech. Theo. Exp.* **2018**, *2018* , 053402. [CrossRef]
61. Garcia, L.M.T.; Roux, A.V.D.; Martins, A.C.R.; Yang, Y.; Florindo, A.A. Exploring the emergence and evolution of population patterns of leisure-time physical activity through agent-based modelling. *Int. J. Behav. Nutr. Phys. Act.* **2018**, *15*, 112. [CrossRef] [PubMed]
62. Tang, T.; Chorus, C.G. Learning opinions by observing actions: Simulation of opinion dynamics using an action-opinion inference model. *J. Artif. Soc. Soc. Simul. (JASSS)* **2019**, *22*, 2. [CrossRef]
63. Martins, A.C.R. Network generation and evolution based on spatial and opinion dynamics components. *Int. J. Mod. Phys. C* **2019**, *30*, 1950077. [CrossRef]
64. Martins, A.C.R. Discrete opinion dynamics with m choices. *Eur. Phys. J. B* **2020**, *93*, 1. [CrossRef]
65. León-Medina, F.J.; Tena-Sánchez, J.; Miguel, F.J. Fakers becoming believers: How opinion dynamics are shaped by preference falsification, impression management and coherence heuristics. *Qual. Quant.* **2020**, *54*, 385–412. [CrossRef]
66. Maciel, M.V.; Martins, A.C.R. Ideologically motivated biases in a multiple issues opinion model. *Phys. A Stat. Mech. Appl.* **2020**, *553*, 124293. [CrossRef]
67. Fang, A.; Yuan, K.; Geng, J.; Wei, X. Opinion dynamics with bayesian learning. *Complexity* **2020**, *2020* , 8261392. [CrossRef]
68. Sun, Z.; Müller, D. A framework for modeling payments for ecosystem services with agent-based models, bayesian belief networks and opinion dynamics models. *Environ. Model. Softw.* **2013**, *45*, 15–28. [CrossRef]
69. Orléan, A. Bayesian interactions and collective dynamics of opinion: Herd behavior and mimetic contagion. *J. Econ. Behav. Organ.* **1995**, *28*, 257–274. [CrossRef]
70. Rabin, M.; Schrag, J.L. First impressions matter: A model of confirmatory bias. *Quart. J. Econ.* **1999**, *114*, 37–82. [CrossRef]
71. Andreoni, J.; Mylovanov, T. Diverging opinions. *Am. Econ. J. Microecon.* **2012**, *4*, 209–232. [CrossRef]
72. Nishi, R.; Masuda, N. Collective opinion formation model under bayesian updating and confirmation bias. *Phys. Rev. E* **2013**, *87*, 062123. [CrossRef]
73. Eguíluz, V.M.; Masuda, N.; Fernández-Gracia, J. Bayesian decision making in human collectives with binary choices. *PLoS ONE* **2015**, *10*, e0121332. [CrossRef]
74. Wang, Y.; Gan, L.; Djurić, P.M. Opinion dynamics in multi-agent systems with binary decision exchanges. In *2016 IEEE International Conference on Acoustics, Speech and Signal Processing (ICASSP)*; Proceedings; The Institute of Electrical and Electronics Engineers, Inc.: New York, NY, USA, 2016; pp. 4588–4592. [CrossRef]
75. Knill, D.C.; Pouget, A. The Bayesian brain: The role of uncertainty in neural coding and computation. *Trends Neurosci.* **2004**, *27*, 712–719. [CrossRef]
76. Martins, A.C.R. Probabilistic biases as Bayesian inference. *Judgm. Decis. Mak.* **2006**, *1*, 108–117. [CrossRef]

77. Tenenbaum, J.B.; Kemp, C.; Shafto, P. Theory-based Bayesian models of inductive reasoning. In *Inductive Reasoning: Experimental, Developmental, and Computational Approaches*; Feeney, A., Heit, E., Eds.; Cambridge University Press: Cambridge, UK, 2007; pp. 167–204. [CrossRef]
78. Tenenbaum, J.B.; Kemp, C.; Griffiths, T.L.; Goodman, N.D. How to grow a mind: Statistics, structure, and abstraction. *Science* **2011**, *331*, 1279–1285. [CrossRef]
79. Martins, A.C.R. *Arguments, Cognition, and Science: Need and Consequences of Probabilistic Induction in Science*; Rowman & Littlefield Publishers: Lanham, MD, USA, 2020.
80. Martins, A.C.R. Embracing undecidability: Cognitive needs and theory evaluation. *arXiv* **2020**, arXiv:2006.02020. [CrossRef]
81. Simon, H.A. Rational choice and the structure of environments. *Psychol. Rev.* **1956**, *63*, 129–138. [CrossRef] [PubMed]
82. Selten, R. What is bounded rationality? In *Bounded Rationality: The Adaptive Toolbox*; Gigerenzer, G., Selte, R., Eds.; The MIT Press: Cambridge, MA, USA, 2001; pp. 147–171. [CrossRef]
83. Cox, R.T. *The Algebra of Probable Inference*; The John Hopkins Press: Baltimore, MD, USA, 1961. Available online: https://bayes.wustl.edu/Manual/cox-algebra.pdf (accessed on 15 june 2024).
84. Jaynes, E.T. *Probability Theory: The Logic of Science*; Cambridge University Press: New York, NY, USA, 2003. [CrossRef]
85. Caticha, A.; Giffin, A. Updating probabilities. *AIP Conf. Proc.* **2006**, 872, 31–42. [CrossRef]
86. Eberhardt, F.; Danks, D. Confirmation in the cognitive sciences: The problematic case of bayesian models. *Minds Mach.* **2011**, *21*, 389–410. [CrossRef]
87. Elqayam, S.; Evans, J.S.B.T. Rationality in the new paradigm: Strict versus soft bayesian approaches. *Think. Reason.* **2013**, *19*, 453–470. [CrossRef]
88. Chater, N.; Felin, T.; Funder, D.C.; Gigerenzer, G.; Koenderink, J.J.; Krueger, J.I.; Noble, D.; Nordli, S.A.; Oaksford, M.; Schwartz, B.; Stanovich, K.E.; Todd, P.M. Mind, rationality, and cognition: An interdisciplinary debate. *Psychon. Bull. Rev.* **2018**, *25*, 793–826. [CrossRef]
89. Watson, P.C.; Johnson-Laird, P. *Psychology of Reasoning: Structure and Content*; Harvard University Press: Cambridge, MA, USA, 1972. Available online: https://archive.org/details/psychologyofreas0000waso (accessed on 15 June 2024).
90. Tversky, A.; Kahneman, D. Extension versus intuituive reasoning: The conjuction fallacy in probability judgement. *Psychol. Rev.* **1983**, *90*, 293–315. [CrossRef]
91. Oskamp, S. Overconfidence in case-study judgments. *J. Consult. Psychol.* **1965**, *29*, 261–265. [CrossRef] [PubMed]
92. Johnson-Laird, P.N.; Legrenzi, P.; Legrenzi, M.S. Reasoning and a sense of reality. *Brit. J. Psychol.* **1972**, *6*, 395–400. [CrossRef]
93. Gigerenzer, G.; Todd, P.M.; the ABC Research Group. *Simple Heuristics That Make Us Smart*; Oxford University Press, Inc.: New York, NY, USA, 1999.
94. Tversky, A.; Kahneman, D. Availability: A heuristic for judging frequency and probability. *Cogn. Psychol.* **1973**, *5*, 207–232. [CrossRef]
95. Gigerenzer, G.; Goldstein, D.G. Reasoning the fast and frugal way: Models of bounded rationality. *Psychol. Rev.* **1996**, *103*, 650–669. [CrossRef] [PubMed]
96. Kahan, D.M. Ideology, motivated reasoning, and cognitive reflection. *Judgm. Decis. Mak.* **2013**, *8*, 407–424. [CrossRef]
97. Kahan, D.M. The expressive rationality of inaccurate perceptions. *Behav. Brain Sci.* **2017**, *40*, e6. [CrossRef] [PubMed]
98. Mercier, H.; Sperber, D. Why do humans reason? arguments for an argumentative theory. *Behav. Brain Sci.* **2011**, *34*, 57–111. [CrossRef] [PubMed]
99. Mercier, H.; Sperber, D. *The Enigma of Reason*; Harvard University Press: Cambridge, MA, USA, 2017. [CrossRef]
100. Kahneman, D.; Tversky, A. Prospect theory: An analysis of decision under risk. *Econometrica* **1979**, *47*, 263–291. [CrossRef]
101. Edwards, W. Conservatism in human information processing. In *Formal Representation of Human Judgment*; Kleinmuntz, B., Ed.; John Wiley & Sons, Inc.: New York, NY, USA, 1968; pp. 359–369. Available online: https://pages.ucsd.edu/~mckenzie/Edwards1968excerpts.pdf (accessed on 15 June 2024).
102. Plous, S. *The Psychology of Judgment and Decision Making*; McGraw-Hill: New York, NY, USA, 1993.
103. Baron, J. *Thinking and Deciding*; Cambridge University Press: New York, NY, USA, 2023. [CrossRef]
104. Fitelson, B.; Thomason, N. Bayesians sometimes cannot ignore even very implausible theories (even ones that have not yet been thought of). *Australas. J. Log.* **2008**, *6*, 25–36. [CrossRef]
105. Galam, S. Contrarian deterministic effect: The hung elections scenario. *Phys. A Stat. Mech. Appl.* **2004**, *333*, 453–460. [CrossRef]
106. O'Hagan, A. *Kendall's Advanced Theory of Statistics. Volume 2B: Bayesian Inference*; Edward Arnold: London, UK, 1994.
107. Jervis, R. *Perception and Misperception in International Politics*; Princeton University Press: Princeton, NJ, USA, 1976. [CrossRef]
108. Clifford, P.; Sudbury, A. A model for spatial conflict. *Biometrika* **1973**, *60*, 581–588. [CrossRef]
109. Holley, R.; Liggett, T.M. Ergodic theorems for weakly interacting systems and the voter model. *Ann. Probab.* **1975**, *3*, 643–663. [CrossRef]
110. Galam, S. Modelling rumors: The no plane pentagon french hoax case. *Phys. A Stat. Mech. Appl.* **2003**, *320*, 571–580. [CrossRef]
111. Galam, S. Opinion dynamics, minority spreading and heterogeneous beliefs. In *Econophysics and Sociophysics: Trends and Perspectives*; Chakrabarti, B.K., Chakraborti, A., Chatterjee, A., Eds.; WILEY-VCH Verlag GmbH & Co. KGaA: Weinheim, Germany, 2006; Chapter 13. [CrossRef]

112. Nickerson, R.S. Confirmation bias: A ubiquitous phenomenon in many guises. *Rev. Gen. Psychol.* **1998**, *2*, 175–220. [CrossRef]
113. The R Development Core Team. *R: A Language and Environment for Statistical Computing. Reference Index. Version 2.7.0*; R Foundation for Statistical Computing: Vienna, Austria, 2008. Available online: https://ringo.ams.stonybrook.edu/images/2/2b/Refman.pdf (accessed on 15 June 2024).

physics

MDPI

Article

Fake News: "No Ban, No Spread—With Sequestration"

Serge Galam

CEVIPOF—Centre for Political Research, Sciences Po—CNRS, 1, Place Saint Thomas d'Aquin, 75007 Paris, France; serge.galam@sciencespo.fr

Abstract: To curb the spread of fake news, I propose an alternative to the current trend of implementing coercive measures. This approach would preserve freedom of speech while neutralizing the social impact of fake news. The proposal relies on creating an environment to naturally sequestrate fake news within quite small networks of people. I illustrate the process using a stylized model of opinion dynamics. In particular, I explore the effect of a simultaneous activation of prejudice tie breaking and contrarian behavior, on the spread of fake news. The results show that indeed most pieces of fake news do not propagate beyond quite small groups of people and thus pose no global threat. However, some peculiar sets of parameters are found to boost fake news so that it "naturally" invades an entire community with no resistance, even if initially shared by only a handful of agents. These findings identify the modifications of the parameters required to reverse the boosting effect into a sequestration effect by an appropriate reshaping of the social geometry of the opinion dynamics landscape. Then, all fake news items become "naturally" trapped inside limited networks of people. No prohibition is required. The next significant challenge is implementing this groundbreaking scheme within social media.

Keywords: fake news; freedom of speech; sequestration; opinion dynamics; prejudices; contrarians; tipping points; attractors; sociophysics

1. Introduction

Fake news has emerged as a significant and critical component of today's information landscape. It plays a pivotal role in the shaping outcomes of critical debates surrounding social, political, and societal matters. Fake news exerts a substantial influence on the formation of public opinion via social media [1,2].

Frequently, fake news is crafted with malicious intent, disseminating false and deceptive information while employing emotional manipulation and exploring many prejudices deeply rooted in our social and political representations. Indeed, the motivation of individual spreaders has been investigated [3,4].

In addition, instances of foreign-originated fake news were observed during the 2016 US and 2022 French presidential elections, in which they interfered in favor of specific candidates. However, investigations have not established a quantifiable impact on the final electoral results, despite evidence pointing to a reinforcement of existing individual opinions [5].

Presently, fake news has evolved into a pervasive tool for distorting public discourse on crucial issues faced by modern societies. Consequently, its associated impact has become a pressing concern for democratic institutions. Policymakers are taking measures to implement various regulations aimed at curbing the spread of fake news.

A series of regulation measures are thus implemented to control and restrict the use of social media [6–8], which in practical terms are likely to end in limitations of individual freedom of speech. In addition, considerable resources and effort are being dedicated to establishing fact-checking platforms to assess the reliability of information disseminated online. Yet, fake news continues to thrive within social media platforms.

Citation: Galam, S. Fake News: "No Ban, No Spread—With Sequestration". *Physics* **2024**, *6*, 859–876. https://doi.org/10.3390/physics6020053

Received: 7 March 2024
Revised: 9 May 2024
Accepted: 14 May 2024
Published: 6 June 2024

Moreover, it is of importance to caution against the relevance of these coercive policies, because such regulatory bodies can be misused for partisan purposes, as proven in the past. Indeed, they could also turn counterproductive by preventing the revelation of real facts hidden by official institutions, providing a framework for dismissing state secrets revealed by whistleblowers as "fake news". Famous examples include the "Watergate Scandal" [9], "Tuskegee Syphilis Study" [10], and "Iran-Contra Affair" [11].

At odd, Elon Musk has launched an opposite controversial governance of his social media X, advocating for freedom of speech [12,13], which in turn has produced a backslash from quite a number of people and institutions accusing him of promoting hate speech in the name of freedom of speech [14].

In this paper, I address the issue of curbing the spread of fake news by exploring another avenue I denote "No Ban, No Spread—with Sequestration". The aim is to unveil a framework to sequestrate fake news posts within quite small networks of agents. My starting hypothesis is to consider that it may not be the content of fake news per se that matters, but rather the interaction of that content with the social and psychological geometry in which fake news emerges and propagates.

While this geometry is given independently of the specific content of a fake news item, it plays a crucial role in facilitating its spread. Given such a layout, my focus is to identify the related parameters capable of reversing the shape of the associated social and psychological geometry to make it block the spread of "fake news" instead of boosting it. Once done, fake news become "naturally" trapped and sequestrated in quite small networks of agents.

To test the soundness of my proposal, I consider a stylized social framework within sociophysics [15–18], to explore the features that control the propagation of fake news beyond its mere content. Sociophysics is a new active field of physics that tackles social and political phenomena adopting a physicist-like approach [19,20]. The challenge is not to substitute social sciences but to create a new hard science by itself [21–35].

Indeed, among the large spectrum of social and political issues covered by sociophysics [36–55], the study of the dynamics of opinion occupies a central place [56–64]. A good deal of papers consider binary variables [65–86] and fewer three or more discrete opinions [87–94]. Currently, the field of sociophysics is attracting growing interest with a rather large number of published papers [95–108].

Here, I extend the model of opinion dynamics I have been developing for a few decades (then, the Galam model) deployed in a multi-dimensional space of parameters, by exploring a novel combination of the heterogeneity of agents with contrarians embedded among floaters when tie-breaking prejudice is activated [109–115].

The study reveals that the combination of contrarians and tie-breaking prejudice is instrumental in shaping the geometrical landscape, which either blocks fake news items or on the contrary unlocks others fostering an overall spreading. It happens that the activation of a quite small proportion of contrarians associated with a favorable tie-breaking prejudice opens the path to a massive spread of fake news, even when it started being believed by only a handful of agents.

The findings could serve as a foundation to design non-restrictive regulations, which could prevent "naturally" any invasion of social media by fake news. The related lever is the setting of a geometrical sequestration of fake news within quite small networks of agents. Then, neither prohibition or restriction is required.

The rest of the paper is organized as follows. Section 2 reviews the Galam model of opinion dynamics, while Section 3 introduces a new combination of contrarians and prejudice tie breaking in local group updates of opinion. The mechanisms locking or unlocking the spread of fake news are unveiled in Section 4. Section 5 identifies new strategies to sequestrate fake news posts without prohibiting them. The Conclusions contains a summary of the main results.

2. The Galam Model of Opinion Dynamics

2.1. Floaters Dynamics and Local Majority

The bare Galam model considers two competing discrete opinions A and B within a homogeneous population of floaters. Floaters are agents holding an opinion. They advocate to promote it. However, floaters listen to opposite arguments in favor of the competing opinion and thus are susceptible to become convinced to shift opinion [111–113].

Given initial proportions p_0 and $(1 - p_0)$ in favor of A and B, a dynamic is implemented by iterating a three-step procedure to update individual opinions. First, agents are distributed randomly in quite small groups of size r. Then, a local majority rule is applied separately to each group where all agents adopt the majority opinion. Third, agents are reshuffled. The update modifies the proportions p_0 and $(1 - p_0)$ to new ones p_1 and $(1 - p_1)$.

The scheme is then repeated n times with $p_0 \to p_1 \to p_2 \to \ldots \to p_n$. The associated update equation is given by

$$p_{r,1} = \left[\sum_{l=\bar{r}+1}^{r} \binom{r}{l} p_0^l (1 - p_0)^{r-l} + \frac{1}{2}\delta\left(\bar{r} - \frac{r}{2}\right) \binom{r}{r/2} p_0^{r/2} (1 - p_0)^{r/2} \right], \tag{1}$$

where $\bar{r} \equiv I[r/2]$, with $I[x]$ denoting the integer part of x and $\delta(x - a)$ is the Dirac delta function. Last term of Equation (1) means that for an even value of r at a tie, agents do not shift opinion.

The full landscape of the dynamics is obtained solving the fixed-point equation $p_{r,1} = p_0$. The equation yields one tipping point $p_t = 1/2$ and two attractors $p_A = 1$ and $p_B = 0$, which are all independent of the value of r. The associated dynamics are thus perfectly balanced with the opinion, which has gathered the majority of individual initial opinions, becoming larger and larger to reach eventually unanimity provided the number of updates n is sufficient.

The majority of agents have convinced individually via local discussing groups, agents holding initially the minority opinion to adopt the initial majority opinion. The dynamics are democratic with $p_0 < p_1 < p_2 < ... < p_n$ when $p_0 > 1/2$ and $p_0 > p_1 > p_2 > \cdots > p_n$ when $p_0 < 1/2$.

2.2. Floaters with Tie-Breaking Prejudice

However, the above ideal picture of democratic opinion dynamics breaks down when a tie breaking is included for evenly sized groups. In this case, at a tie, the group gets trapped into a collective doubt, with both opinions being equally acceptable. Since rationality cannot help to decide, everyone selects one of the two opinions at random, like tossing a coin. Individual choices are made by chance.

But contrary to the above rationale of random choices, the Galam model hypothesizes that indeed the "coin" is biased. For each agent, the state of doubting puts unconsciously some prejudice in control of the choice. The decision-making bias is monitored by the prejudice, which has been activated by the issue at stake. The decision is not made in the name of prejudice, but in the name of chance. To account for a distribution of different prejudices among agents, at a tie, opinion A is selected with probability k and opinion B with probability $(1 - k)$.

Accordingly, for $r = 4$, Equation (1) reduces to

$$p_{4,k} = p_0^4 + 4p_0^3(1 - p_0) + 6kp_0^2(1 - p_0)^2, \tag{2}$$

which still has the two attractors $p_A = 1$ and $p_B = 0$. But now, the attractors are separated by a tipping point located at

$$p_{t,k} = \frac{(6k - 1) - \sqrt{13 - 36k + 36k^2}}{6(2k - 1)}, \tag{3}$$

instead of $p_t = 1/2$. For $k = 0, 1/2$, and 1, Equation (3) yields, respectively, $p_{t,0} = (1+\sqrt{13})/6 \approx 0.77$, $p_{t,1/2} = 1/2$, and $p_{t,1} = (5-\sqrt{13})/6 \approx 0.23$. Accordingly, $1/2 \leq p_{t,k} \leq 0.77$ when $0 \leq k \leq 1/2$ and $0.23 \leq p_{t,k} \leq 1/2$ when $1/2 \leq k \leq 1$. (Note some misprints in Equation (3) and the following paragraph in Ref. [115]).

With $p_t \approx 0.23$, instead of 0.50, the case $k = 1$ illustrates the phenomenon of minority spreading. With opinion A being favored by the group prejudice, it needs to gather at minimum initial minority support of only 0.23% to convince the initial majority of agents, who are sharing opinion B, to adopt instead opinion A via local and open-minded discussions.

The previous democratic character of the opinion dynamics has been broken naturally and unconsciously without notice in favor of the choice in tune with the prejudices of the group. No coercion has been used. However, the initial support of A must be larger than 0.23. Otherwise, the minority opinion does lose support.

2.3. Floaters and Contrarians

Like floaters, contrarians are agents having an opinion, arguing for it, and listening to opposite arguments. However, unlike floaters, instead of following the local majority in a discussing group (fake news is true, fake news is false), they automatically adopt the opposite opinion (fake news is false, fake news is true) whatever the majority is [114]. They do it also in case of initial unanimity.

On this basis, while local majority rule favors the strengthening of an initial majority between two competing opinions, contrarians on the other hand favor a minority stand, which in turn reduces the gap between the respective proportions of the two competing opinions.

As expected, contrarians prevent the disappearance of the minority opinion even when many successive updates have been implemented. For instance, for discussing groups of size 4, update Equation (2) becomes,

$$p_{1,x} = (1-2x)\left[p_0^4 + 4p_0^3(1-p_0) + 3p_0^2(1-p_0)^2\right] + x, \tag{4}$$

where $k = 1/2$ (no tie breaking), x is the proportion of contrarians, and $(1-x)$ is the proportion of floaters.

The associated fixed-point equation $p_{1,x} = p_0$ yields $p_{t,x} = 1/2$ and

$$p_{A,x;B,x} = \frac{1 - 2x \pm \sqrt{1 - 8x + 12x^2}}{2(1-2x)}, \tag{5}$$

provided $0 \leq x \leq x_c$ with $x_c = 1/6 \approx 0.167$.

While the tie-prejudice effect preserves the floater attractors $p_A = 1$ and $p_B = 0$, shifting the tipping point away from $p_t = 1/2$, contrarians on the other hand preserve the tipping point $p_t = 1/2$ but shift both attractors, which in turn stabilize a coexistence of a large majority and a quite small minority with $p_{A,x} < 1$ and $p_{B,x} > 0$.

It is worth noticing that at $x = x_c = 1/6$, $p_A = p_B = 1/2$. Accordingly, for $x > x_c$, the dynamics is driven by a single attractor located at $1/2$. Any initial condition ends up at a perfectly balanced support for A and B. Here, I assume that $x \leq 1/2$, which is sound given the definition of a contrarian.

2.4. From Minority Opinion to Fake News

The Galam model deals with competing opinions A and B given some initial proportions p_0 and $(1-p_0)$ of respective supports, making by definition one minority and the other one majority. Applying the model to fake news is performed naturally, noticing a restriction of the proportions. Denoting A fa piece of fake news implies by nature of the phenomenon to have A being ultra minority with $p_0 \ll 1/2$. The competing opinion B then denotes the dismissing of A for being false and always starts with overwhelming support.

In addition, it is worth stressing that each fake news post activates specific prejudices. Given the content, within a social and political context, a fake news item produces contrarians in different proportions.

3. Contrarians with Tie Prejudice Breaking

In Section 2, I reviewed some main results obtained from the Galam model of opinion dynamics. I now investigate for the first time the impact of having contrarians in a community of floaters with active tie prejudice breaking. I restrict the study to discussing groups of size 4. Combining Equations (2) and (4) leads to the update equation

$$p_{4,k,x} = (1-2x)\left[p_0^4 + 4p_0^3(1-p_0) + 6kp_0^2(1-p_0)^2\right] + x. \quad (6)$$

Despite of a simplicity of Equation (6), the related fixed-point equation $p_{4,k,x} = p_0$ cannot be solved analytically. A numerical treatment is required to determine the fixed points $p_{A,k,x}, p_{B,k,x}, p_{t,k,x}$, with their domains of existence within the full parameter space $0 \leq k \leq 1$ and $0 \leq x \leq 1$.

Before starting the investigation, it is worth reminding the effect on the opinion dynamics of having contrarians among a community of floaters without tie-breaking prejudice ($k = 1/2$) as shown in Figure 1 and discussed in Section 2.3. With contrarians opposing the local majorities, as expected, they prevent reaching unanimity, securing always a resilient proportion of agents sharing the minority opinion. The more contrarians, the larger the stable minority with both attractors moving towards the tipping point, which remains located at 50%.

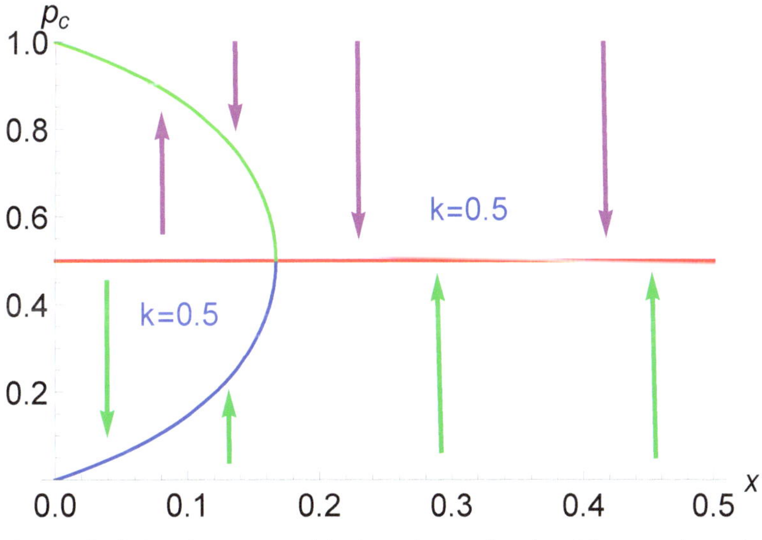

Figure 1. Evolution of attractors and tipping points as a function of the proportion x of contrarians for probability $k = 1/2$ (no tie breaking). The lower curve represents the attractor $p_{B,0.5,x}$ (in blue), the upper curve the attractor $p_{A,0.5,x}$ (in green), and the first part of middle line the tipping point $p_{t,0.5,x} = 0.5$ (in red). At $x_c = 0.167$, $p_{A,0.5,x_c} = p_{B,0.5,x_c} = p_{t,0.5,x_c} = 0.5$. For $x > x_c$, the unique attractor is located at precisely $1/2$ (the second part of the red line). The violet and green arrows show the different directions of opinion dynamics. See text for more details.

Nevertheless, despite being smooth, the shift of attractors brings them to the tipping point for the small proportion of contrarians $x_c = \approx 0.167$. Then, for $x > x_c$, the dynamics are turned upside down, becoming a single 50% attractor dynamic. Beyond x_c, contrarians erase totally any initial difference in the proportions of support for A and B. The whole related dynamics is and stays symmetric ensuring a democratic balance.

However, as soon as $k \neq 0$, the symmetry between A and B is broken. To identify the consequences associated with the activation of contrarians, I choose arbitrarily to start from $k = 1$ to study the range $0 \leq k \leq 1$.

3.1. Fake News in Tune with Active Prejudices

Figure 2 exhibits the dynamics landscape of opinion as a function of x for $k = 1$. At the corner $(k = 1, x = 0)$, $p_{A,1,0} = 1$, $p_{B,1,0} = 0$ and $p_{t,1,0} \approx 0.23$. From there, slightly increasing the proportion x of contrarians shifts $p_{A,1,0}$, and $p_{B,1,0}$ to, respectively, lower and higher values $p_{A,1,x}$ and $p_{B,1,x}$, as one would expect and is shown in the Figure 2.

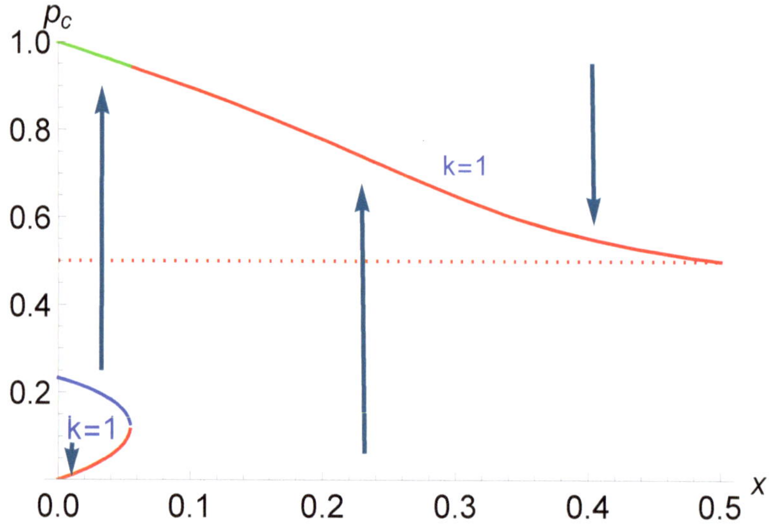

Figure 2. Evolution of attractors and tipping points as a function of the proportion x of contrarians for $k = 1$. The lower curve represents the attractor $p_{B,1,x}$ (in red), the upper curve the attractor $p_{A,1,x}$ (in green and red), and the middle curve the tipping point $p_{t,1,x}$ (in blue). At $x_c = 0.055$, $p_{A,1,x_c} = 0.944$ and $p_{A,1,0.20} = 0.778$, and $p_{A,1,0.30} = 0.649$. The red dotted line represents the majority threshold. The arrows show the different directions of opinion dynamics.

However, the value of the tipping point $p_{t,1,x}$ is seen to decrease, which is not expected. Indeed, being in a region where p is less than 0.23, a lower value for the tipping point means that A will keep increasing even when above 50%. That is not expected since contrarians are expected to decrease any majority as seen with the two attractors $p_{A,1,0} = 1$ and $p_{B,1,0} = 0$ being shifted to, respectively, lower and high values.

Moreover, contrary to the symmetric case where the three fixed points merge to result in one unique attractor located at 0.50, here, only the two fixed points $p_{B,1,x}$ and $p_{t,1,x}$ merge and disappear at a small value $x_c = 0.055$, leaving $p_{A,1,x}$ as the unique attractor driving the dynamics with $p_{A,1,x_c} = 0.944$. It is noticeable that when $x > x_c = 0.055$, $p_{A,1,x}$ is always located above 50%.

In addition to being counter-intuitive, the above results show a unexpected and alarming reality about the spread of fake news. Given a fake news item totally in tune $(k = 1)$ with the leading prejudice of a community, as soon as the proportion of active contrarians is larger than a few percent $(x > x_c = 0.055)$, a handful of agents sharing initially the fake news are sufficient to have it spread inexorably and invade large parts of this community, as seen with $p_{A,1,x_c} = 0.944$.

Although increasing the proportion of contrarians decreases the value $p_{A,1,x}$ towards equal probability as expected, the fake news can still reach more than the majority of the community, as seen with $p_{A,1,0.20} = 0.778$ and $p_{A,1,0.30} = 0.649$ in Figure 2.

In case the prejudices are heterogeneous with respect to the fake news post, the above phenomenon persists, requiring only a few more active contrarians. But the fake news post still requires only a handful of initial support agents to spread over and become the majority, for instance, with $k = 0.60$, $x_c = 0.114$, and $p_{A,0.60,x_c} = 0.843$, as seen in Figure 3.

For $x = 0.20$, $p_{A,0.60,0.20} = 0.648$, which indicates that a large minority $1 - p_{A,0.60,0.20} = 0.352$ remain dismissing the piece of fake news. For $x = 0.30$, $p_{A,0.60,0.30} = 0.537$, with the opposed minority reaching $1 - p_{A,0.60,0.30} = 0.463$.

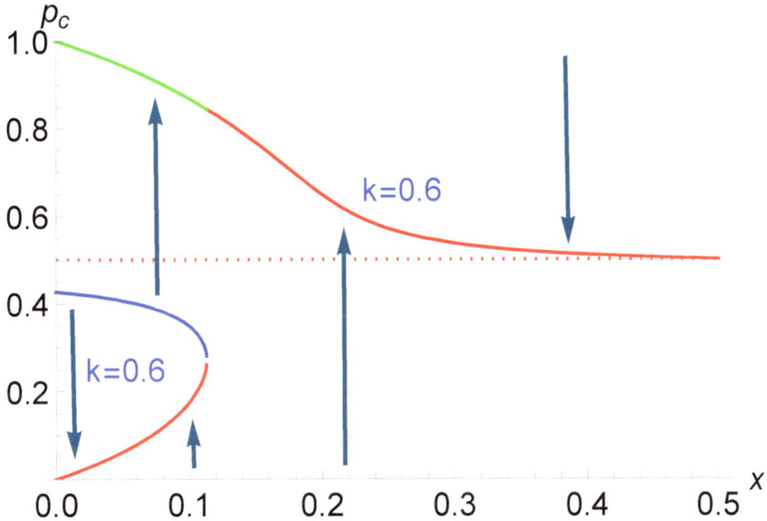

Figure 3. Evolution of attractors and tipping point as a function of the proportion x of contrarians for $k = 0.60$. The lower curve represents the attractor $p_{B,0.60,x}$ (in red), the upper curve the attractor $p_{A,0.60,x}$ (in green and red), and the middle curve the tipping point $p_{t,0.60,x}$ (in blue). At $x_c = 0.114$, $p_{A,0.60,x_c} = 0.843$, $p_{A,0.60,0.20} = 0.648$, and $p_{A,0.60,0.30} = 0.537$. The red dotted line represents the majority threshold. The arrows show the different directions of opinion dynamics.

Figure 4 exhibits both $k = 0.53$ and $k = 0.501$ cases, which show how the symmetrical case $k = 0.50$ is recovered (see Figure 2). Figure 2, left, has $x_c = 0.142$, $p_{A,0.53,x_c} = 0.753$, $p_{A,0.53,0.20} = 0.562$, and $p_{A,0.53,0.30} = 0.511$. Figure 2, right, has $x_c = 0.164$, $p_{A,0.501,x_c} = 0.589$, $p_{A,0.501,0.20} = 0.502$, and $p_{A,0.501,0.30} = 0.500$.

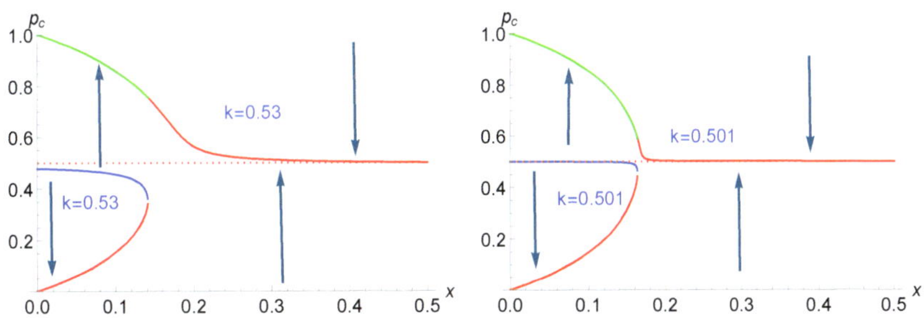

Figure 4. Evolution of attractors and tipping points as a function of the proportion x of contrarians for $k = 0.53$, $x_c = 0.142$, $p_{A,0.53,x_c} = 0.753$, $p_{A,0.53,0.20} = 0.562$, and $p_{A,0.53,0.30} = 0.511$ (**left**) and $k = 0.501$, $x_c = 0.164$, $p_{A,0.501,x_c} = 0.589$, $p_{A,0.501,0.20} = 0.502$, and $p_{A,0.501,0.30} = 0.500$ (**right**). The lower curves represent the attractor $p_{B,k,x}$ (in red), the upper curves the attractor $p_{A,k,x}$ (in green and red), and the middle curves the tipping point $p_{t,k,x}$ (in blue). The red dotted line represents the majority threshold. The arrows show the different directions of opinion dynamics.

3.2. Fake News at Odds with Active Prejudices

Let us look at the reverse situation with a fake news item at odds with the leading prejudices of the community. The extreme case $k = 0$ dynamics landscape is shown in

Figure 5. Comparing Figures 2 and 5 shows that the situation is anti-symmetrical from $k = 1$, where being in tune with the leading prejudices fosters drastically the spread of fake news even when initially shared by only a handful of agents.

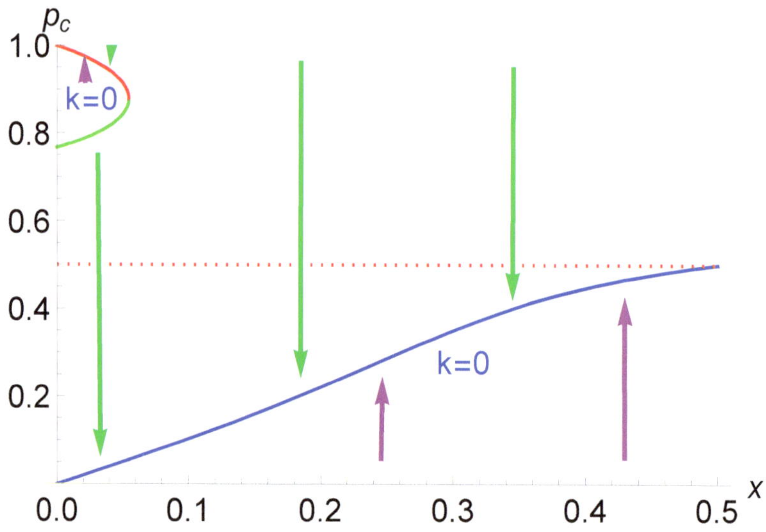

Figure 5. Evolution of attractors and tipping points as a function of the proportion x of contrarians for $k = 0$. The lower curve represents the attractor $p_{B,0,x}$ (in blue), the upper curve the attractor $p_{A,0,x}$ (in red), and the middle curve the tipping point $p_{t,0,x}$ (in green). At $x_c = 0.055$, $p_{B,0,x_c} = 0.056$, $p_{B,0,0.20} = 0.222$, and $p_{B,0,0.30} = 0.351$. The red dotted line represents the majority threshold. The violet and green arrows show the different directions of opinion dynamics.

When the fake news item is at odds with the active prejudices, even if a large majority of agents have initially believed it is true, the repeated discussions between agents in quite small groups reduce drastically their proportion. Figure 5 exhibits the dynamics landscape as a function of x for $k = 0$. At the corner $(k = 0, x = 0)$, $p_{A,0,0} = 1, p_{B,0,0} = 0$ and $p_{t,0,0} \approx 0.77$. From there, slightly increasing the proportion of contrarians shifts $p_{A,0,0}$ and $p_{B,0,0}$ to, respectively, lower and higher values, as would be expected.

However, here, contrary to $p_{t,1,x}$, the value of the tipping point $p_{t,0,x}$ increases with x, which is expected since a higher value for the tipping point makes it more complicated for a majority to hold its status with contrarians reducing the gap between the majority and the minority. However, as just above, even when A turns into the minority, it keeps losing support.

Now, the two fixed points $p_{A,0,x}$ and $p_{t,0,x}$ instead of $p_{B,1,x}$ and $p_{t,1,x}$ merge and disappear, but still at the same small value $x_c = 0.055$, leaving $p_{B,0,x}$ to be the unique attractor driving the dynamics with $p_{B,0,x_c} = 0.056$.

While the results just above were alarming and unexpected, here, the results are reassuring with respect to the spontaneous curbing of fake news diffusion. Given a fake news item totally at odds $(k = 0)$ with the leading prejudice of a community, as soon as the proportion of active contrarians is larger than a few percent $(x > x_c = 0.055)$, even a particularly majority of agents sharing initially the fake news item will eventually shift opinion, making their proportion to shrink inexorably down to quite low values of believers as seen with $p_{B,0,x_c} = 0.056$.

Moreover, increasing the proportion of contrarians increases the value $p_{B,0,x_c}$ towards equal probability as expected but keeps it lower than 0.50. Indeed, when the fake news supporters are the majority, the dynamics turn them down to a minority, as seen with $p_{B,0,0.20} = 0.222$ and $p_{B,0,0.30} = 0.351$ in Figure 5.

In case the prejudices are heterogeneous with respect to the fake news item, the above phenomenon persists, requiring only a few more active contrarians. The fake news item still ends up in the minority. For instance with $k = 0.40$, $x_c = 0.114$, and $p_{B,0.40,x_c} = 0.157$, as seen in Figure 6. Nevertheless, Figure 6 shows that a large minority remains concerning the fake news item with $p_{B,0.40,0.20} = 0.352$ and $p_{B,0.40,0.30} = 0.463$ for, respectively, $x = 0.20$ and $x = 0.30$.

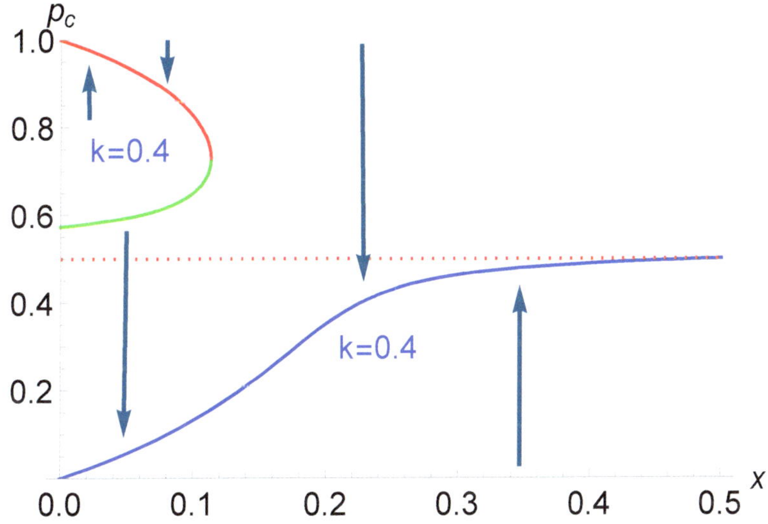

Figure 6. Evolution of attractors and tipping points as a function of the proportion x of contrarians for $k = 0.40$. The lower curve represents the attractor $p_{B,0.40,x}$ (in blue), the upper curve the attractor $p_{A,0.40,x}$ (in red), and the middle curve the tipping point $p_{t,0.40,x}$ (in green). At $x_c = 0.114$, $p_{B,0.40,x_c} = 0.157$, $p_{B,0.40,0.20} = 0.352$, and $p_{B,0.40,0.30} = 0.463$. The red dotted line represents the majority threshold. The arrows show the different directions of opinion dynamics.

Figure 7 exhibits both $k = 0.47$ and $k = 0.499$ cases, showing how the symmetrical case $k = 0.50$ is recovered from a lower value of k (see Figure 2). Figure 7, left, has $x_c = 0.142$ and $p_{B,0.47,x_c} = 0.247$. For, respectively, $x = 0.20$ and $x = 0.30$, $p_{B,0.47,0.20} = 0.438$ and $p_{B,0.47,0.30} = 0.489$. Figure 7, right, has $x_c = 0.164$ and $p_{B,0.499,x_c} = 0.411$ with $p_{B,0.499,0.20} = 0.498$ and $p_{B,0.499,0.30} = 0.500$.

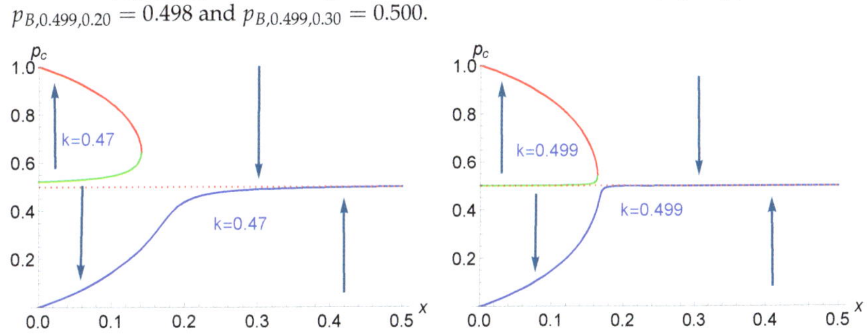

Figure 7. Evolution of attractors and tipping points as a function of the proportion x of contrarians for $k = 0.47$, $x_c = 0.142$, $p_{B,0.47,x_c} = 0.247$, $p_{B,0.47,0.20} = 0.438$, and $p_{B,0.47,0.30} = 0.489$ (**left**) and $k = 0.499$, $x_c = 0.164$, $p_{B,0.499,x_c} = 0.411$, $p_{B,0.499,0.20} = 0.498$, and $p_{B,0.499,0.30} = 0.500$ (**right**). The lower curves represent the attractor $p_{B,k,x}$ (in blue), the upper curves the attractor $p_{A,k,x}$ (in red), and the middle curves the tipping point $p_{t,k,x}$ (in green). The red dotted line represents the majority threshold. The arrows show the different directions of opinion dynamics.

4. Unlocking or Locking Fake News

The analyses in Section 3 have highlighted the critical impact of a few percent of contrarians on the dynamics of spreading fake news, once tie breaking by prejudice is activated. In addition, the direction of the contrarian impact is set by the overlap between the content of fake news and the leading prejudices prevailing in the social community.

It is useful to reemphasize that the selection of the prejudice being activated by the fake news is performed unconsciously when a discussing group gets trapped in a local doubt, determining the fake news validity. Either the fake news item benefits from the leading activated prejudices with $k > 1/2$ or it is impeded with $k < 1/2$. The respective effects on the dynamics of spreading are drastically different. But in both cases, only a few percent of contrarians are required to implement the drastic bias of the dynamics, as shown in Figure 8 and Table 1.

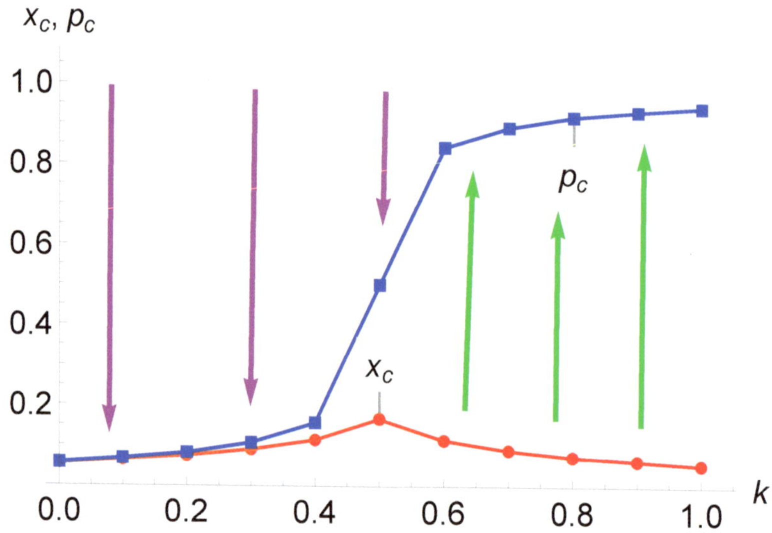

Figure 8. The values of x_c as a function of k. p_c denotes the values p_{B,k,x_c} for $0 \le k \le 0.5$ and p_{A,k,x_c} for $0.5 \le k \le 1$. The violet and green arrows show the different directions of opinion dynamics. See text for details.

Table 1. The values of x_c as a function of k with the values p_{B,k,x_c} for $0 \le k \le 0.5$ and p_{A,k,x_c} for $0.5 \le k \le 1$. See text for details.

k	0, 1	0.1, 0.9	0.2, 0.8	0.3, 0.7	0.4, 0.6	0.5
x_c	0.055	0.064	0.074	0.09	0.114	0.167
$p_{B,k\le 0.5,x_c}$	0.056	0.067	0.0815	0.107	0.157	0.5
$p_{A,k\ge 0.5,x_c}$	0.944	0.933	0.919	0.893	0.843	0.5

Moreover, the values of x_c are identical for k and $(1-k)$ in the range $0 \le k \le 1/2$. The associated values of the unique attractor at x_c satisfy $p_{B,k\le 0.5,x_c} + p_{A,k\ge 0.5,x_c} = 1$, as seen in Table 1. In addition, it is worth noticing that the values of the unique attractors at x_c stay either very low ($0 \le k < 0.4$) or very high ($0.6 < k \le 1$) beside in the range $0.4 < k < 0.6$ where the values, respectively, fall towards 0.5. Three very different regimes are thus obtained as a function of k:

Regime 1: $0 \le k < 0.4$. When the activated prejudices are mainly detrimental to fake news, the unique attractor $p_{B,k,x}$ is always much lower than $1/2$, as seen in Table 1. This means that even if a fake news item is first believed to be true by an overwhelming majority of the agents, the subsequent informal discussions among quite small groups of agents will eventually turn most of them to reject the fake news item as being false.

Regime 2: $0.4 < k < 0.6$. When the activated prejudices are almost equally distributed with respect to fake news, whatever initial conditions, the fake news post ends up being shared by almost half of the agents. It is less than half when $0.4 < k < 0.5$ and more than half for $0.5 < k < 0.6$. In both cases, a substantial part of the community believes the fake news is true, while another substantial part believes it is false. The society is polarized with respect to the validity of the piece of fake news.

Regime 3: $0.6 < k \leq 1$. When the activated prejudices are mainly at the benefit of the fake news item, the unique attractor is $p_{A,k,x}$ is always much larger than $1/2$, as seen in Table 1 and Figure 8. It means that even if fake news is first believed by only a handful of agents, the informal discussions among them will inexorably increase the proportion of believers to end up with an overwhelming part of the community. Such case is most concerned with the fake news item reaching a stable status of being "true" within the related community.

To grasp the whole landscape of the various types of dynamics, I have aggregated the cases $k = 1, 0.6, 0.501$, and $k = 0.499, 0.4, 0$ shown in Figure 9, upper left and upper right, respectively. Figure 9, lower, includes both series.

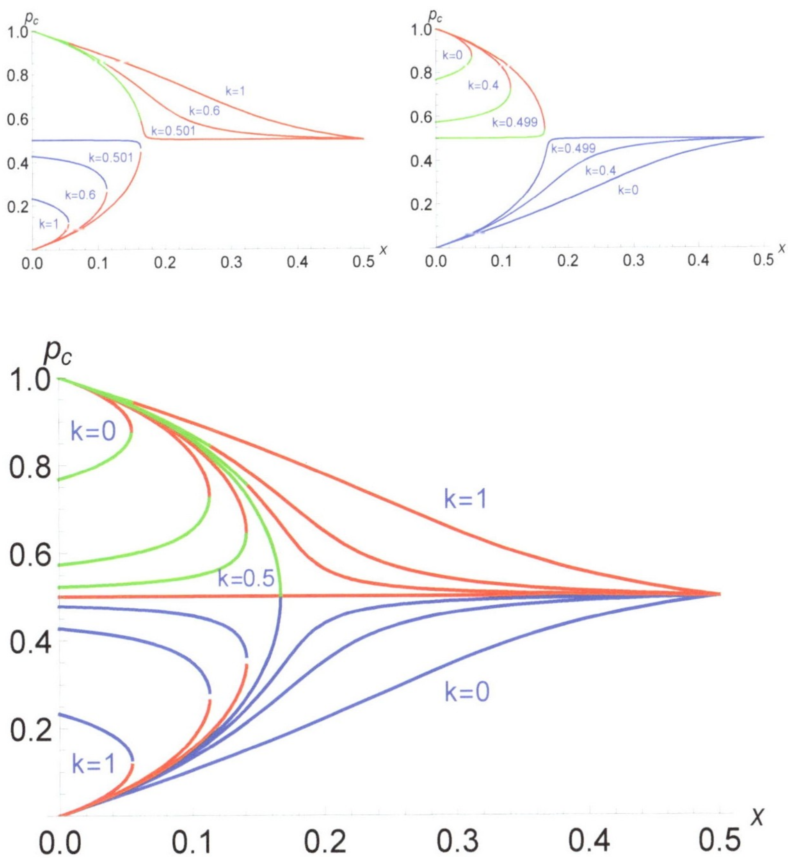

Figure 9. Attractors and tipping point as a function of x for $k = 1, 0.6, 0.501$ (**upper left**), $k = 0.499, 0.4, 0$ (**upper right**), and for both series (**lower**) for the attractors $p_{B,k,x}$ (in blue), $p_{A,k,x}$ (in red), and the tipping point $p_{t,k,x}$ (in green)—all denoted by p_c.

5. Strategies to Sequestrate Fake News without Prohibiting Them

The series of results obtained here highlight the critical roles played by contrarians and activated prejudices in shaping the fate of a given piece of fake news. It is worth stressing that their respective roles are implemented in parallel and without interaction between them. The findings set the frame to identify new avenues for designing strategies to curb the prominent spread of fake news and sequestrate it naturally into quite small networks of agents without the need for prohibition. The key findings are as follows.

(i) A few percent of contrarians are enough to modify drastically the full landscape of the associated dynamics of spreading or shrinking as exhibited in Figure 9. Contrarians have always the same impact, which is transforming the tipping-point dynamic into that of a single attractor. The consequence is that any initial support for any fake news ends up at this unique attractor whose value is independent of that of the initial support.
(ii) The location of the unique attractor of the dynamics is either above or below $1/2$ depending on the distribution of prejudices activated by the fake news item. It is also worth emphasizing that the single attractor is located mostly at either considerably low or considerably high values as a function of the distribution of activated prejudices.
(iii) The heterogeneities of the activated prejudices depend on the sociocultural composition of each community. Moreover, the prejudices that are activated spontaneously are selected by the content of the piece of fake news. As a result, some identical fake news may spread in some communities and shrink in others.

5.1. Most Fake News Do not Spread

While contrarians are instrumental to the drastic reshaping of the geometry of the dynamics landscape, I notice that they need to reach a proportion ranging between 6% and 20% to turn the tipping-point dynamics into a single attractor dynamics (see Table 1). More than 10% is a significant figure, which in turn indicates that it is unlikely that much fake news would generate such proportions of contrarians.

(i) When the contrarians are few in number, even in the case of beneficial (prejudices $k = 1$), fake news needs to start with a rather high proportion of individual believers and also rather hard to reach, as seen in Figure 9. For instance, with $x = 0$, the initial proportion must be higher than 23%. However, there are quite rare exceptions, such as the false claim that Israel bombed a hospital in Gaza in October 2023, which reached an impressive number of believers around the world in a matter of hours [116].
(ii) When the fake news item is at odds with the prejudices, the challenge becomes out of reach. In the case $k = 0$ and $x = 0$, the initial support must be higher than 77% (Figure 9). A particularly large figure is impossible to reach in most cases.
(iii) In cases where the fake news item does generate proportions of contrarians about 10%, substantial proportions of initial believers are still necessary for both $k < 1/2$ and $k > 1/2$, as seen in Figure 9,

All these observations indicate that most fake news does not spread and thus does not pose a threat to democratic unbiased public debates.

5.2. Some Rare Fake News Turn Spontaneously Invasive

When the fake news item is able to simultaneously generate more than 15% of contrarians and be in tune with most activated prejudices, only a handful of initial believers is sufficient to launch an invasive dynamics of opinion. Then, the proportion of supporters grows unseen till reaching eventually a majority of agents despite being initially dismissed as false by the majority of the community as shown in Figure 9,

It may also happen, although quite rarely, that a fake news post is believed at once as true by almost everyone [116]. When that happens, if the fake news item is in tune with the activated prejudices, it stays widely believed through local discussions and settled as "true" within the population.

5.3. Novel Strategies to Curb Invasive Fake News

Indeed, the above findings and results indicate that to prevent malicious fake news from thwarting the outcomes of balanced democratic public debates about central societal and political issues. The focus could be not on banning the making of fake news but on either preventing their spread or driving their downsizing.

As long as a fake news item stays confined to a quite small group of agents it does not pose a threat. Therefore, instead of limiting the freedom of speech with heavy control of social media, it may be more efficient to address the geometry of the landscape of opinion, which drives the propagation of fake news. The subsequent goal becomes to identify the features that keep most fake news confined and then apply them to the ones that do spread over. Accordingly, I advocate setting the conditions for sequestrating invasive fake news within quite small networks of agents. In the rare cases where the initial proportion of believers is quite high, the aim is to reduce the initial support to quite small values and to keep it there.

Within the extended Galam model of opinion dynamics, two parameters can fulfill this goal of "No ban, No spread—with Sequestration". The first one is k, the spectrum of prejudices activated spontaneously when groups of discussing agents become trapped at a tie, with the two choices, true and fake, being equally acceptable. The second one is x, the proportion of contrarian agents, which is produced by both the content of the fake news item and the social environment of the agents.

The values of these two parameters determine the geometry of the landscape in which the fake news dynamics are deployed. When the geometry boosted a fake news propagation, its reshaping requests to modify these values to set them at quite low values to activate the sequestration of fake news with no possibility of spreading out. More specifically, the proportion of contrarians must be reduced, which implies individual changes of attitude. In addition, tie-breaking cases must be tuned against the fake news item, which requires modifying the associated prejudices. In the case of an initial high proportion of believers, turning k from quite high to quite low values triggers a shrinking dynamic.

Once this avenue is selected, the follow-up instrumental question is how to implement such a reshaping scheme within the real world. That requires elaborating appropriate novel tools to modify k and x in order to reshape the social geometry of the landscape of the dynamics of fake news.

However, defining the appropriate corresponding tools is beyond my skills and expertise. At this stage, I have shown that the scheme of "No Ban, No Spread—with Sequestration" is feasible within a stylized model of opinion dynamics. To transpose the related findings to the real world requires an interdisciplinary collaboration with scientists from computer, psychological, cognitive, and behavioral sciences. I hope this paper will stimulate in these communities an interest along this point of view.

6. Conclusions

In this paper, I addressed the issue of how to curb the spread of fake news by exploring a new avenue denoted "No Ban, No Spread—with Sequestration", which aims both to preserve full freedom of speech and neutralize the impact of fake news.

To illustrate my proposal, I used the sociophysics Galam model of opinion dynamics focusing on the impact of having simultaneously contrarians when prejudice tie breaking is activated. These two parameters are independent of each other. Solving the update equations has revealed the existence of a critical value for the proportion of contrarians, above which the dynamics landscape is turned upside down. There, the dynamics shift abruptly from tipping-point dynamics to single attractor dynamics. Remarkably, the related contrarian critical value is small with values between 6% and 20%.

In parallel, the associated single attractor is found to be located at either relatively high or quite low proportions of fake news believers. The selection depends on the fake news item being in tune or at odds with the main activated prejudices. A high value means the fake news item does eventually invade the social space, whatever its initial support, even with only a handful of agents. On the contrary, a quite low value means the fake ends up sequestrated into a quite small proportion of believers, even if it starts from an initial high support.

These results are instrumental in determining the fate of fake news. Unveiling the social geometry of the opinion landscape has shown that most pieces of fake news never reach significant proportions of believers and stay confined to quite small numbers of agents. Therefore, they pose no threat.

There is, however, a subgroup of fake news that simultaneously generates sufficient contrarians and is in tune with most activated prejudices. Those items of fake news do spread and invade large parts of a social community, even when initially shared by only a handful of agents. The phenomenon is counter-intuitive and unexpected, highlighting an alternative understanding of the "natural" spread of fake news.

At this point, it is worth noticing that while my study is performed using groups of size 4, the main results are still valid when accounting for a combination of group sizes, including larger sizes (weaker tie-breaking effect) and pairs of people (stronger tie-breaking effect) [109]. Moreover, although social media groups may be quite large when considering a given thread of discussion, most contributors are likely to read only a few earlier comments, keeping quite small the effective size of the group. For future work, investigating the effect of networks and the influence of external factors like competence could be of interest.

The above findings allow envisioning new targeted paths to tackle only those threatening pieces of invasive fake news by sequestrating them via a modification of the underlying social geometry to prevent their otherwise propagation towards a majority of believers. Neither ban nor control of the net is required, allowing both full freedom of speech and the guarantee of getting the threatening fake news items naturally sequestrated into limited social networks involving only quite small numbers of agents. However, it is of importance to stress that allowing the existence of fake news per se does not mean allowing speech of hate, which is forbidden by law and thus must be prosecuted and condemned.

To conclude, I would like to stress that I am aware that there is a significant gap between my findings in the context of a stylized model and the implementation of these findings in the real world of social media. My purpose is to open up a new hypothetical direction for curbing the spread of fake news, hoping that experts working on the ground could use the proposal to eventually design new appropriate protocols. In particular, a discussion within the sociophysics community to tackle the challenge of possible real-life implementation would be particularly useful.

Funding: This research received no external funding.

Data Availability Statement: No data have been used in this study.

Acknowledgments: I thank Kevin Arceneaux for valuable comments about the earlier draft of this paper.

Conflicts of Interest: The author declares no conflicts of interest.

References

1. Pennycook, G.; Rand, D.G. The implied truth effect: Attaching warnings to a subset of fake news stories increases perceived accuracy of stories without warnings. *Manag. Sci.* **2018**, *67*, 4944–4957.
2. Lewandowsky, S.; Ecker, U.K.H.; Cook, J. Beyond misinformation: Understanding and coping with the post-truth era. *J. Appl. Res. Mem. Cogn.* **2017**, *6*, 353–369. [CrossRef]
3. Petersen, M.B.; Osmundsen, M.; Arceneaux, K. The "Need for Chaos" and motivations to share hostile political rumors. *Am. Political Sci. Rev.* **2023**, *117*, 1486–1505. [CrossRef]
4. Arceneaux, K.; Gravelle, T.B.; Osmundsen, M.; Petersen, M.B.; Reifler, J.; Scotto, T.J. Some people just want to watch the world burn: The prevalence, psychology and politics of the "Need for Chaos". *Philos. Trans. R. Soc. B* **2021**, *376*, 20200147. [CrossRef] [PubMed]
5. Grinberg, N.; Joseph, K.; Friedland, L.; Swire-Thompson, B.; Lazer, D. Fake news on twitter during the 2016 U.S. presidential election. *Science* **2019**, *363*, 374–378. [CrossRef] [PubMed]

6. Carter-Ruck. Insights Hub. Fake News, Authentic Views. 2019. Carter-Ruck: London, UK, 2019. Available online: https://www.carter-ruck.com/insight/fakes-news-authentic-views/ (accessed on 11 May 2024).
7. Edwards, L. How to Regulate Misinformation. *The Royal Society. Blog*, 25 January 2022. Available online: https://royalsociety.org/blog/2022/01/how-to-regulate-misinformation/ (accessed on 11 May 2024).
8. Tan, C. Regulating disinformation on Twitter and Facebook. *Griffith Law Rev.* **2022**, *31*, 513–536. [CrossRef]
9. Woodward, B.; Bernstein, C. *All the President's Men*; Simon and Schuster: New York, NY, USA, 1974. Available online: https://archive.org/details/allpresidentsmen0000bern (accessed on 11 May 2024).
10. Jones, J.H. *Bad Blood: The Tuskegee Syphilis Experiment*; The Free Press/Macmillan Publishing Co, Inc.: New York, NY, USA, 1981. Available online: https://archive.org/details/badbloodtuskegee00jone (accessed on 11 May 2024).
11. Tower, J.; Muskie, E.; Scowcroft, B. *The Tower Commission Report: The Full Text of the President's Special Review Board*; Bantham Books, Inc.; Times Books, Inc.: New York, NY, 1987. Available online: https://archive.org/details/towercommission00unit/ (accessed on 11 May 2024).
12. Sullivan, M. Elon Musk's hypocrisy about free speech hits a new low. *The Guardian*, 7 September 2023. Available online: https://www.theguardian.com/commentisfree/2023/sep/07/elon-musks-hypocrisy-about-free-speech-hits-a-new-low (accessed on 11 May 2024).
13. Nover, S. Elon Musk is finally fighting a genuine free speech battle. *Quartz*, 13 September 2023. Available online: https://qz.com/elon-musk-is-finally-fighting-a-genuine-free-speech-bat-1850833829 (accessed on 11 May 2024).
14. Milmo, D. Anti-hate speech group accuses Elon Musk's X Corp of intimidation over legal threat. *The Guardian*, 31 July 2023. Available online: https://www.theguardian.com/technology/2023/jul/31/anti-hate-speech-group-accuses-elon-musk-x-corp-intimidation (accessed on 11 May 2024).
15. Brazil, R. The physics of public opinion. *Phys. World* **2020**, 33, 24–28. [CrossRef]
16. Schweitzer, F. Sociophysics. *Phys. Today* **2018**, *71*, 40–47. [CrossRef]
17. Galam, S. *Sociophysics: A Physicist's Modeling of Psycho-Political Phenomena*. Springer Science + Business Media, LLC: New York, NY, USA, 2012. [CrossRef]
18. Chakrabarti, B.K.; Chakraborti, A.; Chatterjee, A. (Eds.) *Econophysics and Sociophysics: Trends and Perspectives*; Wiley-VCH Verlag GmbH & Co. KGaA: Weinheim, Germany, 2006.
19. Galam, S. Physicists, non physical topics, and interdisciplinarity. *Front. Phys.* **2022**, *10*, 986782. [CrossRef]
20. da Luz, M.G.E.; Anteneodo, C.; Crokidakis, N.; Perc, M. Sociophysics: Social collective behavior from the physics point of view. *Chaos Solitons Fractals* **2023**, *170*, 113379. [CrossRef]
21. Mobilia, M. Polarization and consensus in a voter model under time-fluctuating influences. *Physics* **2023**, *5*, 517–536. [CrossRef]
22. Zheng, S.; Jiang, N.; Li, X.; Xiao, M.; Chen, Q. Faculty hiring network reveals possible decision-making mechanism. *Physics* **2023**, *5*, 851–861. [CrossRef]
23. Filho, E.A.; Lima, F.W.; Alves, T.F.A.; Alves, G.d.; Plascak, J.A. Opinion dynamics systems via Biswas–Chatterjee–Sen model on Solomon networks. *Physics* **2023**, *5*, 873–882. [CrossRef]
24. Oestereich, A.L.; Pires, M.A.; Queirós, S.M.D.; Crokidakis, N. Phase transition in the Galam's majority-rule model with information-mediated independence. *Physics* **2023**, *5*, 911–922. [CrossRef]
25. Ellero, A.; Fasano, G.; Favaretto, D. Mathematical programming for the dynamics of opinion diffusion. *Physics* **2023**, *5*, 936–951 [CrossRef]
26. Malarz, K.; Masłyk, T. Phase diagram for social impact theory in initially fully differentiated society. *Physics* **2023**, *5*, 1031–1047 [CrossRef]
27. Li, S.; Zehmakan, A.N. Graph-based generalization of Galam model: Convergence time and influential nodes. *Physics* **2023**, *5*, 1094–1108. [CrossRef]
28. Ghosh, A.; Chakrabarti, B.K. Do successful researchers reach the self-organized critical point? *Physics* **2024**, *6*, 46–59. [CrossRef]
29. Mori, S.; Nakamura, S.; Nakayama, K.; Hisakado, M. Phase transition in ant colony optimization. *Physics* **2024**, *6*, 123–137. [CrossRef]
30. Kaufman, M.; Kaufman, S.; Diep, H.T. Social depolarization: Blume–Capel model. *Physics* **2024**, *6*, 138–147. [CrossRef]
31. Ausloos, M.; Rotundo, G.; Cerqueti, R. A Theory of best choice selection through objective arguments grounded in linear response theory concepts. *Physics* **2024**, *6*, 468–482. [CrossRef]
32. Llabrès, J.; Oliver-Bonafoux, S.; Anteneodo, C.; Toral, R. Aging in some opinion formation models: A comparative study. *Physics* **2024**, *6*, 515–528. [CrossRef]
33. Caticha, N.; Calsaverini, R.S.; Vicente, R. Statistical mechanics of social hierarchies: A mathematical model for the evolution of human societal structures. *Physics* **2024**, *6*, 629–644. [CrossRef]
34. Merlone, U.; Forno, A.D. The influence of lobbies: Analyzing group consensus from a physics approach. *Physics* **2024**, *6*, 659–673. [CrossRef]
35. Deffuant, G. Complex transitions of the bounded confidence model from an odd number of clusters to the next. *Physics* **2024**, *6*, 742–759. [CrossRef]
36. Demming, A. The laws of division. *Phys. World* **2024**, *37*, 37–41. [CrossRef]
37. Vilone, D.; Polizzi, E. Modeling opinion misperception and the emergence of silence in online social system. *PLoS ONE* **2024**, *19*, e0296075. [CrossRef]

38. Cui, P.-B. Exploring the foundation of social diversity and coherence with a novel attraction-repulsion model framework. *Physica A* **2023**, *618*, 128714. [CrossRef]
39. Liu, W.; Wang, J.; Wang, F.; Qi, K.; Di, Z. The precursor of the critical transitions in majority vote model with the noise feedback from the vote layer. *arXiv* **2023**, arXiv:2307.11398. [CrossRef]
40. Banisch, S.; Shamon, H. Validating argument-based opinion dynamics with survey experiments. *J. Artif. Soc. Soc. Simul. (JASSS)* **2024**, *27*, 17. [CrossRef]
41. Pal, R.; Kumar, A.; Santhanam, M.S. Depolarization of opinions on social networks through random nudges. *Phys. Rev. E* **2023**, *108*, 034307. [CrossRef] [PubMed]
42. Bagarello, F. Phase transitions, KMS condition and decision making: An introductory model. *Philos. Trans. R. Soc. A Math. Phys. Engin. Sci.* **2023**, *381*, 20220377. [CrossRef] [PubMed]
43. Galam, S.; Mauger, A. On reducing terrorism power: A hint from physics. *Phys. A Stat. Mech. Appl.* **2003**, *323*, 695–704. [CrossRef]
44. Crokidakis, N. Radicalization phenomena: Phase transitions, extinction processes and control of violent activities. *Int. J. Mod. Phys. C* **2023**, *34*, 2350100. [CrossRef]
45. Coquidé, C.; Lages, J.; Shepelyansky, L.D. Prospects of BRICS currency dominance in international trade. *Appl. Netw. Sci.* **2023**, *8*, 65. [CrossRef]
46. Mulyaa, D.A.; Muslim, R. Phase transition and universality of the majority-rule model on complex networks. *Int. J. Mod. Phys. C* **2024**, *in print*. [CrossRef]
47. Neirotti, J.; Caticha, N. Rebellions and impeachments in a neural network society. *arXiv* **2023**, arXiv:2309.11188. [CrossRef]
48. Shen, C.; Guo, H.; Hu, S.; Shi, L.; Wang, Z.; Tanimoto, J. How committed individuals shape social dynamics: A survey on coordination games and social dilemma games. *Eur. Phys. Lett. (EPL)* **2023**, *144*, 11002. [CrossRef]
49. Grabisch, M.; Li, F. Anti-conformism in the threshold model of collective behavior. *Dyn. Games Appl.* **2020**, *10*, 444–477. [CrossRef]
50. Forgerini, F.L.; Crokidakis, N.; Carvalho, M.A.V. Directed propaganda in the majority-rule model. *Int. J. Mod. Phys. C* **2024**, *in print*. [CrossRef]
51. Crokidakis, N. Recent violent political extremist events in Brazil and epidemic modeling: The role of a SIS-like model on the understanding of spreading and control of radicalism. *Int. J. Mod. Phys. C* **2024**, *35*, 2450015. [CrossRef]
52. Huang, C.; Bian, H.; Han, W. Breaking the symmetry neutralizes the extremization under the repulsion and higher order interactions. *Chaos Solitons Fractals* **2024**, *180*, 114544. [CrossRef]
53. Azhari, A.; Muslim, R.; Mulya, D.A.; Indrayani, H.; Wicaksana, C.A.; Rizky, A. Independence role in the generalized Sznajd model. *arXiv* **2023**, arXiv:2309.13309. [CrossRef]
54. Naumisa, G.G.; Samaniego-Steta, F.; del Castillo-Mussot, M.; Vázquez, G.J. Three-body interactions in sociophysics and their role in coalition forming. *Phys. A* **2007**, *379*, 226–234. [CrossRef]
55. Galam, S. The invisible hand and the rational agent are behind bubbles and crashes. *Chaos Solitons Fractals* **2016**, *88*, 209–217. [CrossRef]
56. Hamann, H. Opinion dynamics with mobile agents: Contrarian effects by spatial correlations. *Front. Robot. AI* **2018**, *5*, 63. [CrossRef] [PubMed]
57. Guo, L.; Cheng, Y.; Luo, Z. Opinion dynamics with the contrarian deterministic effect and human mobility on lattice. *Complexity* **2015**, *20*, 43–49. [CrossRef]
58. Tiwari, M.; Yang, X.; Sen, S. Modeling the nonlinear effects of opinion kinematics in elections: A simple Ising model with random field based study. *Phys. A* **2021**, *582*, 126287. [CrossRef]
59. Chacoma, A.; Zanette, D.H. Critical phenomena in the spreading of opinion consensus and disagreement. *Pap. Phys.* **2014**, *6*, 060003. [CrossRef]
60. Cheon, T.; Morimoto, J. Balancer effects in opinion dynamics. *Phys. Lett. A* **2016**, *380*, 429–434. [CrossRef]
61. Landry, N.W.; Restrepo, J.G. Opinion disparity in hypergraphs with community structure. *Phys. Rev. E* **2023**, *108*, 034311. [CrossRef]
62. Mulyaa, D.A.; Muslim, R. Destructive social noise effects on homogeneous and heterogeneous networks: Induced-phases in the majority-rule model. *Int. J. Mod. Phys. C* **2024**, *in print*. [CrossRef]
63. Mihara, A.; Ferreira, A.A.; Martins, A.C.R.; Ferreira, F.F. Critical exponents of master-node network model. *Phys. Rev. E* **2023**, *108*, 054303. [CrossRef] [PubMed]
64. Gärtner, B.; Zehmakan, A.N. Threshold behavior of democratic opinion dynamics. *J. Stat. Phys.* **2020**, *178*, 1442–1466. [CrossRef]
65. Toth, G.; Galam, S. Deviations from the majority: A local flip model. *Chaos Solitons Fractals* **2022**, *159*, 112130. [CrossRef]
66. Kowalska-Styczeń, A.; Malarz, K. Noise induced unanimity and disorder in opinion formation. *PLoS ONE* **2020**, *15*, e0235313. [CrossRef] [PubMed]
67. N. Crokidakis, Nonequilibrium phase transitions and absorbing states in a model for the dynamics of religious affiliation. *Phys. A Stat. Mech. Appl.* **2024**, 643, 129820. [CrossRef]
68. Redner, S. Reality-inspired voter models: A mini-review. *C. R. Phys.* **2019**, *20*, 275–292. [CrossRef]
69. Galam, S.; Moscovici, S. Towards a theory of collective phenomena. III: Conflicts and forms of power. *Eur. J. Soc. Psychol.* **1995**, *25*, 217–229. [CrossRef]
70. Jedrzejewski, A.; Marcjasz, G.; Nail, P.R. Sznajd-Weron, K. Think then act or act then think? *PLoS ONE* **2018**, *13*, e0206166. [CrossRef]

71. Singh, P.; Sreenivasan, S.; Szymanski, B.K.; Korniss, G. Competing effects of social balance and influence. *Phys. Rev. E* **2016**, *93*, 042306. [CrossRef] [PubMed]
72. Bagnoli, F.; Rechtman, R. Bifurcations in models of a society of reasonable contrarians and conformists. *Phys. Rev. E* **2015**, *92*, 042913. [CrossRef] [PubMed]
73. Carbone, G.; Giannoccaro, I. Model of human collective decision-making in complex environments. *Eur. Phys. J. B* **2015**, *88*, 339. [CrossRef]
74. Sznajd-Weron, K.; Szwabiński, J.; Weron, R. Is the Person-situation debate important for agent-based modeling and vice-versa? *PLoS ONE* **2014**, *9*, e112203. [CrossRef] [PubMed]
75. Florian, R.; Galam, S. Optimizing conflicts in the formation of strategic alliances. *Eur. Phys. J. B* **2000**, *16*, 189–194. [CrossRef]
76. Javarone, M.A. Networks strategies in election campaigns. *J. Stat. Mech.* **2014**, *2014*, P08013. [CrossRef]
77. Javarone, M.A.; Singh, S.P. Strategy revision phase with payoff threshold in the public goods game. *J. Stat. Mech.* **2024**, *2024*, 023404. [CrossRef]
78. Goncalves, S.; Laguna, M.F.; Iglesias, J.R. Why, when, and how fast innovations are adopted. *Eur. Phys. J. B* **2012**, *85*, 192. [CrossRef]
79. Ellero, A.; Fasano, G.; Sorato, A. A modified Galam's model for word-of-mouth information exchange. *Phys. A Stat. Mech Appl.* **2009**, *388*, 3901–3910. [CrossRef]
80. Gimenez, M.C.; Reinaudi, L.; Vazquez, F. Contrarian voter model under the influence of an oscillating propaganda: Consensus, bimodal behavior and stochastic resonance. *Entropy* **2022**, *24*, 1140. [CrossRef]
81. Iacominia, E.; Vellucci, P. Contrarian effect in opinion forming: Insights from Greta Thunberg phenomenon. *J. Math. Sociol.* **2023**, *47*, 123–169. [CrossRef]
82. Brugnoli, E.; Delmastro, M. Dynamics of (mis)information flow and engaging power of narratives. *arXiv* **2022**, arXiv:2207.12264. [CrossRef]
83. Sobkowicz, P.; Now, W. Opinion modelers? *Front. Phys.* **2020**, *8*, 587009 . [CrossRef]
84. Weron, T.; Nyczka, P.; Szwabiński, J. Composition of the influence group in the *q*-voter model and its impact on the dynamics of opinions. *Entropy* **2024**, *26*, 132. [CrossRef]
85. Gsänger, M.; Hösel, V.; Mohamad-Klotzbach, C.; Müller, J. Opinion models, data, and politics. *Entropy* **2024**, *26*, 212. [CrossRef] [PubMed]
86. Shang, Y. Hybrid consensus for averager-copier-voter networks with non-rational agents. *Chaos Solitons Fractals* **2018**, *110*, 244–251. [CrossRef]
87. Mobilia, M.; Fixation and polarization in a three-species opinion dynamics model. *Eur. Phys. Lett.* **2011**, *95*, 50002. [CrossRef]
88. Calvaõ, A.M.; Ramos, M.; Anteneodo, C.; Role of the plurality rule in multiple choices. *J. Stat. Mech.* **2016**, *2016*, 023405. [CrossRef]
89. Maciel, M.V.; Martins, A.C.R.; Ideologically motivated biases in a multiple issues opinion model. *Phys. A Stat. Mech. Appl.* **2020**, *553*, 124293. [CrossRef]
90. Dworak, M.; Malarz, K. Vanishing opinions in Latané model of opinion formation. *Entropy* **2023**, *25*, 58. [CrossRef]
91. Galam, S.The drastic outcomes from voting alliances in three-party democratic voting (1990–2013). *J. Stat. Phys.* **2013**, *151*, 46–68. [CrossRef]
92. Ferri, I.; Gaya-Ávila, A.; Guilera, A.D. Three-state opinion model with mobile agents. *Chaos* **2023**, *33*, 093121. [CrossRef] [PubMed]
93. Tsintsaris, D.; Tsompanoglou, M.; Ioannidis, E. Dynamics of social influence and knowledge in networks: Sociophysics models and applications in social trading, behavioral finance and business. *Mathematics*, 2024, *12*, 1141. [CrossRef]
94. Coquidé, C.; Lages, J.; Shepelyansky, D.L. Opinion formation in the world trade network. *Entropy* **2024**, *26*, 141. [CrossRef] [PubMed]
95. Oliveira, I.V.G.; Wang, C.; Dong, G.; Du, R.; Fiore, C.E.; Vilela, A.L.M.; Stanley, H.E. Entropy production on cooperative opinion dynamics. *Chaos Solitons Fractals* **2024**, *181*, 114694. [CrossRef]
96. Segovia-Martin, J.; Rivero, R. Cross-border political competition. *PLoS ONE*, **2024**, *19*, e0297731 [CrossRef] [PubMed]
97. Zimmaro, F.; Galam, S.; Javarone, M.A. Asymmetric games on networks: Mapping to Ising models and bounded rationality. *Chaos Solitons Fractals*, **2024**, *181*, 114666. [CrossRef]
98. Macias, R.C.; Vera; Vera, J.M.R. Dynamics of opinion polarization in a population. *Math. Soc. Sci.* **2024**, *128*, 31–40. [CrossRef]
99. Battiston F.; Cairoli A.; Nicosia V.; Baule, A.; Latora V. Interplay between consensus and coherence in a model of interacting opinions. *Phys. D Nonlinear Phenom.* **2016**, *323–324*, 12–19. [CrossRef]
100. Ghosh, S.; Bhattacharya, S.; Mukherjee, S.; Chakravarty, S. Promote to protect: Data-driven computational model of peer influence for vaccine perception. *Sci. Rep.* **2024**, *14*, 306. [CrossRef]
101. Götz, T.; Krüger, T.; Niedzielewski, K.; Pestow, R.; Schäfer, M.; Schneider, J. Chaos in opinion-driven disease dynamics. *Entropy* **2024**, *26*, 298. [CrossRef] [PubMed]
102. Soares, J.P.M.; Fontanari, J.F. *N*-player game formulation of the majority-vote model of opinion dynamics. *Phys. A Stat. Mech. Appl.* **2024**, *643*, 129829. [CrossRef]
103. Zhang, H.-B.; Tang, D.-P. Effects of group size and noise on cooperation in population evolution of dynamic groups. *arXiv* **2024**, arXiv:2402.16317. [CrossRef]
104. Kononovicius, A.; Astrauskas, R.; Radavičius, M.; Ivanauskas, F. Delayed interactions in the noisy voter model through the periodic polling mechanism. *arXiv* **2024**, arXiv:2403.10277. [CrossRef]

105. Cao, W.; Zhang, H.; Kou, G.; Zhang, B. Discrete opinion dynamics in social networks with stubborn agents and limited information. *Inf. Fusion* **2024**, *109*, 102410. [CrossRef]
106. Maksymov, I.S.; Pogrebna, G. Quantum-mechanical modelling of asymmetric opinion polarisation in social networks. *Information* **2024**, *15*, 170. [CrossRef]
107. Behrens, F.; Hudcová, B.; Zdeborová, L. Dynamical phase transitions in graph cellular automata. *Phys. Rev. E* 2024, 109, 044312. [CrossRef] [PubMed]
108. Martins, A.C.R. Discrete opinion dynamics with *M* choices. *Eur. Phys. J. B* **2020**, *93*, 1. [CrossRef]
109. Galam, S.; Cheon, T. Tipping points in opinion dynamics: A universal formula in five dimensions. *Front. Phys.* **2020**, *8*, 446. [CrossRef]
110. Galam, S. Majority rule, hierarchical structures, and democratic totalitarianism: A statistical approach. *J. Math. Psychol.* **1986**, *30*, 426–434. [CrossRef]
111. Galam, S.; Chopard, B.; Masselot, A.; Droz, M. Competing species dynamics: Qualitative advantage versus geography. *Eur. Phys. J. B* **1998**, *4*, 529–531. [CrossRef]
112. Galam, S. Minority opinion spreading in random geometry. *Eur. Phys. J. B* **2002**, *25*, 403–406. . [CrossRef]
113. Galam, S. Heterogeneous beliefs, segregation, and extremism in the making of public opinions. *Phys. Rev. E* **2005**, *71*, 046123. [CrossRef] [PubMed]
114. Galam, S. Contrarian deterministic effects on opinion dynamics: "The hung elections scenario". *Phys. A Stat. Mech. Appl.* **2004**, *333*, 453–460. [CrossRef]
115. Galam, S. Unanimity, coexistence, and rigidity: Three sides of polarization. *Entropy* **2023**, 25, 622. [CrossRef]
116. Greenblatt, J.A. ADL: Media helped spread blood libel against Israel, jews in Gaza hospital news coverage. *USA Today Opinion*, 22 October 2023. Available online: https://eu.usatoday.com/story/opinion/2023/10/22/gaza-hospital-explosion-hamas-israel-war-misinformation/71246499007/ (accessed on 11 May 2024).

physics MDPI

Article

Complex Transitions of the Bounded Confidence Model from an Odd Number of Clusters to the Next

Guillaume Deffuant [1,2]

[1] LISC, Inrae, Université Clermont-Auvergne, 63178 Aubière, France; guillaume.deffuant@inrae.fr
[2] LAPSCO, Université Clermont-Auvergne, 63000 Clermont-Ferrand, France

Abstract: The bounded confidence model assumes simple continuous opinion dynamics in which agents ignore opinions which are too far from their own. The two initial variants—Hegselmann–Krause (HK) and Deffuant–Weisbuch (DW) of the model have attracted significant attention since the early 2000s. This paper revisits the version of the HK model applied to a probability distribution, earlier studied by Jan Lorenz. It shows that the bifurcation diagram depends on the parity of the size of the discretisation and that adding a small noise to the initial conditions leads to complex transitions involving several phases.

Keywords: bounded confidence model; agent and distribution models; bifurcation diagram

Citation: Deffuant, G. Complex Transitions of the Bounded Confidence Model from an Odd Number of Clusters to the Next. *Physics* **2024**, *6*, 742–760. https://doi.org/10.3390/physics6020046

Received: 8 January 2024
Revised: 19 February 2024
Accepted: 4 March 2024
Published: 8 May 2024

1. Introduction

The field of sociophysics produces a wide variety of models and invests a lot of effort into their study. These models are new conceptual objects lying at the boundary between mathematics, physics and computer science. From the early models of James Sakoda [1] and Thomas Schelling [2], the aim was to instantiate the complex interactions between individual and collective dynamics or between psychology and sociology [3], that physicists tend to call the interactions between micro and macro behaviours. Serge Galam illustrates this point with his extensive production of models about diverse social dynamics such as pyramidal vote, minority opinion diffusion, the diffusion of terrorist activities and the formation of coalitions and uses a variety of tools from physics to study them (for instance, see a review in Ref. [4]). The models used in the field indeed often require sustained theoretical efforts to fully formalise the micro–macro dynamics (see for instance the voter model [5,6] or the Axelrod model of culture [7,8] or the CODA opinion and action model [9]). This paper presents such an effort to understand the details of the collective dynamics emerging from specific models that can be related to sociophysics: bounded confidence models of opinions.

The idea the bounded confidence model, namely of agents that are influenced by others' opinions only when these are close enough to their own, appeared almost simultaneously in two variants in the early 2000s. The Hegselmann–Krause (HK) variant [10], inaugurating the term bounded confidence to designate the model, was motivated by the dynamics of opinions within groups of experts, and assumes that the agents consider the opinions of all of the others when revising their own. The Deffuant–Weisbuch (DW) variant [11,12] was initially aimed at modelling exchanges among farmers and assumes that agents revise their attitude when discussing with a single randomly chosen other.

Running these models with agents that hold the same confidence bound leads to one or more separate clusters of opinions, depending on the value of the confidence bound. A number of papers are devoted to studying these models and their variants [13]. Early papers study the model in which all agents have the same confidence bound, showing the details of the bifurcation diagrams when model parameters change. Other researchers investigate populations with different confidence bounds. Several studies focus on including so-called extremists agents, whose opinion is at the border of the opinion interval [14–16]. Others consider agents with confidences drawn in a given interval [17]. Many papers assess the effect of different types of networks of interactions [18–20], and in some of them, the topology changes with rules depending on the opinion dynamics [21,22]. Introducing noise into these models also significantly modifies their qualitative behaviour [23–25]. Some reviews are fully or partly devoted to these models and their variations [13,26,27].

This paper focuses on a specific theoretical approach, initially proposed in Ref. [28] to study the DW variant and then by Jan Lorenz [29] to study the HK variant. The approach uses a probability distribution of the opinions instead of a discrete set of opinions. Therefore, it approximates the behaviour of the model starting from a perfectly uniform initial distribution and an infinity of agents. Reference [28] identifies several features of the model that were not well apparent when running it in its agent version. The main feature is the periodicity of the model which is made apparent by changing the commonly used variables of representation. Another feature is the systematic presence of smaller clusters, called minor or secondary clusters, in addition to the usual major or primary clusters.

Following Lorenz [29], this paper studies the HK bounded confidence in distribution. As discussed by Lorenz, the utility of a distribution version of the HK model is not straightforward as the discretised initial uniform probability distribution is equivalent to starting with initially equally spaced agents, a case which has been the subject of several papers [10,30]. Moreover, the discretisation introduces approximation errors, whereas a computation with rational numbers yields an exact result. However, the distribution approach is more efficient computationally, making systematic simulation studies easier. With this approach, Reference [29] shows that, with the HK model, the number of clusters does not change monotonously when the bound decreases. Indeed, generally, a central cluster reappears instead of two clusters equally distant from the center. Lorenz notices long convergence times, indicating metastable states in this case.

This study replicates the approach of Ref. [29] with several differences: first, this study uses the same axes of representation as in Ref. [28], in order to make more apparent a possible periodicity. Second, this study uses both odd and even size of discretisation of the opinion axis. The results indeed show a striking difference in the behaviour of the model when the parity of the discretisation size changes. Finally, this paper investigates the effect of adding a small noise to the initial distribution. This leads to complex transitions between an odd number of clusters and the next, involving different phases.

2. The Models and Their Approximation with Distributions

2.1. The Two Variants of the BC Model

Let us consider the initial models, assuming a total mixing of the agents all sharing the same confidence bound ϵ. The two variants of the Bounded Confidence model differ in the schedule they use for agents to update their opinions. In the HK variant, all the agents update their opinions simultaneously, considering all of the agents with close enough opinions (see Algorithm 1). Importantly, the HK variant is deterministic. By contract, in the DW variant, at each time step, a couple of agents' opinions (a_i, a_j) is randomly chosen, and if the distance between opinions is below the confidence bound ϵ, then the opinion a_i is updated by its influence on a_j (see Algorithm 2). Note that, in the original version [11], both agents update their opinions. This change is actually not much consequential, as it does not change the general behaviour of the model. In both cases, we assume that the opinions are updated until a convergence criterion is reached. Indeed, in both cases the agents ultimately form clusters that are at more than ϵ-distance one from another.

A noticeable difference between the algorithms is that an iteration of the HK model updates all the opinions while an iteration of the DW model updates only a randomly chosen couple of opinions.

Algorithm 1: `runHK`($[a_1, \ldots, a_n]$ (initial opinions), ϵ (confidence bound))

```
while convergence ([a_1,...,a_n]) = FALSE do
    for i ← 1 to n do
        â_i ← average({a_j, |a_i − a_j| < ε})          // average of close enough
          opinions;
    end
    for i ← 1 to n do
        a_i ← â_i            // all opinions change simultaneously;
    end
end
```

Algorithm 2: `runDW`($[a_1, \ldots, a_n]$ (initial opinions), ϵ (confidence bound))

```
while convergence ([a_1,...,a_n]) = FALSE do
    (a_i, a_j) ← random(2, [a_1,...,a_n])        // random distinct couple;
    if |a_i − a_j| < ε then
        a_i ← (a_i + a_j)/2
    end
end
```

2.2. Distribution Version

Let us consider a probability distribution of agents' opinions of density ρ on the opinion range $[0,1]$, or more conveniently $[-1,1]$. Theoretically this density is continuously defined over this range. Computers cannot manipulate such an object and we approximate the continuous density by a vector of size n_g on a discrete subset of the opinion range. We start with a uniform distribution, hence all the values in the vector ρ_0 are equal to $1/n_g$.

In general, physicists determine the evolution of the distribution over time by writing a master equation, which expresses the balance between the density flows coming and leaving at each discrete location of the discrete grid. Here, I use a method (already used in Ref. [15]) that is equivalent but can be seen as more straightforward. Indeed, it applies the model directly on the discrete sites, stores all the results in a buffer vector Δ and once all the sites are completed, it updates distribution ρ by adding Δ to it. Actually, in the end, Δ includes the results of the master equation at the considered time step. In the case of the HK model, instead of computing a difference, we directly compute the new distribution $\widehat{\rho}$ and then replace ρ with $\widehat{\rho}$.

As the opinion axis is discretised, the method always produces some errors, particularly when the new opinion value coming from the interaction of opinions is not located exactly at a discrete value. In particular, with the DW version, the method should transfer amounts of density from site i on the axis to the theoretical position $(i+j)/2$. When an integer k exists, such that $i+j=2k$, the whole δ is transferred to site k. However, if $i+j=2k+1$, then it would be arbitrary to transfer the whole δ to site k or to site $k+1$. Therefore, $\delta/2$ is transferred to each site k and $k+1$. Generalised to the HK model, the method defines the new position a where the amount of density δ should be transferred as a weighted average that can be located anywhere in a segment $[k,k+1]$. Then the method affects δ to sites k and $k+1$ proportionally to the proximity of a to the site. $a-k$ is the proximity of a to site $k+1$ and $k+1-a$ is the proximity of a to site k. Lorenz uses the same approximation in his distribution version of the HK model [29].

Then the investigation process includes, for different values of the confidence bound ϵ, running the model approximation until the distribution stabilises and then determining the positions and amplitude of the final clusters. With the chosen approximation, the convergence is ensured when there are only either a single or a couple of isolated sites with non-zero density; isolated meaning that there is no other non-zero site at a distance lower than n_ϵ. Algorithms 3 and 4 display the pseudo code of the method for the two variants of the BC model.

As noticed in Ref. [28], it is convenient to use an opinion range $[-1, 1]$ to vary $1/\epsilon$ instead of ϵ and a/ϵ instead of the opinion a because the DW model then shows a periodic behaviour in the vicinity of $a/\epsilon = 0$. Taking this into account, and following [28], we also define n_g as an integer close to $2n_\epsilon/\epsilon$ (number of discrete values describing the probability distribution ρ_t), where t is the number of iterations.

Algorithm 3: distributionRunDW(ϵ (confidence bound), n_ϵ (ϵ discretisation size))

```
n_g ← round(2n_e/ε)          // n_g opinion range discretisation size;
ρ ← [1/n_g, ..., 1/n_g]              // ρ vector of size n_g;
while convergence (ρ) = FALSE do
    Δ ← [0,...,0]                    // Δ vector of size n_g;
    for i ∈ {1,...,n_g} do
        m ← max(1, i − n_ε)       // minimum location interacting with i;
        M ← min(n_g, i + n_ε)     // maximum location interacting with i;
        for j ← m to M do
            if i + j = 2k then
                Δ[k] ← Δ[k] + ρ[i]ρ[j]
            end
            if i + j = 2k + 1 then
                Δ[k] ← Δ[k] + ρ[i]ρ[j]/2 ;
                Δ[k + 1] ← Δ[k + 1] + ρ[i]ρ[j]/2 ;
            end
            Δ[i] ← Δ[i] − ρ[i]ρ[j] ;
        end
    end
    ρ ← ρ + Δ
end
```

Algorithm 4: distributionHK(ϵ (confidence bound), n_ϵ (ϵ discretisation size))

```
n_g ← round(2n_e/ε)          // n_g opinion range discretisation size;
ρ ← [1/n_g, ..., 1/n_g]              // ρ vector of size n_g;
while convergence (ρ) = FALSE do
    ρ̂ ← [0,...,0]                    // ρ̂ vector of size n_g;
    for i ∈ {1,...,n_g} do
        m ← max(1, i − n_ε)       // minimum location interacting with i;
        M ← min(n_g, i + n_ε)     // maximum location interacting with i;
        a ← (Σ_{j=m}^{M} jρ[j]) / (Σ_{j=m}^{M} ρ_t[j])      // Average weighted by ρ_t;
        k ← int(a)                // int (a) integer part of a ;
        ρ̂[k] ← ρ̂[k] + (k + 1 − a)ρ[i] ;
        ρ̂[k + 1] ← ρ̂[k + 1] + (a − k)ρ[i] ;
    end
    ρ ← ρ̂
end
```

In order to evaluate the importance of a cluster, its mass (total distribution in the cluster) is divided by 2ϵ. If the result is higher than 0.5, the cluster is major. If it is lower than 0.01, it is minor. Reference [28] explains the emergence of minor clusters in the DW model qualitatively. Their explanation suggests that the HK model should not produce minor clusters.

2.3. *Dealing with Computational Instabilities: Forcing Symmetry and Adding Noise*

Lorenz notices [29] cases of computational instabilities in the HK distribution model for some values of the parameters, which are revealed by asymmetric results. Indeed, as the initial state and the rules of change are all symmetric, these asymmetries necessarily come from computational errors.

This study deals with these potential computational instabilities in two ways. Firstly, the model runs on only half of the distribution and then it copies these results by symmetry to the other half. In this way, computational instabilities lead to different outcomes that are always symmetric. Secondly, in order to better understand the behaviour of the model in these zones of computational instability, I compare the results to the ones obtained when adding a small amount of noise to the initial distribution. In my tests, the initial mass $1/n_g$ of each discrete site in the distribution becomes $(1+\delta)/n_g$, where δ is uniformly drawn in $[-0.001, 0.001]$. I only apply this noise in the segment $[-\epsilon, \epsilon]$, as the instabilities only occur in this part of the space.

3. Simulation Results

3.1. *DW Model*

Figure 1 shows the final clusters of the DW model for opinions in range $[-1, 1]$ and $1/\epsilon \in [2, 9]$. The colour code of site i is determined by $\rho(i)/2\epsilon$ where $\rho(i)$ is the mass at the site. For the major clusters, this value varies from 0.5 (green dots) to 1 and higher (red dots). The smaller values are represented as blue dots, shown by the plus symbols for quite low ones (lower than 0.01), representing the minor clusters.

Figure 1 replicates the results from Ref. [28] (also replicated by Lorenz [29], but not using the same axes) with systematic minor clusters and a clear periodicity. I find a discontinuity in the presence of the minor clusters that does not appear in Ref. [28], and their positions are slightly different. These differences are small and could come from a different convergence criterion (the simulations stop when the absolute value of the change vector is lower than 10^{-5}).

Figure 1 was obtained with only odd values of n_g, the size of the total discretisation. However, taking even values instead yields exactly the same results. Let us recall these results in order to contrast them with the ones obtained with the HK model, which are the focus of this paper.

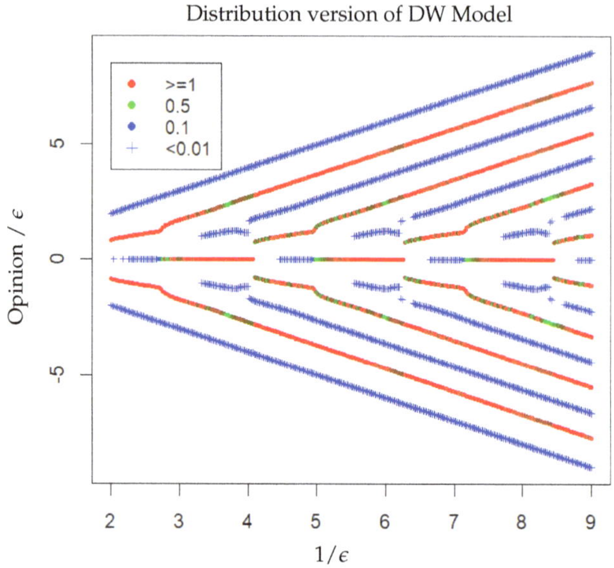

Figure 1. *Cont.*

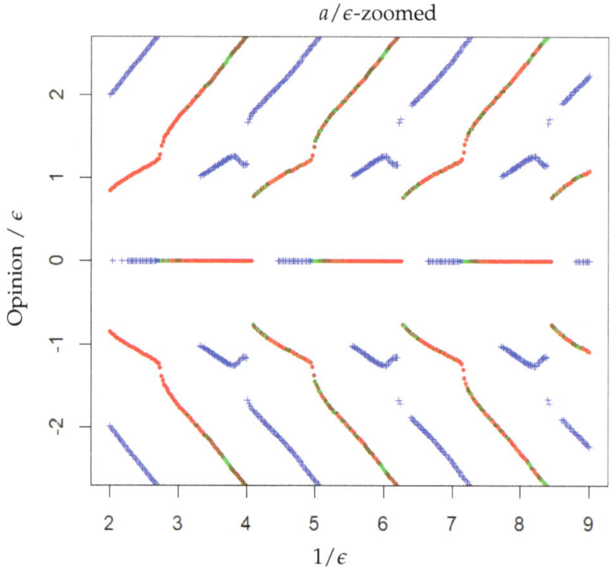

Figure 1. Positions of the clusters (a/ϵ) in the full (**upper**) and zoomed (**lower**) a/ϵ ranges for the opinion, a, range $[-1,1]$. The discretisation of ϵ is $n_\epsilon = 101$. The colors and symbols represent the values of the total mass of the cluster divided by 2ϵ as indicated. See text for details.

3.2. HK Model with Odd or Even Discretisation Size

Figure 2 shows the positions and types of clusters obtained with the HK distribution model, with $1/\epsilon$ varying from 2 to 13 (with an opinion range of $[-1,1]$), with odd (top) and even (bottom) discretisation sizes n_g. In both cases, the discretisation of ϵ is $n_\epsilon = 1001$. The results look periodic for both odd and even values of n_g, but they are quite different. Moreover, unlike the DW model, in which, periodically, one new central cluster or two new central clusters are created and remain stable, it looks that the HK model goes through periodic complex transitions with instabilities until two new major clusters stabilise, at about 2ϵ from the centre.

In the case of odd values of n_g (Figure 2, upper) in each transition, when $1/\epsilon$ increases, two clusters around a minor cluster in the middle appear briefly, replacing a central major cluster, and then there is a central cluster which is alone again, and finally two clusters around the central one, at a distance of about ϵ. This phenomenon was already observed by Lorenz [29], and highlighted in the title of his paper ("consensus strikes back"). However, the periodicity was less apparent in Lorenz's representation. The details of the three first transitions shown in Figure 3, left, suggest that this periodicity is not perfect. In the first and the third transitions, there are a few more values of $1/\epsilon$ leading to a couple of major clusters around the centre, where the equivalent values lead to a central cluster in the second transition. Moreover, the slope of the positions of the clusters appearing at the beginning of the transition looks to be smaller in the second transition than in the first, and even smaller in the third.

In the case of even values of n_g (Figure 2, lower), the transitions include more cases of convergence towards a couple of clusters around the centre. Moreover, the transitions are more complicated as can be seen in more details in Figure 3. The central cluster separates into two clusters which remain for more values of $1/\epsilon$ than when n_g is odd, and then the positions of these clusters move closer to the central cluster as $1/\epsilon$ increases. Moreover, the results then look unstable as there are cases where the central cluster reappears. Again, the transitions are not exactly the same at each period and the instability of the results may not be sufficient to explain all the differences. Indeed, there looks to be significantly more convergences to the central cluster in the third transition.

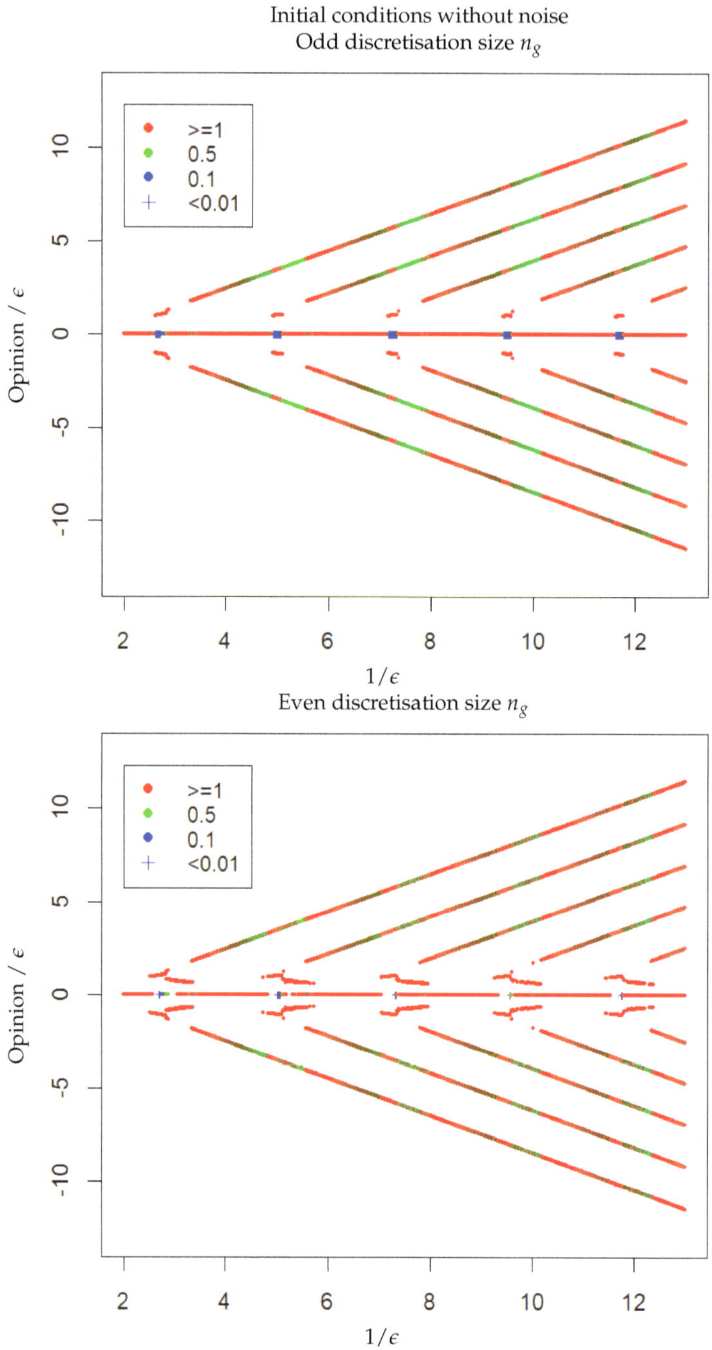

Figure 2. Positions of the clusters (a/ϵ) for the opinion range $[-1, 1]$) for the odd (**upper**) and bottm (**lower**) opinion range discretisation, n_g. ϵ discretisation size, $n_\epsilon = 1001$. The colors and symbols represent the values of the total mass of the cluster divided by 2ϵ as indicated.

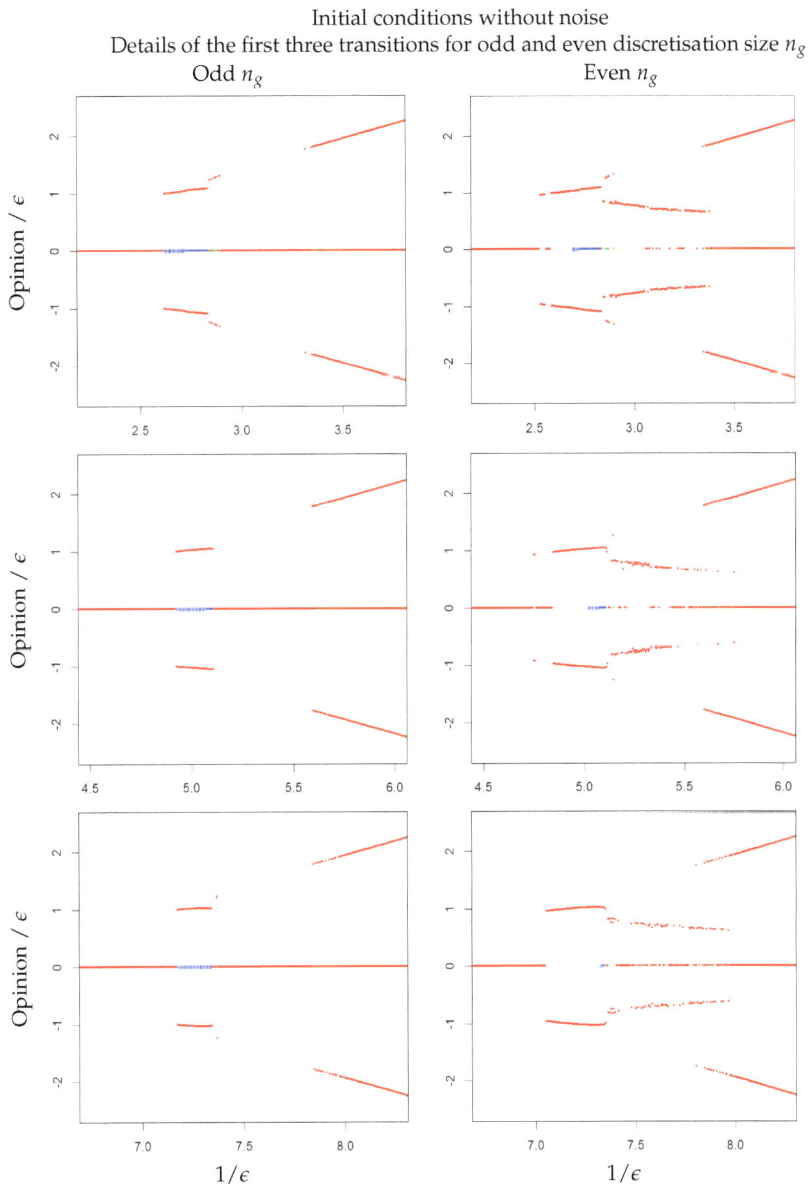

Figure 3. A closer look at the first three transitions for odd (**left**) and even (**right**) opinion discretisation sizes shown in Figure 2

The difference in behaviour between the odd and even discretization sizes is illustrated by examples of simulation with odd and even values of n_g in Figure 4. The simulations use a smaller discretisation size, $n_\epsilon = 15$ of ϵ, for $1/\epsilon = 3$, which is in the first transition, in order to make all the sites in the distribution visible, with an odd $n_g = 91$ in Figure 4, upper, and an even $n_g = 90$ in Figure 4, lower.

In both cases, a central major cluster, two major clusters at positions larger than $\pm\epsilon$ and two minor clusters located just below $\pm\epsilon$ appear after a few iterations. Then, the minor clusters quite slowly attract the two major clusters towards the centre, until the major clusters arrive at ϵ from the central cluster.

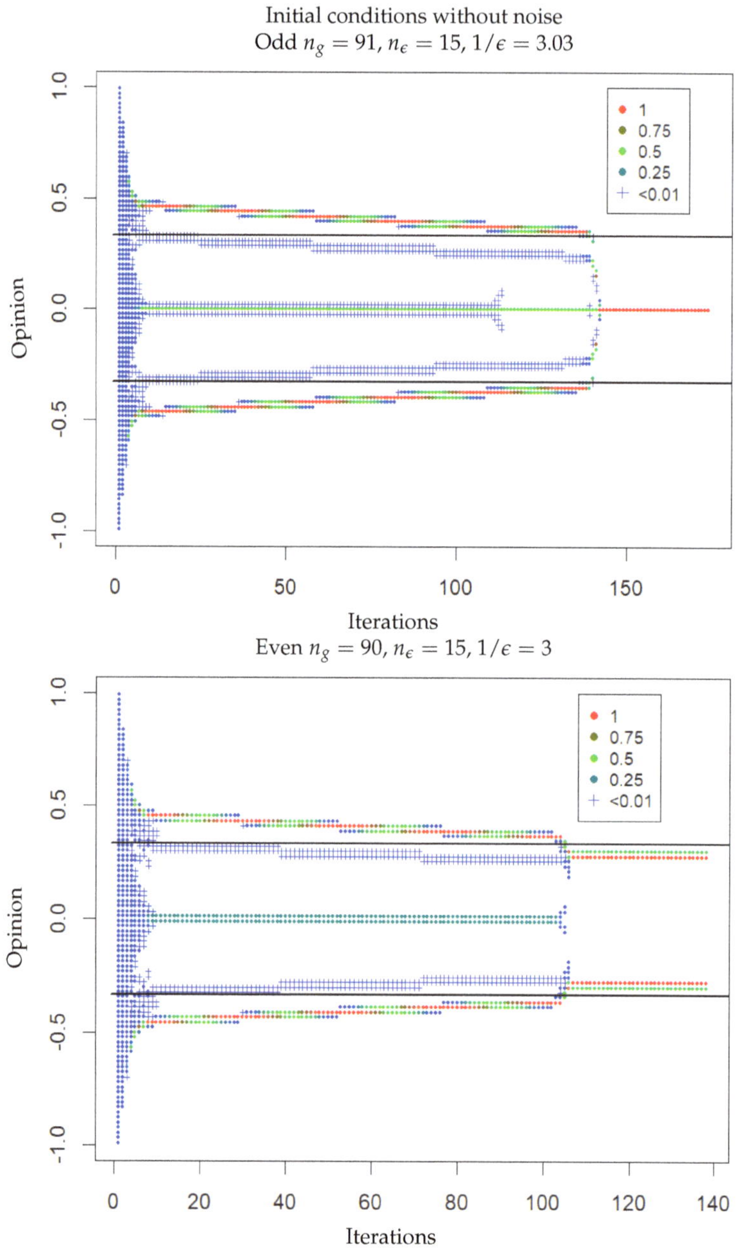

Figure 4. Examples of runs of HK distribution model for $1/\epsilon \approx 3$ and $n_\epsilon = 15$ when the total discretisation, n_g, is odd (**upper**) or even (**lower**). The two black lines are at $\pm\epsilon$. The colors and symbols represent the values of the total mass of the cluster divided by $2n_\epsilon$, as indicated.

Then, the process depends on the parity of n_g. If n_g is odd, the central cluster involves only one discrete site with a significant mass (with possibly very small residual masses in the sites just below and above). Then, the two major clusters at $\pm\epsilon$ are both attracted by the central cluster, yielding a single major central cluster. If n_g is even, the central cluster is made of two sites with a significant mass. Each of them is attracted by the closest of the two clusters situated at $\pm\epsilon$ and the final result is two major clusters at the same distance, smaller than ϵ, from the centre.

These examples clarify why central clusters are more likely with odd n_g and couples of clusters around the centre are more likely with even n_g.

It can be argued that a central cluster including only a one site with a significant mass is a better representation of the agent model (taking initially regularly spaced agents). Indeed, starting with two sites with significant masses defining the central cluster, the agent algorithm would progressively merge them. This justifies Lorenz's choice to privilege the odd discretisation size.

However, the situation becomes more complicated than the two simple cases shown in Figure 4 suggest. Indeed, these cases hide potential instabilities that can take place, which are more likely when the discretisation size n_g is larger. In order obtain a better view of these instabilities, let us now run the model when adding a small noise to the initial state.

3.3. HK Model with Noise on the Initial Conditions

Figure 5 shows the results of the HK model in distribution when adding a small noise to the initial distribution, as specified in Section 2.3. In this case, the unstable part of the transition appears clearly. The odd (Figure 5, upper) and even (Figure 5, lower) discretisation sizes n_g yield rather similar results.

In order to obtain a more precise view of the model's behaviour at the transition, I run the model for $n_\epsilon = 500$ for 1501 values of $1/\epsilon \in [2.25, 3.75]$, an interval in which the first transition takes place, and n_g takes all of the integer values such that $2250 \leq n_g \leq 3750$. The values of n_g are thus alternatively odd and even. Indeed, in this case, the value the discretisation size n_g is $1000/\epsilon$. Figure 6 shows the results, with the cluster positions in Figure 6, upper, and the convergence times in Figure 6, bottom.

One can identify three main phases in this transition (delimited by dotted lines on Figure 6):

- From $1/\epsilon \approx 2.53$ to $1/\epsilon \approx 2.62$, for an odd n_g, one central cluster is reached after a few hundred iterations, and for an even n_g, two clusters, at distance roughly $\pm\epsilon$ from the centre (slowly increasing with $1/\epsilon$), are reached after about 10 iterations.
- From $1/\epsilon \approx 2.62$ to $1/\epsilon \approx 2.85$, two clusters at a distance of roughly $\pm\epsilon$ from the centre (slowly increasing with $1/\epsilon$) with one minor cluster at the centre are reached after a dozen iterations,
- From $1/\epsilon \approx 2.85$ to $1/\epsilon \approx 3.35$, there are two clusters in asymmetric positions and of different masses. The cluster of smaller mass is located at $\pm\epsilon$ and the one of bigger mass is closer to the centre on the other side, and moves slowly closer to the centre when $1/\epsilon$ increases. The convergence is reached after a large number of iterations (several thousands), except at the beginning and in the second half the phase where some cases of fast convergence also appear.

Figure 7 shows two examples of simulations in the first phase of the transition, one for an odd discretisation size $n_g = 2531$ ($1/\epsilon = 2.351$; Figure 7, top) the other for an even $n_g = 2532$ ($1/\epsilon = 2.352$; Figure 7, middle), and one example in the third phase for $n_g = 3001$ ($1/\epsilon = 3.001$; Figure 7, bottom).

In Figure 7, top, for $n_g = 2531$, there is a quite small cluster remaining at the centre (blue plus symbols) which slowly attracts both clusters initially located at a bit less than ϵ from the centre until they merge. In Figure 7, middle, for $n_g = 2532$, the initial evolution of the distribution is quite similar, except that there is no minor cluster at the centre. Therefore, the two major clusters remain in their positions and the simulation converges in about ten iterations. In this first phase, the convergence is the same with or without noise.

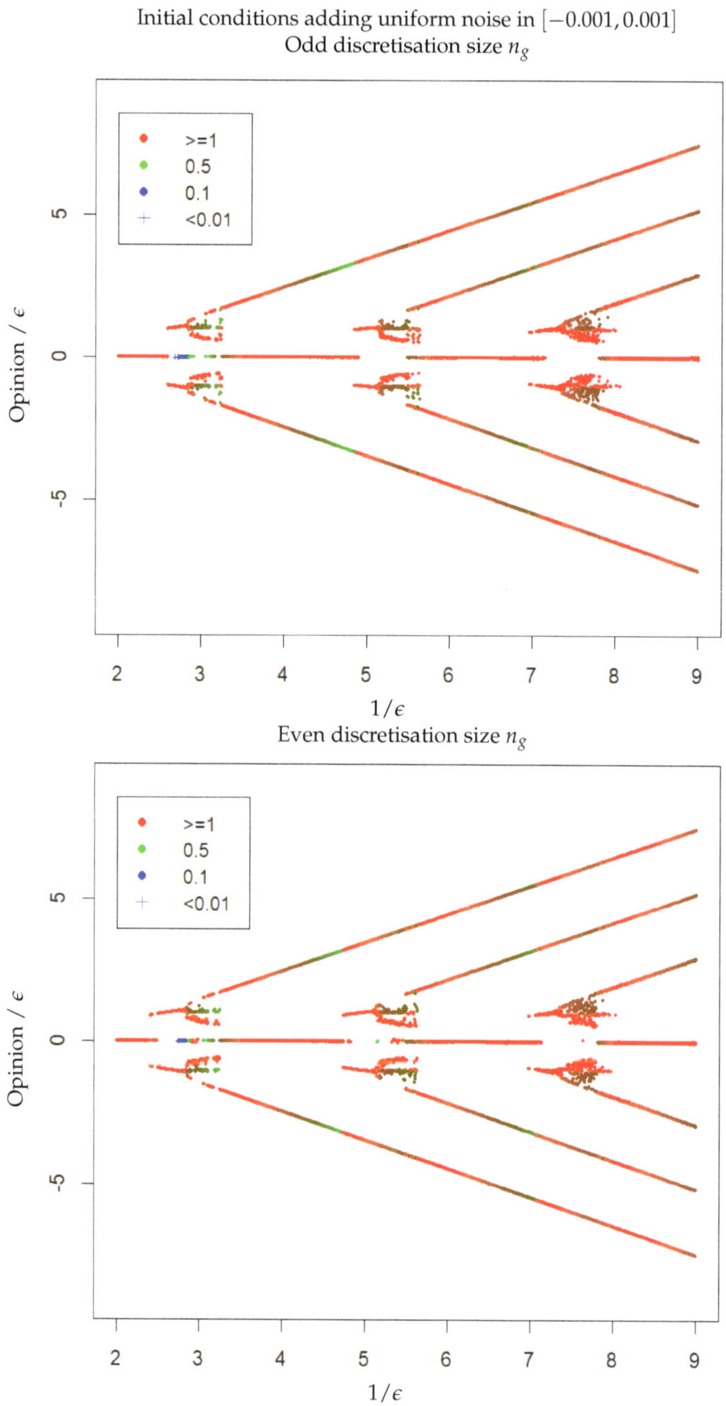

Figure 5. Positions of the clusters (a/ϵ) (opinion range $[-1, 1]$) for odd (**upper**) and even (**lower**) opinion range discretisation, n_g. ϵ discretisation size $n_\epsilon = 101$. The colors and symbols represent the values of the total mass of the cluster divided by $2n_\epsilon$, as indicated.

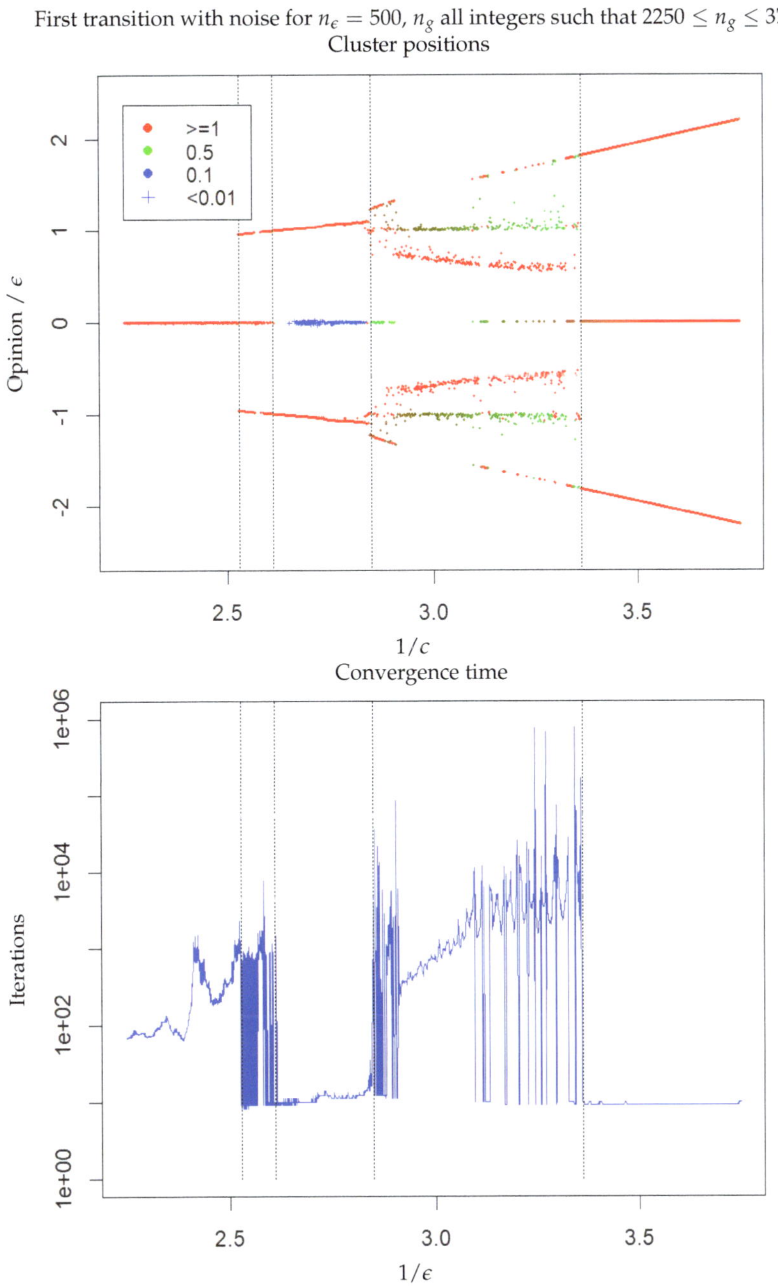

Figure 6. Positions of the clusters (a/ϵ) (**upper**) and convergence time (**lower**) (opinion range $[-1, 1]$). Discretisation n_g is odd or even. ϵ discretisation size $n_\epsilon = 500$. Upper: the colors and symbols represent the values of the total mass of the cluster divided by 2ϵ, as indicated.

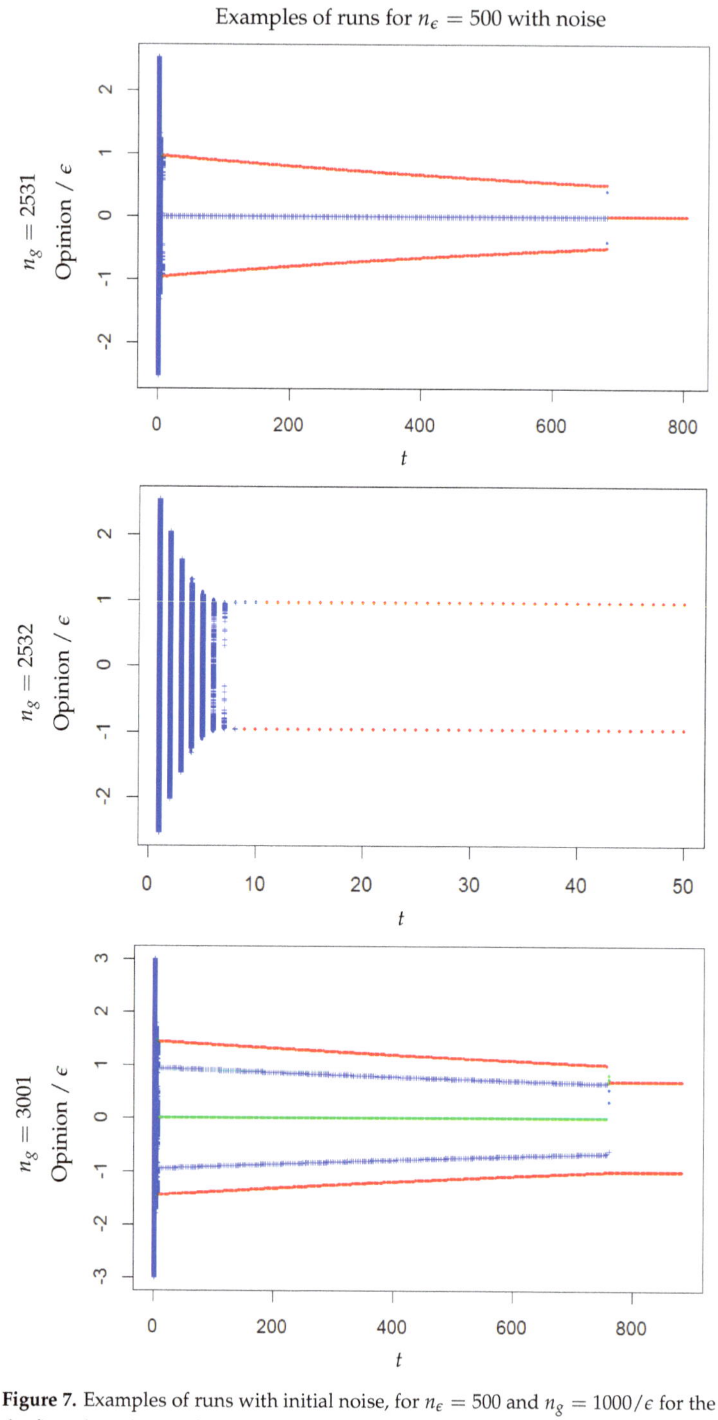

Figure 7. Examples of runs with initial noise, for $n_\epsilon = 500$ and $n_g = 1000/\epsilon$ for the simulations in the first phase (**top** and **middle**) and in the third phase (**bottom**) of the transition. The colors and symbols are as in Figure 1.

Figure 7, bottom, represents an example of a simulation in the third phase of the transition. In this case, the noise strongly influences the result. The configuration emerging after a dozen time steps includes one major central cluster (in green), two minor clusters located at about $\pm\epsilon$ (in blue) and two external major clusters symmetrically located a bit further (in red). Then, the minor clusters attract the two further major clusters and also move progressively closer to the centre. As a consequence of the noise, the configuration is slightly asymmetric. The major external cluster located on the positive side becomes closer than ϵ to the central cluster before the other, and at this moment, it merges with the central cluster and the minor cluster into a major cluster which is closer than ϵ to the centre. The other major external cluster merges with the minor cluster on its side and remains located quite close to $-\epsilon$. Certainly, depending on the noise, the first major cluster to merge can also be the one located on the negative side.

This last example explains the shape of the cluster positions in Figure 6, upper, showing two main branches on each side of the centre: the one in red corresponds to the external cluster that arrives closer to the centre first and merges with the central cluster, the other in green is the one that arrives after and only merges with the minor cluster on its side, and remains close to $\pm\epsilon$.

Note that, at the beginning and in the second half of the third phase, there are many cases where the minor clusters are located slightly further than ϵ and they directly merge with the external major clusters, with the whole distribution converging to three clusters in a dozen iterations. This explains the frequent oscillations in the convergence time, at the beginning and in the second half of the third phase of the transition (these differences could also define more specific phases in the transition).

Figure 8 shows the positions of the clusters during the transition from three to five clusters (for $1/\epsilon \in [4.5, 6]$), focusing on opinions a such that $a/\epsilon \in [-2.5, 2.5]$, where the transition takes place.

The second transition (between three and five clusters) is similar to first one (between one and three clusters). Three phases can also be identified as having broadly similar characteristics. Some differences are noticeable in the third phase though. Indeed, in the first transition, one underlines that, in the second half of the third phase, the results can be either an asymmetric configuration of two clusters or a symmetric configuration of three clusters. By contrast, during the whole third phase of the second transition, only the asymmetric configuration of two clusters takes place (here we omit to count the two symmetric clusters that are located further than 2.5ϵ and that do not appear in Figure 8, upper). Moreover, a fourth phase can then be identified, in which the results alternate between a new asymmetric configuration, with one major cluster around $\pm\epsilon$ and the other around $\pm 2\epsilon$, and the standard symmetric configuration of three clusters. The limits of the four phases appear as vertical dotted lines in Figure 8.

An example of a simulation yielding this new configuration is represented on Figure 9. In this case, there is a single minor cluster that is formed between the central major cluster and the first cluster located in its negative side. The minor cluster slowly attracts both major clusters which finally merge. In other simulations, the minor cluster can be formed between the central cluster and the first cluster on the positive side (located around 2ϵ).

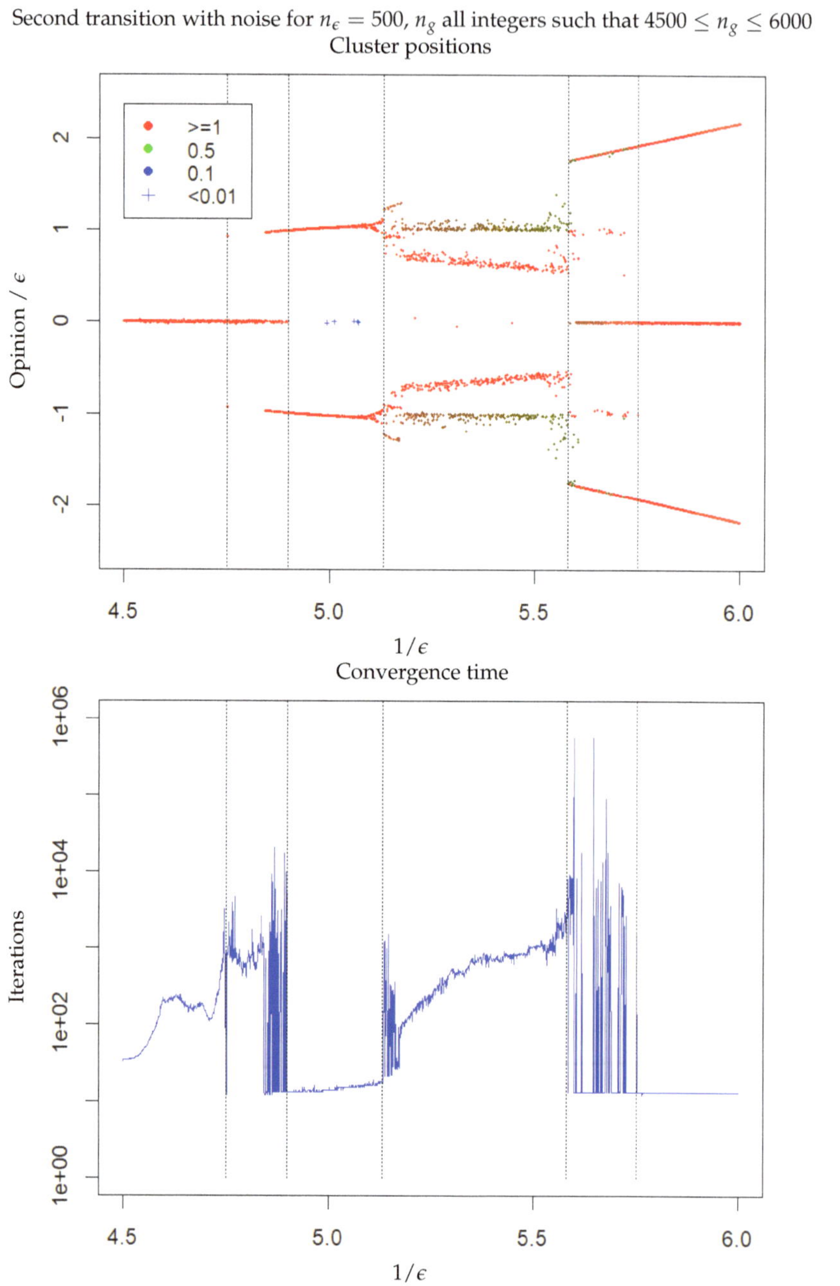

Figure 8. Positions of the clusters (a/ϵ) (**upper**) and convergence time (**lower**) (opinion range $[-1,1]$). The vertical dotted lines delimit the phases of the transition. Upper: the colors and symbols represent the values of the total mass of the cluster divided by 2ϵ, as indicated.

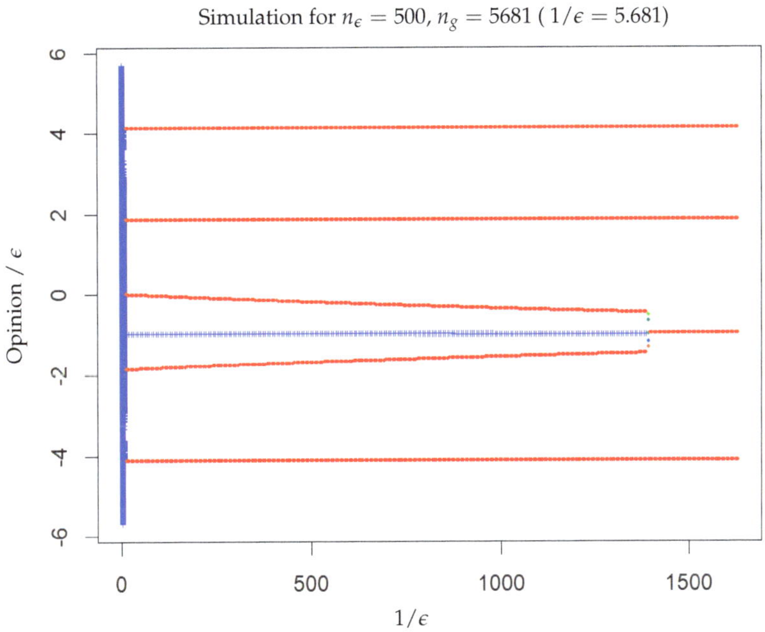

Figure 9. Example of simulation in the fourth phase of the second transition. The colors and symbols are as in Figure 1.

4. Discussion

Starting from a slightly noisy uniform distribution, the HK model of distribution shows surprisingly complex transitions from an odd number of clusters to the next, when the confidence bound, ϵ, decreases. Indeed, the first transition (from one to three clusters) includes three different phases. In the first phase, the result depends on the parity of the discretisation. In the third phase, the result is sensitive to small initial asymmetries, because of the presence of persisting minor clusters that slowly attract major clusters.

The second transition (from three to five clusters) shows strong similarities with the first one, but also striking differences. In particular, a fourth phase appears, in which a single persisting minor cluster appears between the central major and one of its neighbours, slowly attracting these clusters until they finally merge. Further investigations are necessary to explain why this fourth phase takes place in the second transition and not in the first one.

We did not investigate the third transition in detail, but strong differences with the two first transitions are clear. These observations suggest that, unlike the DW model, the HK model is not perfectly periodic. Indeed, the differences between the transitions look too systematic to be due to the noise only.

These results suggest that the HK version of the bounded confidence is more different from the DW version than suggested by the comparison made by Lorenz in 2006 [29]. Indeed, the DW version shows a clear periodicity and no sensitivity to the parity of the discretisation or to a small amount of initial noise.

However, before making definite conclusions, it is strongly advisable to closely compare the distribution version of the HK model with its agent version. Indeed, a distribution model is particularly relevant when it averages some randomness in the agent model. This is the case for the distribution version of the DW model, which averages the randomness of the pair interactions of the agent version. As the HK model is deterministic, the distribution version introduces approximations without averaging, hence its relevance is rather questionable. Moreover, as shown by Rainer Hegselmann [31], when starting from regularly spaced agents on the opinion axis, it is possible to compute the exact result of the HK model for a rational confidence bound ϵ and to find all the values of ϵ leading to different outcomes.

However, the results presented in this paper could promote new investigations into the agent model and new interpretations of its results. Therefore, the merit of the distribution version could simply be to provide a different perspective on the HK model, hopefully leading to a better understanding of its properties.

Overall, although the HK version looks simpler in its principle than the DW version, as it is deterministic, the collective behaviours that it generates in the case of initially regularly spaced agents do not look fully understood more than twenty years after its first publications [10,11] about this model. This looks to us a typical example of the general endeavour of sociophysics to understand precisely how collective regularities emerge from individual dynamics.

Funding: The research has been partly funded by the TED4LAT (Twinning in Environmental Data and Dynamical Systems Modelling for Latvia) European project (Grant agreement 101079206).

Data Availability Statement: The program in R can be downloaded at https://github.com/guillaumeDeffuant/DistributionHKModel.

Acknowledgments: I warmly thank Serge Galam for his suggestions of improvements on an earlier version of the article.

Conflicts of Interest: The author declares no conflicts of interest.

References

1. Sakoda, J.M. The checkerboard model of social interaction. *J. Math. Sociol.* **1971**, *1*, 119–132. [CrossRef]
2. Schelling, T.C. Dynamic models of segregation. *J. Math. Sociol.* **1971**, *1*, 143–186. [CrossRef]
3. Hegselmann, R. Thomas C. Schelling and James M. Sakoda: The intellectual, technical, and social history of a model. *J. Artif. Soc. Soc. Simul. (JASSS)* **2017**, *20*, 15.
4. Galam, S. Sociophysics: A review of Galam models. *Int. J. Mod. Phys. C* **2008**, *19*, 409–440. [CrossRef]
5. Cox, B. What is Hopi gossip about? Information management and Hopi factions. *Man* **1970**, *5*, 88–98. [CrossRef]
6. Schweitzer, F.; Beberra, L. Nonlinear voter models: The transition from invasion to coexistence. *Eur. Phys. J. B* **2009**, *67*, 301–318. [CrossRef]
7. Axelrod, R. The dissemination of culture: A model with local convergence and global polarization. *J. Confl. Resolut.* **1997**, *41*, 203–226. [CrossRef]
8. Klemm, K.; Eguíluz, V.M.; Toral, R.; San Miguel, M. Global culture: A noise-induced transition in finite systems. *Phys. Rev. E* **2003**, *67*, 045101(R). [CrossRef] [PubMed]
9. Martins, A.C.R. Continuous opinions and discrete actions in opinion dynamics problems. *Int. J. Mod. Phys. C* **2008**, *19*, 617–624. [CrossRef]
10. Hegselmann, R.; Krause, U. Opinion dynamics and bounded confidence: Models, analysis and simulation. *J. Artif. Soc. Soc. Simul. (JASSS)* **2002**, *5*, 2. Available online: https://www.jasss.org/5/3/2.html (accessed on 1 March 2024).
11. Deffuant, G.; Neau, D.; Amblard, F.; Weisbuch, G. Mixing beliefs among interacting agents. *Adv. Complex Syst.* **2000**, *3*, 87–98. [CrossRef]
12. Weisbuch, G. Bounded confidence and social networks. *Eur. Phys. J. B* **2004**, *38*, 339–343. [CrossRef]
13. Lorenz, J. Continuous opinion dynamics under bounded confidence: A survey. *Int. J. Mod. Phys. C* **2007**, *18*, 1819–1838. [CrossRef]
14. Deffuant, G.; Weisbuch, F.A.G.; Faure, T. How can extremism prevail? A study based on the relative agreement interaction model. *J. Artif. Soc. Soc. Simul. (JASSS)* **2002**, *5*, 1. Available online: https://www.jasss.org/5/4/1.html (accessed on 1 March 2024).
15. Deffuant, G. Comparing extremism propagation patterns in continuous opinion models. *J. Artif. Soc. Soc. Simul. (JASSS)* **2006**, *9*, 8. Available online: https://www.jasss.org/9/3/8.html (accessed on 1 March 2024).
16. Mathias, J.-D.; Huet, S.; Deffuant, G. Bounded confidence model with fixed uncertainties and extremists: The opinions can keep fluctuating indefinitely. *J. Artif. Soc. Soc. Simul. (JASSS)* **2016**, *19*, 6. [CrossRef]
17. Schawe, H.; Fontaine, S.; Hernández, L. When network bridges foster consensus. Bounded confidence models in networked societies. *Phys. Rev. Res.* **2021**, *3*, 023208. [CrossRef]
18. Lorenz, J.; Urbig, D. About the power to enforce or prevent consensus by manipulating communication rules. *Adv. Complex Syst.* **2007**, *10*, 251–269. [CrossRef]
19. Stauffer, D.; Meyer-Ortmanns, H. Simulation of consensus model of Deffuant et al. on a Barabási–Albert network. *Int. J. Mod. Phys. C* **2004**, *15*, 241–246. [CrossRef]
20. Gargiulo, F.; Gandica, Y. The role of homophily in the emergence of opinion controversies. *J. Artif. Soc. Soc. Simul. (JASSS)* **2017**, *20*, 8. [CrossRef]

21. Carletti, T.; Righi, S.; Fanelli, D. Emerging structures in social networks guided by opinions' exchange. *Adv. Complex Syst.* **2011**, *14*, 13–30. [CrossRef]
22. Kan, U.; Feng, M.; Porter, M.A. An adaptive bounded-confidence model of opinion dynamics on networks. *J. Complex Netw.* **2022**, *11*, cnac055. [CrossRef]
23. Pineda, M.; Toral, R.; Hernández-García, E. Noisy continuous-opinion dynamics. *J. Stat. Mech. Theory Exp.* **2009**, *2009*, P08001. [CrossRef]
24. Pineda, M.; Toral, R.; Hernádez-Garciía, E. Diffusing opinions in bounded confidence processes. *Eur. Phys. J. D* **2011**, *62*, 109–117. [CrossRef]
25. Flache, A.; Macy, M.W. Small worlds and cultural polarization. *J. Math. Sociol.* **2011**, *35*, 146–176. [CrossRef]
26. Castellano, C.; Fortunato, S.; Loreto, V. Statistical physics of social dynamics. *Rev. Mod. Phys.* **2007**, *81*, 591–646. [CrossRef]
27. Flache, A.; Mäs, M.; Feliciani, T.; Chattoe-Brown, E.; Deffuant, G.; Huet, S.; Lorenz, J. Models of social influence: Towards the next frontiers. *J. Artif. Soc. Soc. Simul. (JASSS)* **2017**, *20*, 4. [CrossRef]
28. Ben-Naim, E.; Krapivsky, P.L.; Redner, S. Bifurcations and patterns in compromise processes. *Phys. D Nonlin. Phenom.* **2003**, *183*, 190–204. [CrossRef]
29. Lorenz, J. Consensus strikes back in the Hegselmann-Krause model of continuous opinion dynamics under bounded confidence. *J. Artif. Soc. Soc. Simul. (JASSS)* **2006**, *9*, 8. Available online: https://www.jasss.org/9/1/8.html (accessed on 1 March 2024).
30. Hegarty, P.; Wedin, E. The Hegselmann-Krause dynamics for equally spaced agents. *J. Differ. Equat. Appl.* **2016**, *22*, 1621–1645. [CrossRef]
31. Hegselmann, R. Bounded confidence revisited: What we overlooked, underestimated, and got wrong. *J. Artif. Soc. Soc. Simul. (JASSS)* **2023**, *26*, 11. [CrossRef]

physics

MDPI

Article

The Influence of Lobbies: Analyzing Group Consensus from a Physics Approach

Ugo Merlone [1,*] and Arianna Dal Forno [2]

[1] Department of Psychology, University of Torino, 10124 Torino, Italy
[2] Department of Economics, University of Molise, 86100 Campobasso, Italy; arianna.dalforno@unimol.it
* Correspondence: ugo.merlone@unito.it

Abstract: In this paper, we study the influence of a small group of agents (i.e., a lobby) that is trying to spread a rumor in a population by using the known model proposed by Serge Galam. In particular, lobbies are modeled as subgroups of individuals who strategically choose their seating in the social space in order to protect their opinions and influence others. We consider different social gatherings and simulate, using finite Markovian chains, opinion dynamics by comparing situations with a lobby to those without a lobby. Our results show how the lobby can influence opinion dynamics in terms of the prevailing opinion and the mean time to reach unanimity. The approach that we take overcomes some of the problems that behavioral economics and psychology have recently struggled with in terms of replicability. This approach is related to the methodological revolution that is slowly changing the dominant perspective in psychology.

Keywords: opinion dynamics; sociophysics; Markovian chain physics models; social influence

1. Introduction

Besides the known incident [1] and the more recent case [2] in behavioral economics, replicability has also been questioned in psychology since at least 2015 [3], and alternative approaches to analyzing the phenomena studied by these disciplines could help to shed light on human behavior in groups.

Social influence is a topic that has attracted great interest in both social psychology [4] and behavioral economics [5]. Social psychology in particular has provided some ground-breaking experiments in which the effect of groups in changing opinions [6] has been investigated. Of additional relevance is the research on the influence of minorities using a simulation approach on lattices [7] and on networks [8,9]; for a discussion of spatial interactions in agent-based models, see [10]. Another interesting example can be found in studies on the influence of small groups on majority consent [11]. These studies analyze the influence of groups at the microscopic level and provide some insights that can be applied in the analysis of political change [12]. Previous research has focused in particular on two different aspects: the influence of the majority (i.e., conformity, Ref. [6]) and the influence of the minority [13]. Studies on conformity show that the most common opinion in a group is able to exert an enormous influence on individuals and leads them to give inaccurate answers [6]. By contrast, studies on minority influence have focused on how people resist group pressure and how individuals are able to change the opinion upheld by the dominant position [14]. Minority studies have been applied to innovation and social change [15] and also to the study of influence on dominant positions [12]; for reviews on the influence of small groups, see [11,16,17]. Further experimental studies have been conducted in order to shed light on the psychological processes of minority influences [18] and the strategies of influencing dominant positions [19]. In the field of social epistemology, both mathematical and computational models of social influence are being developed to compare the ability to secure consensus in different types of social arrangements [20] and

Citation: Merlone, U.; Dal Forno, A. The Influence of Lobbies: Analyzing Group Consensus from a Physics Approach. *Physics* **2024**, *6*, 659–673. https://doi.org/10.3390/physics6020043

Received: 23 December 2023
Revised: 31 January 2024
Accepted: 8 February 2024
Published: 1 May 2024

how agents may fail to converge to the truth [21]. Another important research branch is devoted to the topic of strategic influence. In particular (see, e.g., [22,23]), strategic influence is applied to voting schemes. In this paper, we consider lobbies as groups of individuals who strategically choose their seats in the social space to protect their opinions and influence other individuals. Finally, some contributions in the literature on group decision making and consensus building can be found in Refs. [24,25].

More recently, the spread of conspiracy theories [26] on various topics [27], but also on science [28,29] and the social processes underlying their spread [30], show how relevant these phenomena can be to the society. Given the problems of replicability mentioned above, analytical approaches or replicable simulation models can be highly useful. Therefore, we introduce minorities in the known model of Serge Galam [31], who analyzed the spread of hoaxes twenty years ago using sociophysics.

The paper is structured as follows. In Section 2, we discuss some possible approaches to modeling opinion spreading. In Section 3, we summarize Galam's model [31] and illustrate how we extend it by introducing lobbies. The results of our simulation are analyzed in Section 4. Finally, Section 5 gives the conclusions and suggests further research.

2. Analytical and Computational Approaches to Opinion Spreading

Although experimental studies provided by social psychology offer important contributions in understanding the proximate mechanisms that influence human behavior, this approach presents some intrinsic limits [13]. One of these limits is that experimental studies cannot investigate the temporal dimensions and the long-run effects of opinion dynamics on a population. Therefore, it is crucial to integrate the knowledge of the proximate processes derived from social psychology with methods that can focus on a more dynamic dimension. In other words, once it is known that a local majority opinion can influence individuals in small groups, how will this effect spread to the population? One attempt to fill this gap has been provided by sociophysics [32]. A sociophysics approach focuses on the dynamic dimension, i.e., the time needed for a minority to reach a consensus in a population. One of the first attempts to integrate social psychological theories with the methods of sociophysics was proposed in Ref. [33]. This study provides a different context to test theories developed in an experimental setting and allows for the validation of these concepts on a broader scale. In particular, this study focuses on the phenomenon of polarization in groups, i.e., the tendency of individuals to hold more radical opinions in the presence of others.

Other approaches that can be integrated with social psychology for the study of opinion dynamics are mathematical modeling [34–36] and agent-based simulations [37,38]. Recently, the integration of other approaches, such as agent-based modeling, has become more popular (see, e.g., [39]), because they make it possible to consider several phenomena in social psychology "as emergent results of dynamically interactive processes taking place in their contexts" [40]. This is part of the quiet methodological revolution referred to in [41], as statistics moves from the mechanical application of a series of procedures to the building and the evaluation of models.

Among other disciplines, sociophysics offers some important insights into the analysis of social influence on opinion dynamics and consensus [37,42,43]. One of the main interesting models is proposed in [31]. Galam's model focuses on the analysis of the propagation or diffusion of hoaxes in public debates. To be more specific, in Galam's model, individuals interact together on a specific topic in groups of different dimensions. The model shows that even if, at the beginning, an opinion represents the minority, after a certain amount of discussion, this opinion can be diffused to the population. The configuration of different tables of different sizes where people discuss is called a *social space* [31,44] for details). An example of interaction in Galam's model can be found in Figure 1. Further research has introduced more complexity in this model [45]. For example, the agents proposed in Ref. [45] hold always the same opinion and have the ability to influence all the agents sitting at the table. In this research, we introduce the presence of lobbies in Galam's model.

Lobbies are modeled as minorities that are able to influence other opinions with a specific strategy. As a matter of fact, members of the lobby have the ability to find and seat, when possible, at a table where they can be the majority and exert their influence more efficiently.

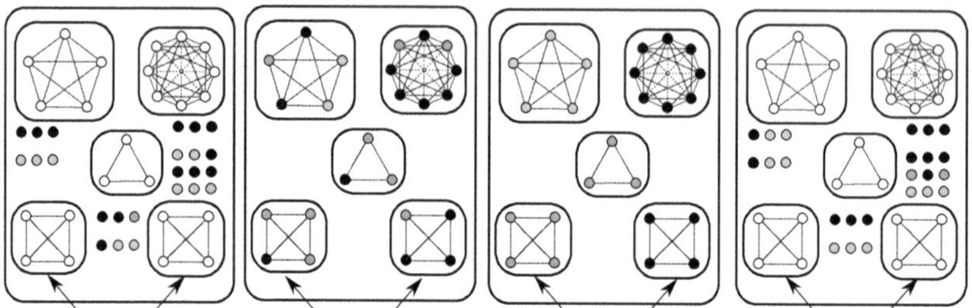

Figure 1. An exemplification of Galam's model: a one-step opinion dynamics. **Left** to **right**: First stage: individuals (black and gray circles) sharing the two opinions are moving around in order to sit at the available seats (white circles) of the tables. Gray circle individuals are "in favor of opinion $\mathcal{X}$", while black circle individuals are "against $\mathcal{X}$". Second stage: individuals are partitioned into groups of various sizes (tables). Before the discussion occurs, there are eleven individual in favor (gray circles) and thirteen against (black circles). Third stage: within each group, the discussion occurs and consensus has been reached. As a result, there are now twelve individuals in favor, and twelve individuals against. Last stage: individuals are again moving around with no discussion.

3. Rumors and Lobbies in Galam's Model

Galam's model can be considered a dynamic application of the conformity influence in different social gatherings. In Ref. [31], individuals have the same characteristics (i.e., they all meet randomly and change their opinion according to the majority) and a consensus can be reached when a specific minority threshold is reached. Therefore, Galam's model does not explain the dynamics in which small organized groups such as lobbies can implement some strategies to influence others' behavior. Lobbies can be modeled in different ways and can have a wide range of strategies, behaviors or interactions. In this paper, we present a model of a lobby capable of influencing the dynamics of opinions at strategic tables. More specifically, as in other studies examining seating positions at specific tables [46–48], lobbies are groups of individuals who always sit at tables where they can influence others. The influence rule is the same as in Galam's model [31] but, in the case considered, one has a small part of the population who strategically sit at the tables where they possibly represent the majority and then persuade others. To summarize, we consider a Galam-like model with a lobby, where the lobby is modeled as a subgroup of individuals who are "against opinion $\mathcal{X}$", and coordinate their sitting at the tables of the social space with the aim of influencing as many agents as possible of the opposite opinion without being overridden. As a matter of fact, in the case here considered, the lobby members are subject to the table majority rule too; in other words, they are different from the opinion leaders considered in [45], who are always able to persuade all agents at their table. The strategic choice of the lobby depends on the social space configuration. As it is impossible to consider all the possible social space configurations for any number of individuals, we consider an example to study opinion dynamics.

In Galam's model [31], gatherings take place at different times, and each individual holds one of two possible opinions, e.g., in favor of $\mathcal{X}$, or against $\mathcal{X}$. When the individuals meet in a specific gathering configuration (a table with $3, 4, \ldots n$ individuals), they change their opinion according to the majority. Then, they will discuss again randomly in different social gatherings and change always their opinion according to this rule.

3.1. The Model without Lobby

We consider a population of N individuals who, at time t, have either opinion $\mathcal{X}$ (in favor) or $\overline{\mathcal{X}}$ (against). At each time t, j denotes the number of individuals with the opinion $\mathcal{X}$ and defines the state of the system; therefore, the set of the possible statesof the system is $J = \{0, 1, \ldots, N\}$, where $j = 0$ indicates a consensus on the opinion $\overline{\mathcal{X}}$ and $j = N$ on the opinion $\mathcal{X}$. In particular, as in Ref. [44], individuals interact in a social space consisting of a set of available tables $T_r \in \mathcal{N} = \{T_1, T_2, \ldots, T_L\}$, whose size is denoted s_r. The sizes of the tables in $\mathcal{N}$ are represented as a vector $\mathbf{s} = (s_1, s_2, \ldots, s_L) \in \mathbb{R}^L$ (the case $L = N$, i.e., all tables of dimension 1, is trivial and therefore is not considered), where $L < N$ and $\sum_{T_r \in \mathcal{N}} s_r = N$. The number of individuals who hold the same opinion at a given table of size s is denoted by $n(\mathcal{X})$ and $n(\overline{\mathcal{X}})$, respectively, so that $n(\mathcal{X}) + n(\overline{\mathcal{X}}) = s$. The distribution of individuals in favor of opinion $\mathcal{X}$ across the L tables is represented by the vector $\mathbf{n}(\mathcal{X}) = (n_1(\mathcal{X}), n_2(\mathcal{X}), \ldots, n_L(\mathcal{X}))$ and $j = \sum_{r=1}^{L} n_r(\mathcal{X})$ is the state of the system.

At each time step t, the dynamics of each individual's opinion is governed by the following set of rules:

- individuals are randomly distributed to the available seats;
- individuals change their opinion according to a local or table majority rule, i.e., at each table, if $n(\mathcal{X}) > n(\overline{\mathcal{X}})$, then all individuals sitting at the table will change to the opinion $\mathcal{X}$; vice versa, if $n(\overline{\mathcal{X}}) \geq n(\mathcal{X})$, then all the individuals sitting at the table will change to opinion $\overline{\mathcal{X}}$.

This approach leads to a stochastic process where the individuals submit to the local majority with certainty; this is different, for instance, from Ref. [49], where individuals may also change opinion, with a certain probability that depends on a propensity parameter.

3.2. The Model with Lobby

We assume that an interest group, with the goal of influencing the population's opinion, is able to strategically choose seats. Organized interests quite often target opponents and undecideds [50]. We therefore assume that, within the population of N individuals, there exists a subset of size $N^* \leq N$—called a *lobby*—that strategically distributes itself across the available tables in order to maximize the number of new allies who share its opinion $\overline{\mathcal{X}}$. The number of lobbyists seated at a table T_r is denoted as n_r^*. The distribution of lobbyists across the L tables is represented by the vector $\mathbf{n}^* = (n_1^*, n_2^*, \ldots, n_L^*)$. Taking into account the table majority rule defined above in Section 3.1, the dynamics rules are as follows:

- the lobbyists sit at the tables, possibly occupying half of the seats available at each table (if s is an even number), or $\lceil s/2 \rceil$ (if the seats are an odd number) (recall that the symbol $\lceil \cdot \rceil$ indicates the ceiling function);
- non-lobbyist individuals are randomly distributed across the remaining seats;
- the table majority rule is applied to both lobbyists and non-lobbyists and the opinions are updated.

3.3. Simulation Model

The opinion dynamics can be modeled as a finite Markov chain [44], i.e., a discrete-time process in which the future state depends only on the present state and not on the past states [51]. As a matter of fact, all possible opinion configurations can be considered as states and the probability of a transition between any couple of states can be computed as shown in Ref. [44]:

$$P_j(\mathbf{n}(\mathcal{X})) = \frac{\binom{j}{n_1(\mathcal{X}), n_2(\mathcal{X}), \ldots, n_L(\mathcal{X})} \binom{N-j}{s_1 - n_1(\mathcal{X}), s_2 - n_2(\mathcal{X}), \ldots, s_L - n_L(\mathcal{X})}}{\binom{N}{s_1, s_2, \ldots, s_L}}, \tag{1}$$

where $\binom{b}{b_1, b_2, \ldots, b_L}$ is the common notation for multinomial coefficients.

When considering lobbies, the previous formula can be adapted as follows, conditional on the lobbyists' position:

$$P_j(\mathbf{n}(\mathcal{X})|\mathbf{n}^*) = \frac{\binom{j}{n_1(\mathcal{X}),n_2(\mathcal{X}),\dots,n_L(\mathcal{X})}\binom{N-N^*-j}{s_1-n_1^*-n_1(\mathcal{X}),s_2-n_2^*-n_2(\mathcal{X}),\dots,s_L-n_L^*-n_L(\mathcal{X})}}{\binom{N-N^*}{s_1-n^*,s_2-n^*,\dots,s_L-n^*}}. \quad (2)$$

In our study, we consider a population with $N = 24$ individuals as it offers a large enough number of social spaces; even in this case, the partition number is too large for a systematic analysis of all the possible spaces since, for $N = 24$, the partition number is 1575 [52] (sequence A000041). Therefore, we consider the special case in which all the tables have the same size. In order to be able to assess the role of the lobby, we limit its size to a maximum of six individuals. The social spaces that we consider, together with the strategic seating choices of the lobby members, are shown in Figure 2. Similar to that in Galam's model, when a table has the same number of contrasting opinions, the bias is in favor of the individuals who are "against $\mathcal{X}$".

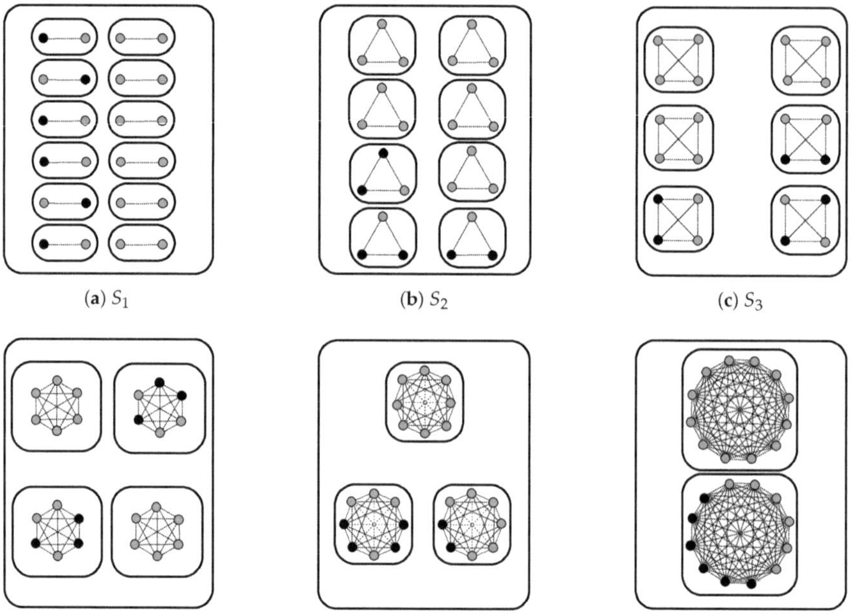

(**a**) S_1 (**b**) S_2 (**c**) S_3 (**d**) S_4 (**e**) S_5 (**f**) S_6

Figure 2. Population of twenty-four people seated in different social spaces. Lobby consists of at most six people (black dots) who are "against $\mathcal{X}$" seating at prefixed strategic tables for different social spaces S_h as follows. (**a**) S_1: twelve tables with two seats each; lobby members sit individually at six tables. (**b**) S_2: eight tables with three seats each; lobby members sit in couples at three tables. (**c**) S_3: six tables with four seats each; here, too, lobby members sit in couples at three tables. (**d**) S_4: four tables with six seats each; lobby members sit in groups of three at two tables. (**e**) S_5: four tables with six seats each; four lobby members sit at one table and are always able to influence, while the two remaining sit at another table. (**f**) S_6: two tables with twelve seats each; lobby members sit together at one table.

As is typical with finite Markov chains, the transition probabilities are arranged in a matrix [53] (p. 374). For each social space S_h (with $h = 1, \cdots, 6$) of Figure 2, we compare the two scenarios, one with and the other without a lobby. The transition matrices can be computed using Formulas (1) and (2) as illustrated in the following example.

Example

We consider $N = 24$ individuals and the social space S_6, which consists of two tables of the same size $s = 12$. The choice of the space S_6 is due to the fact that the multinomial coefficients in the transition probabilities defined in Equations (1) and (2) become binomial coefficients and the notation is simpler. At the end of each time period, the opinions are homogeneous; therefore, only the states 0, 12 and 24 have positive probabilities. From Equation (1), one can derive the positive entries of the matrix $\mathbf{M}_6$ for the scenario without a lobby:

$$
\begin{aligned}
p_{j,0} &= 1 && \text{with } 0 \le j \le 6,\\
p_{j,0} &= \left[\sum_{k=j-s/2}^{s/2} \binom{j}{k}\binom{N-j}{s-k}\right] / \binom{N}{s} && \text{with } 7 \le j \le 12,\\
p_{j,12} &= 2\left[\sum_{k=s/2+1}^{j} \binom{j}{k}\binom{N-j}{s-k}\right] / \binom{N}{s} && \text{with } 7 \le j \le 12,\\
p_{j,12} &= 1 && \text{with } j = 13,\\
p_{j,12} &= 2\left[\sum_{k=s/2}^{N-j} \binom{j}{s/2-k}\binom{N-j}{k}\right] / \binom{N}{s} && \text{with } 14 \le j \le 18,\\
p_{j,24} &= \left[\sum_{k=s/2+1}^{j} \binom{j}{k}\binom{N-j}{s-k}\right] / \binom{N}{s} && \text{with } 14 \le j \le 18,\\
p_{j,24} &= 1 && \text{with } 19 \le j \le 24,
\end{aligned}
$$

that is,

$$
\mathbf{M}_6 = \begin{pmatrix}
1 & 0\,0\,0\,0\,0\,0\,0\,0\,0\,0\,0 & 0 & 0\,0\,0\,0\,0\,0\,0\,0\,0\,0\,0 & 0\\
1 & 0\,0\,0\,0\,0\,0\,0\,0\,0\,0\,0 & 0 & 0\,0\,0\,0\,0\,0\,0\,0\,0\,0\,0 & 0\\
1 & 0\,0\,0\,0\,0\,0\,0\,0\,0\,0\,0 & 0 & 0\,0\,0\,0\,0\,0\,0\,0\,0\,0\,0 & 0\\
1 & 0\,0\,0\,0\,0\,0\,0\,0\,0\,0\,0 & 0 & 0\,0\,0\,0\,0\,0\,0\,0\,0\,0\,0 & 0\\
1 & 0\,0\,0\,0\,0\,0\,0\,0\,0\,0\,0 & 0 & 0\,0\,0\,0\,0\,0\,0\,0\,0\,0\,0 & 0\\
1 & 0\,0\,0\,0\,0\,0\,0\,0\,0\,0\,0 & 0 & 0\,0\,0\,0\,0\,0\,0\,0\,0\,0\,0 & 0\\
1 & 0\,0\,0\,0\,0\,0\,0\,0\,0\,0\,0 & 0 & 0\,0\,0\,0\,0\,0\,0\,0\,0\,0\,0 & 0\\
\frac{\sum_{k=1}^{6}\binom{7}{k}\binom{17}{12-k}}{\binom{24}{12}} & 0\,0\,0\,0\,0\,0\,0\,0\,0\,0\,0 & \frac{2\sum_{k=7}^{7}\binom{7}{k}\binom{17}{12-k}}{\binom{24}{12}} & 0\,0\,0\,0\,0\,0\,0\,0\,0\,0\,0 & 0\\
\frac{\sum_{k=2}^{6}\binom{8}{k}\binom{16}{12-k}}{\binom{24}{12}} & 0\,0\,0\,0\,0\,0\,0\,0\,0\,0\,0 & \frac{2\sum_{k=7}^{8}\binom{8}{k}\binom{16}{12-k}}{\binom{24}{12}} & 0\,0\,0\,0\,0\,0\,0\,0\,0\,0\,0 & 0\\
\frac{\sum_{k=3}^{6}\binom{9}{k}\binom{15}{12-k}}{\binom{24}{12}} & 0\,0\,0\,0\,0\,0\,0\,0\,0\,0\,0 & \frac{2\sum_{k=7}^{9}\binom{9}{k}\binom{15}{12-k}}{\binom{24}{12}} & 0\,0\,0\,0\,0\,0\,0\,0\,0\,0\,0 & 0\\
\frac{\sum_{k=4}^{6}\binom{10}{k}\binom{14}{12-k}}{\binom{24}{12}} & 0\,0\,0\,0\,0\,0\,0\,0\,0\,0\,0 & \frac{2\sum_{k=7}^{10}\binom{10}{k}\binom{14}{12-k}}{\binom{24}{12}} & 0\,0\,0\,0\,0\,0\,0\,0\,0\,0\,0 & 0\\
\frac{\sum_{k=5}^{6}\binom{11}{k}\binom{13}{12-k}}{\binom{24}{12}} & 0\,0\,0\,0\,0\,0\,0\,0\,0\,0\,0 & \frac{2\sum_{k=7}^{11}\binom{11}{k}\binom{13}{12-k}}{\binom{24}{12}} & 0\,0\,0\,0\,0\,0\,0\,0\,0\,0\,0 & 0\\
\frac{\sum_{k=6}^{6}\binom{12}{k}\binom{12}{12-k}}{\binom{24}{12}} & 0\,0\,0\,0\,0\,0\,0\,0\,0\,0\,0 & \frac{2\sum_{k=7}^{12}\binom{12}{k}\binom{12}{12-k}}{\binom{24}{12}} & 0\,0\,0\,0\,0\,0\,0\,0\,0\,0\,0 & 0\\
0 & 0\,0\,0\,0\,0\,0\,0\,0\,0\,0\,0 & 1 & 0\,0\,0\,0\,0\,0\,0\,0\,0\,0\,0 & 0\\
0 & 0\,0\,0\,0\,0\,0\,0\,0\,0\,0\,0 & \frac{2\sum_{k=6}^{10}\binom{14}{12-k}\binom{10}{k}}{\binom{24}{12}} & 0\,0\,0\,0\,0\,0\,0\,0\,0\,0\,0 & \frac{\sum_{k=7}^{7}\binom{14}{k}\binom{10}{12-k}}{\binom{24}{12}}\\
0 & 0\,0\,0\,0\,0\,0\,0\,0\,0\,0\,0 & \frac{2\sum_{k=6}^{9}\binom{15}{12-k}\binom{9}{k}}{\binom{24}{12}} & 0\,0\,0\,0\,0\,0\,0\,0\,0\,0\,0 & \frac{\sum_{k=7}^{8}\binom{15}{k}\binom{9}{12-k}}{\binom{24}{12}}\\
0 & 0\,0\,0\,0\,0\,0\,0\,0\,0\,0\,0 & \frac{2\sum_{k=6}^{8}\binom{16}{12-k}\binom{8}{k}}{\binom{24}{12}} & 0\,0\,0\,0\,0\,0\,0\,0\,0\,0\,0 & \frac{\sum_{k=7}^{9}\binom{16}{k}\binom{8}{12-k}}{\binom{24}{12}}\\
0 & 0\,0\,0\,0\,0\,0\,0\,0\,0\,0\,0 & \frac{2\sum_{k=6}^{7}\binom{17}{12-k}\binom{7}{k}}{\binom{24}{12}} & 0\,0\,0\,0\,0\,0\,0\,0\,0\,0\,0 & \frac{\sum_{k=7}^{10}\binom{17}{k}\binom{7}{12-k}}{\binom{24}{12}}\\
0 & 0\,0\,0\,0\,0\,0\,0\,0\,0\,0\,0 & \frac{2\sum_{k=6}^{6}\binom{18}{12-k}\binom{6}{k}}{\binom{24}{12}} & 0\,0\,0\,0\,0\,0\,0\,0\,0\,0\,0 & \frac{\sum_{i=7}^{11}\binom{18}{k}\binom{6}{12-k}}{\binom{24}{12}}\\
0 & 0\,0\,0\,0\,0\,0\,0\,0\,0\,0\,0 & 0 & 0\,0\,0\,0\,0\,0\,0\,0\,0\,0\,0 & 1\\
0 & 0\,0\,0\,0\,0\,0\,0\,0\,0\,0\,0 & 0 & 0\,0\,0\,0\,0\,0\,0\,0\,0\,0\,0 & 1\\
0 & 0\,0\,0\,0\,0\,0\,0\,0\,0\,0\,0 & 0 & 0\,0\,0\,0\,0\,0\,0\,0\,0\,0\,0 & 1\\
0 & 0\,0\,0\,0\,0\,0\,0\,0\,0\,0\,0 & 0 & 0\,0\,0\,0\,0\,0\,0\,0\,0\,0\,0 & 1\\
0 & 0\,0\,0\,0\,0\,0\,0\,0\,0\,0\,0 & 0 & 0\,0\,0\,0\,0\,0\,0\,0\,0\,0\,0 & 1\\
0 & 0\,0\,0\,0\,0\,0\,0\,0\,0\,0\,0 & 0 & 0\,0\,0\,0\,0\,0\,0\,0\,0\,0\,0 & 1
\end{pmatrix}.
$$

From Equation (2), the entries for the matrix $\overline{\mathbf{M}}_6$ with a lobby of size $N^* = 6$ are

$$
\begin{aligned}
p_{j,0} &= 1 && \text{with } 0 \le j \le 6, \\
p_{j,0} &= \left[\sum_{k=j-s/2}^{s/2} \binom{j}{k} \binom{N-N^*-j}{j-s/2} \right] / \binom{N-N^*}{s-N^*} && \text{with } 7 \le j \le 12, \\
p_{j,12} &= \left[\sum_{k=0}^{j-s/2-1} \binom{j}{k} \binom{N-N^*-j}{s-N^*-k} \right] / \binom{N-N^*}{s-N^*} && \text{with } 7 \le j \le 12, \\
p_{j,12} &= 1 && \text{with } 13 \le j \le 18, \\
p_{j,24} &= 1 && \text{with } 19 \le j \le 24,
\end{aligned}
$$

and one has

$$
\overline{\mathbf{M}}_6 = \left(\begin{array}{cccc}
1 & 0\,0\,0\,0\,0\,0\,0\,0\,0\,0\,0 & 0 & 0\,0\,0\,0\,0\,0\,0\,0\,0\,0\,0\,0 \\
1 & 0\,0\,0\,0\,0\,0\,0\,0\,0\,0\,0 & 0 & 0\,0\,0\,0\,0\,0\,0\,0\,0\,0\,0\,0 \\
1 & 0\,0\,0\,0\,0\,0\,0\,0\,0\,0\,0 & 0 & 0\,0\,0\,0\,0\,0\,0\,0\,0\,0\,0\,0 \\
1 & 0\,0\,0\,0\,0\,0\,0\,0\,0\,0\,0 & 0 & 0\,0\,0\,0\,0\,0\,0\,0\,0\,0\,0\,0 \\
1 & 0\,0\,0\,0\,0\,0\,0\,0\,0\,0\,0 & 0 & 0\,0\,0\,0\,0\,0\,0\,0\,0\,0\,0\,0 \\
1 & 0\,0\,0\,0\,0\,0\,0\,0\,0\,0\,0 & 0 & 0\,0\,0\,0\,0\,0\,0\,0\,0\,0\,0\,0 \\
1 & 0\,0\,0\,0\,0\,0\,0\,0\,0\,0\,0 & 0 & 0\,0\,0\,0\,0\,0\,0\,0\,0\,0\,0\,0 \\
\frac{\sum_{k=1}^{6}\binom{7}{k}\binom{11}{6-k}}{\binom{18}{6}} & 0\,0\,0\,0\,0\,0\,0\,0\,0\,0\,0 & \frac{\sum_{k=0}^{0}\binom{7}{k}\binom{11}{6-k}}{\binom{18}{6}} & 0\,0\,0\,0\,0\,0\,0\,0\,0\,0\,0\,0 \\
\frac{\sum_{k=2}^{6}\binom{8}{k}\binom{10}{6-k}}{\binom{18}{6}} & 0\,0\,0\,0\,0\,0\,0\,0\,0\,0\,0 & \frac{\sum_{k=0}^{1}\binom{8}{k}\binom{10}{6-k}}{\binom{18}{6}} & 0\,0\,0\,0\,0\,0\,0\,0\,0\,0\,0\,0 \\
\frac{\sum_{k=3}^{6}\binom{9}{k}\binom{9}{6-k}}{\binom{18}{6}} & 0\,0\,0\,0\,0\,0\,0\,0\,0\,0\,0 & \frac{\sum_{k=0}^{2}\binom{9}{k}\binom{9}{6-k}}{\binom{18}{6}} & 0\,0\,0\,0\,0\,0\,0\,0\,0\,0\,0\,0 \\
\frac{\sum_{k=4}^{6}\binom{10}{k}\binom{8}{6-k}}{\binom{18}{6}} & 0\,0\,0\,0\,0\,0\,0\,0\,0\,0\,0 & \frac{\sum_{k=0}^{3}\binom{10}{k}\binom{8}{6-k}}{\binom{18}{6}} & 0\,0\,0\,0\,0\,0\,0\,0\,0\,0\,0\,0 \\
\frac{\sum_{k=5}^{6}\binom{11}{k}\binom{7}{6-k}}{\binom{18}{6}} & 0\,0\,0\,0\,0\,0\,0\,0\,0\,0\,0 & \frac{\sum_{k=0}^{4}\binom{11}{k}\binom{7}{6-k}}{\binom{18}{6}} & 0\,0\,0\,0\,0\,0\,0\,0\,0\,0\,0\,0 \\
\frac{\sum_{k=6}^{6}\binom{12}{k}\binom{6}{6-k}}{\binom{18}{6}} & 0\,0\,0\,0\,0\,0\,0\,0\,0\,0\,0 & \frac{\sum_{k=0}^{5}\binom{12}{k}\binom{6}{6-k}}{\binom{18}{6}} & 0\,0\,0\,0\,0\,0\,0\,0\,0\,0\,0\,0 \\
0 & 0\,0\,0\,0\,0\,0\,0\,0\,0\,0\,0 & 1 & 0\,0\,0\,0\,0\,0\,0\,0\,0\,0\,0\,0 \\
0 & 0\,0\,0\,0\,0\,0\,0\,0\,0\,0\,0 & 1 & 0\,0\,0\,0\,0\,0\,0\,0\,0\,0\,0\,0 \\
0 & 0\,0\,0\,0\,0\,0\,0\,0\,0\,0\,0 & 1 & 0\,0\,0\,0\,0\,0\,0\,0\,0\,0\,0\,0 \\
0 & 0\,0\,0\,0\,0\,0\,0\,0\,0\,0\,0 & 1 & 0\,0\,0\,0\,0\,0\,0\,0\,0\,0\,0\,0 \\
0 & 0\,0\,0\,0\,0\,0\,0\,0\,0\,0\,0 & 1 & 0\,0\,0\,0\,0\,0\,0\,0\,0\,0\,0\,0 \\
0 & 0\,0\,0\,0\,0\,0\,0\,0\,0\,0\,0 & 1 & 0\,0\,0\,0\,0\,0\,0\,0\,0\,0\,0\,0 \\
0 & 0\,0\,0\,0\,0\,0\,0\,0\,0\,0\,0 & 0 & 0\,0\,0\,0\,0\,0\,0\,0\,0\,0\,0\,1 \\
0 & 0\,0\,0\,0\,0\,0\,0\,0\,0\,0\,0 & 0 & 0\,0\,0\,0\,0\,0\,0\,0\,0\,0\,0\,1 \\
0 & 0\,0\,0\,0\,0\,0\,0\,0\,0\,0\,0 & 0 & 0\,0\,0\,0\,0\,0\,0\,0\,0\,0\,0\,1 \\
0 & 0\,0\,0\,0\,0\,0\,0\,0\,0\,0\,0 & 0 & 0\,0\,0\,0\,0\,0\,0\,0\,0\,0\,0\,1 \\
0 & 0\,0\,0\,0\,0\,0\,0\,0\,0\,0\,0 & 0 & 0\,0\,0\,0\,0\,0\,0\,0\,0\,0\,0\,1 \\
0 & 0\,0\,0\,0\,0\,0\,0\,0\,0\,0\,0 & 0 & 0\,0\,0\,0\,0\,0\,0\,0\,0\,0\,0\,1
\end{array} \right).
$$

Although it is possible to calculate the transition matrices for all social spaces, this process is not straightforward and is prone to errors. Therefore, we do not formally derive the transition matrices, but determine them through simulations.

As mentioned above in this Section, for each social space S_h, we compare the two scenarios, one with and the other without a lobby. For each scenario, $1,000,000$ simulations are run to determine the transition matrix.

From the transition matrices, we compute the probabilities of converging to the absorbing states as illustrated in Ref. [44]; those probabilities are graphically represented, together with the average time of convergence to the absorbing states, in Figures 3–8. For each social space S_h ($h = 1, \ldots, 6$), Figures 3–8, left, represent the probabilities of convergence to the respective absorbing states from any initial states, $j = 0, \ldots, 24$. The graphs in Figures 3–8, right, represent the mean time, t_j, of convergence towards the absorbing states, starting from any initial state j. Since we are considering discrete states, the lines connecting the dots are for illustrative purposes only. As expected, when $j \geq N/2$, opinion $\overline{\mathcal{X}}$ is no longer a minority opinion. In Appendix A, we provide the pseudocode for the computation of the transition matrices $\mathbf{M}_h$ without a lobby and the transition matrices $\overline{\mathbf{M}}_h$ of the population with the lobby (the R code and the libraries to replicate our results are available from the authors upon request). As j increases to values larger than 18, we correspondingly reduce the number of the lobby members; for example, when $j = 20$, the number of the lobby members are four, and, in the social space S_2, they would only sit at two tables instead of three.

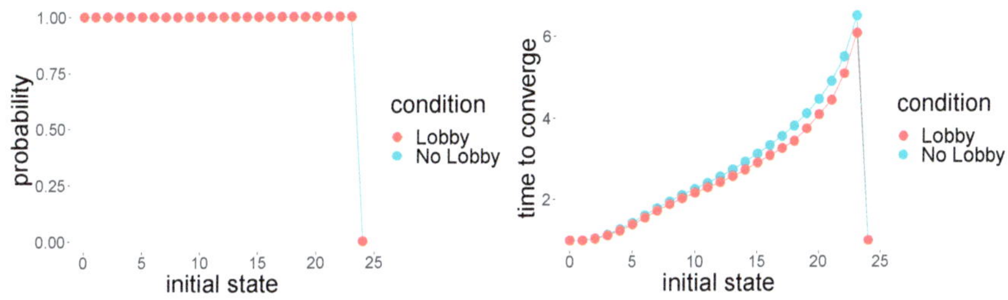

(**a**) (**b**)

Figure 3. Social space S_1: twelve tables with two seats each and the members of the lobby sitting strategically, as shown in Figure 2a. The probability of convergence (**a**) to the unanimous opinion $\overline{\mathcal{X}}$ and the absorption time (**b**) to any unanimous opinion as functions of the initial number of individuals with opinion $\mathcal{X}$ for cases with (red) and without (blue) a lobby.

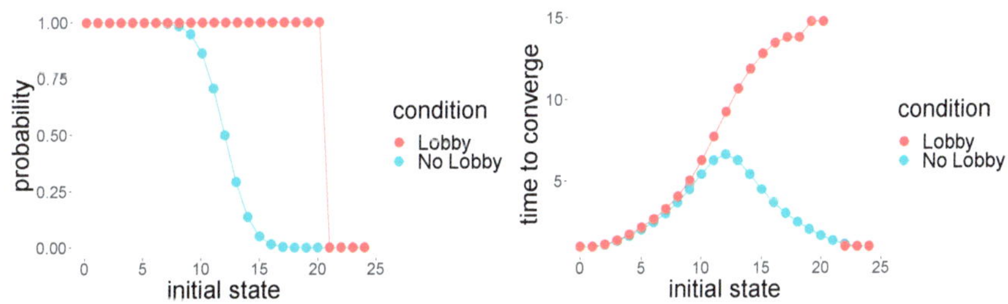

(**a**) (**b**)

Figure 4. Social space S_2: eight tables with three seats each and the members of the lobby sitting strategically, as shown in Figure 2b. The probability of convergence (**a**) to the unanimous opinion $\overline{\mathcal{X}}$ and the absorption time (**b**) to any unanimous opinion as functions of the initial number of individuals with opinion $\mathcal{X}$ for cases with (red) and without (blue) a lobby.

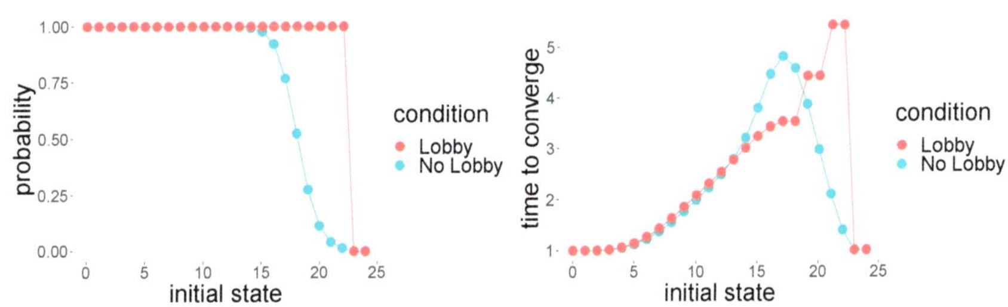

(**a**) (**b**)

Figure 5. Social space S_3: six tables with four seats each and the members of the lobby sitting strategically, as shown in Figure 2c. The probability of convergence (**a**) to the unanimous opinion $\overline{\mathcal{X}}$ and the absorption time (**b**) to any unanimous opinion as functions of the initial number of individuals with opinion $\mathcal{X}$ for cases with (red) and without (blue) a lobby.

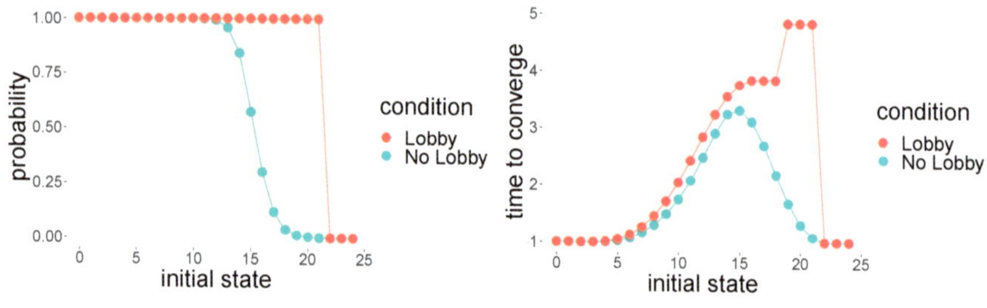

(a) (b)

Figure 6. Social space S_4: four tables with six seats each and the members of the lobby sitting strategically, as shown in Figure 2d. The probability of convergence (**a**) to the unanimous opinion $\overline{\mathcal{X}}$ and the absorption time (**b**) to any unanimous opinion as functions of the initial number of individuals with opinion $\mathcal{X}$ for cases with (red) and without (blue) a lobby.

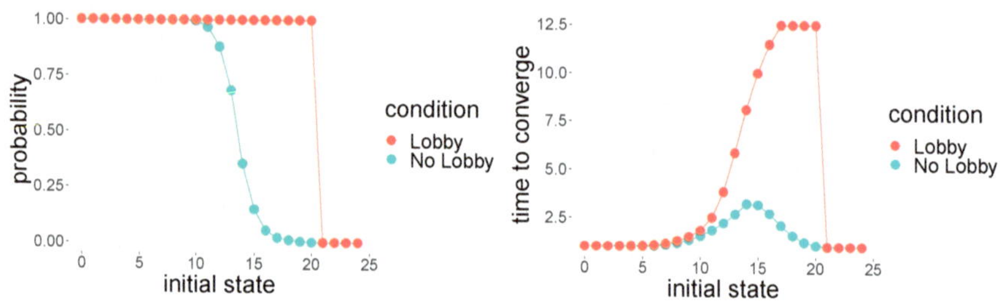

(a) (b)

Figure 7. Social space S_5: three tables with eight seats each and the members of the lobby sitting strategically, as shown in Figure 2e. The probability of convergence (**a**) to the unanimous opinion $\overline{\mathcal{X}}$ and the absorption time (**b**) to any unanimous opinion as functions of the initial number of individuals with opinion $\mathcal{X}$ for cases with (red) and without (blue) a lobby.

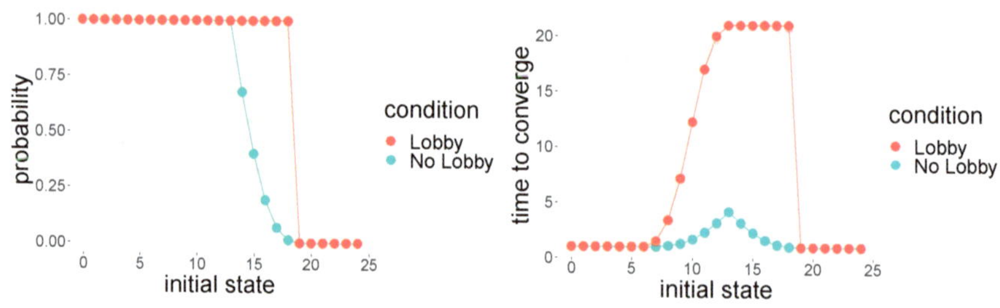

(a) (b)

Figure 8. Social space S_6: two tables with thwelve seats each and the members of the lobby sitting strategically, as shown in Figure 2f. The probability of convergence (**a**) to the unanimous opinion $\overline{\mathcal{X}}$ and the absorption time (**b**) to any unanimous opinion as functions of the initial number of individuals with opinion $\mathcal{X}$ for cases with (red) and without (blue) a lobby.

4. Results and Discussion

Formulas (1) and (2) are general and can be written for finite populations and any social space; moreover, the simulation approach—which is used to avoid the tedious calculation of all possible cases—can also be used in general.

The results show that, across all the social spaces considered, lobbies are able to influence populations with increased sample sizes. This is true for all social spaces except for the simplest one, where there are no differences in dynamics between the situation with and that without a lobby. Moreover, if one considers the time of convergence in each of Figures 3–8, i.e., the time to converge to the opinion held by the lobby ("against opinion $\mathcal{X}$"), the lobbies increase rather than significantly decrease the average number of steps to convergence to an absorbing state, i.e., to a state that cannot be left once entered [51] (p. 35). This means that even if strategic lobbies are able to influence a larger part of the population, their presence can shorten the time to change the public opinion on a given issue. From these two findings, one can conclude that the presence of lobbies can ensure influence on a larger scale, but at the cost of time. In Figures 3–8, we summarize the dynamics with and without lobbies for each of the social spaces considered.

In the social space S_1, the results for convergence are identical for both models with and without lobbies; however, the average convergence time decreases when there are lobby members (Figure 3).

The social space S_2 exhibits a new absorbing state that can be deduced from Figure 2b when the initial state of the population has a lobby with three members. Two of these members strategically sit at the same table, leaving the other member alone at a different table. According to the table majority rule, the pair of lobbyists will persuade the third person at the table in favor of opinion $\overline{\mathcal{X}}$, while the lone lobby member will be persuaded in favor of opinion $\mathcal{X}$, thus maintaining the size of the lobby as originally three members; this is an impasse situation similar to that discussed in Ref. [44]. However, it must be noted that, in this social space, the impasse can only occur when the lobby is active. In Figure 4b, the mean time of convergence to this new absorbing state is not reported as it is not a unanimous state. According to Galam's definition (see [31] (p. 577) for details), a killing point is the unstable fixed point, producing the flow and its direction, determined by the stochastic dynamical system describing the time evolution of the hoax support. In S_2, one can see in Figure 4a that the state $j = 12$ is the killing point when there is no lobby; in contrast, in this social space, when the lobby is present, there is no killing point and this case suggests a bang-bang result [54]. The mean absorption time of unanimity is higher around the killing point; the process is even slower when generated by the lobby, as shown in Figure 4b.

The social space S_3 (Figure 5) has a killing point at state $j = 18$, i.e., at the initial state with eighteen individuals with opinion $\mathcal{X}$ against a lobby of six members. As for space S_2, the mean time of absorption to a unanimous opinion increases when the initial number of individuals in favor of $\mathcal{X}$ is near the killing point. In contrast to space S_2, however, the process triggered by lobbyists near the killing point is faster than in the case without a lobby. Only a further reduction in the size of the lobby leads to a slowdown in the process of convergence to unanimity. This is because the lobbyists' strategy favors the preservation of their opinion over its diffusion in initial states that are smaller than the killing point. That is, one has a type of *echo chamber*, defined as "environments in which the opinion, political leaning, or belief of users about a topic gets reinforced due to repeated interactions with peers or sources having similar tendencies and attitudes" [55]. Expectedly, these cases of social influence generate opinion cascades [56] that reduce individual noise at the expense of group noise [57].

In the social spaces S_4, S_5, and S_6, the average absorption times in the case with a lobby are uniformly higher than without; that is, the process of absorption is slower overall regardless of the killing points (see Figures 6–8). In other words, except for the particular case of the social space S_1, the lobby generally makes a hoax $\overline{\mathcal{X}}$ prevalent even in the cases where it would be impossible; however, this process takes longer on average.

In all the social spaces that we analyzed, except S_1, which is trivial, the killing point is lower in the case without a lobby than in the case with a lobby. This means that with the lobby, in order for the hoax $\overline{\mathcal{X}}$ not to spread, many more people are needed who do not believe it. This is not cost-less; we observe an increase in the mean time of absorption to a unanimous opinion in almost every case. Without lobbying, if the number of individuals in favor of the opinion $\mathcal{X}$ is greater than the number at the killing point, the entire population would also be in favor; with lobbying, however, the attractor changes and the population would end up believing the opinion $\overline{\mathcal{X}}$ but at a slower rate of convergence. This is always the case, except in the social space S_3, as we have already discussed just above. To summarize, we have seen that the lobby increases the killing point and transforms the opinion dynamics into a form of *bang-bang* outcome. In general, however, the lobby increases the mean time to unanimity, as the strategic seating arrangement is primarily aimed at preserving the lobbyists' opinion rather than riskily spreading it. This becomes clearer when analyzing some particular examples. For example, let us consider the social space S_6 and compare the transition probabilities to the states $j = 0$, $j = 12$ and $j = 24$ with a lobby and without a lobby. One can see from Figure 9 that for $7 \leq j \leq 12$, even if the lobby is not present, a unanimous opinion in favor of the opinion $\mathcal{X}$ cannot be reached and, at the same time, a unanimous opinion against $\mathcal{X}$ is more likely without a lobby. In these cases, the lobby strategy is too conservative and slows down the achievement of its goal. If, on the other hand, $14 \leq j \leq 18$, the lobby is effective in avoiding reaching a unanimous opinion in favor of opinion $\mathcal{X}$.

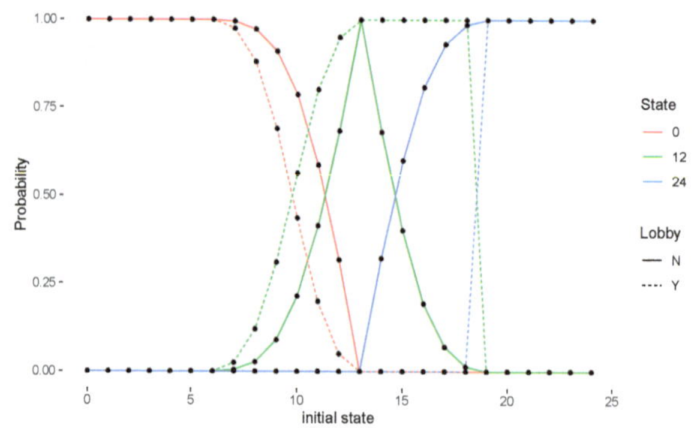

Figure 9. Transition probabilities to the states $j = 0$, $j = 12$ and $j = 24$ with (dashed lines) or without (solid lines) lobby in the social space S_6 versus the number of individuals who are in favor of opinion $\mathcal{X}$.

The same reasoning applies to the other social spaces, even if the number of states with positive probabilities makes the analysis less immediate; see, Figure 10 for the social space S_5 as an example.

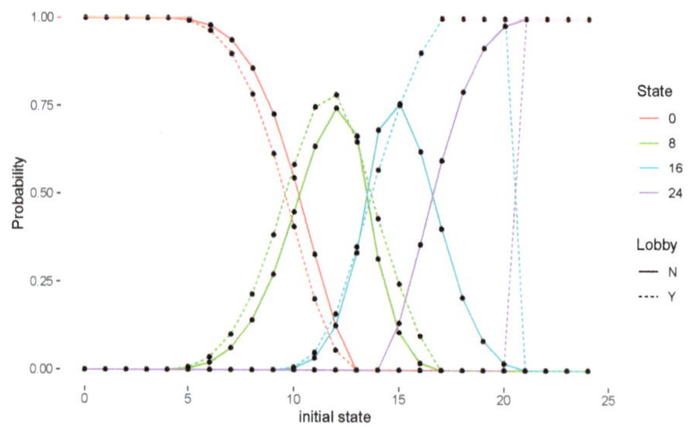

Figure 10. Transition probabilities to the states $j = 0$, $j = 12$ and $j = 24$ with (dashed lines) or without (solid lines) lobby in the social space S_5 versus the number of individuals who are in favor of opinion $\mathcal{X}$.

To avoid the slowing effect of the lobby that occurs in some social spaces, the strategic behavior of the lobby should add what [58] calls the *third dimension* in negotiation—that is, being able to determine the proper sequence of tables to be seated at, depending on the state; see also [59] for features that help groups in generating knowledge.

Finally, as we have seen in some special cases, such as social space S_2, the lobby can generate some impasses that do not occur without lobbies.

5. Conclusions

In this paper, we addressed the influence of lobbies on opinion dynamics considering a population of individuals in different social gatherings. In particular, we considered a lobby that can interact in strategic positions. Extending a known sociophysics model of opinion spreading [31], we introduced lobbies to understand when and how they can be effective in different social spaces. In this case, opinion dynamics can also be analyzed as a Markov chain. Since the calculation of transition probabilities can become tedious, these were calculated using simulations. The results show that lobbies with the investigated strategy are able to overturn the majority opinion, even if this requires a certain amount of time to reach a consensus.

The limitations of the results that we obtained highlight the known impossibility of a theory of social behavior to be simultaneously general, accurate and simple [60]; in particular, in order to have a more accurate model of opinion diffusion in a group, agents should be modeled as heterogeneous entities rather than atoms [32] (p. 29). However, in view of the "quiet methodological revolution" discussed in Ref. [41], the sociophysics perspective [32] should be considered as another possible way to overcome the problem of replicability in psychology, as mentioned in the Introduction.

The model and the analysis of the influence of lobbies can be extended in several ways. First, we have looked at a particular strategy of a lobby and a particular decision rule (i.e., the local majority). Future research would need to examine alternative strategies and decision rules to better understand how opinions can spread in social spaces. Second, we considered different twenty-four -individual social spaces consisting of separate tables; these are special types of networks in which each component is a clique. In future research, we plan to consider more general social networks similar to what was studied in Ref. [61]. In particular, it is interesting to investigate whether the influence of lobbies remains the same and how the time of convergence to unanimity is affected by their presence.

Author Contributions: The authors contributed equally to this work: Conceptualization, U.M. and A.D.F.; methodology, U.M. and A.D.F.; software, U.M.; validation, U.M. and A.D.F.; formal analysis, A.D.F.; investigation, U.M. and A.D.F.; writing—original draft preparation, U.M. and A.D.F.; writing—review and editing, U.M. and A.D.F.; visualization, U.M. and A.D.F.; supervision, A.D.F. All authors have read and agreed to the published version of the manuscript.

Funding: This research received no external funding.

Data Availability Statement: The simulated transition matrices $\mathbf{M}_h$, $\overline{\mathbf{M}}_h$, the R code and the libraries for the replication of our results are available from the authors upon request.

Conflicts of Interest: The authors declare no conflicts of interest.

Appendix A

The pseudocode to compute the transition matrices $\mathbf{M}_h$ without a lobby is reported in Algorithm A1 and the pseudocode to compute the transition matrices $\overline{\mathbf{M}}_h$ of the population with the lobby is reported in Algorithm A2, for each social space S_h.

The parameters that we considered for the simulations in Section 3.3 and the results discussed in Section 4 are the following:

- NumAgents := $N = 24$;
- LobbySize := $N^* = 6$;
- NumberSimulations := 1,000,000.

Algorithm A1 Compute Transition Frequencies with No Lobby

```
procedure TRANSITION
    M ← 0
    for InitialState:=0 to NumAgents do
        repeat
            randomly choose InitialState agents
            set their opinion 'in favor of X'
            set remaining agents opinion to 'against X'
            agents take seats
            agents discuss according to the local majority rule
            compute new state
            update M
        until NumberSimulations
```

Algorithm A2 Compute Transition Frequencies with Lobby

```
procedure TRANSITIONLOBBY
    M̄ ← 0
    for InitialState:=0 to NumAgents do
        repeat
            if InitialState < NumAgents − LobbySize then
                set lobby opinions to 'against X'
                randomly choose InitialState − LobbySize agents
                set their opinion 'in favor of X'
                set remaining agents opinion to 'against X'
            else
                set first NumAgents − InitialState lobby opinions to 'against X'
                set remaining lobby opinions to 'in favor of X'
                set remaining opinions to 'in favor of X'
            lobby agents take seats
            other agents take seats
            all agents discuss according to the local majority rule
            compute new state
            update M̄
        until NumberSimulations
```

References

1. Dominus, S. When the Revolution Came for Amy Cuddy. *The New York Times*, 18 October 2017. Available online: https://www.nytimes.com/2017/10/18/magazine/when-the-revolution-came-for-amy-cuddy.html (accessed on 5 February 2024).
2. Jack, A.; Hill, A. Harvard Fraud Claims Fuel Doubts over Science of Behaviour. *Financial Times*, 30 June 2023. Available online: https://www.ft.com/content/846cc7a5-12ee-4a44-830e-11ad00f224f9 (accessed on 5 February 2024).
3. Aarts, A.A. et al. [Open Science Collaboration]. Estimating the reproducibility of psychological science. *Science* **2015**, *349*, aac4716. [CrossRef] [PubMed]
4. Cialdini, R.B.; Goldstein, N.J. Social influence: Compliance and conformity. *Annu. Rev. Psychol.* **2004**, *55*, 591–621. [CrossRef] [PubMed]
5. Sunstein, C.R. What's available—Social influences and behavioral economics empirical legal realism: A new social scientific assessment of law and human behavior. *Northwest. Univ. Law Rev.* **2002**, *97*, 1295–1314. Available online: https://chicagounbound.uchicago.edu/journal_articles/8621/ (accessed on 5 February 2024).
6. Asch, S.E. Studies of independence and conformity: I. A minority of one against a unanimous majority. *Psychol. Monogr. Gen. Appl.* **1956**, *70*, 1–70. [CrossRef]
7. Jung, J.; Bramson, A. A recipe for social change: Indirect minority influence and cognitive rebalancing. In Proceedings of the Annual Conference of the Computational Social Science Society of the Americas: CSSSA 2016, Santa Fe, NM, USA, 17–20 November 2016. Available online: https://computationalsocialscience.org/wp-content/uploads/2016/11/CSSSA_2016_paper_5-1.pdf (accessed on 5 February 2024).

8. Jung, J.; Page, S.E.; Miller, J.H.; Bramson, A.; Crano, W.D. The impact of indirect minority influence on diversity of opinion and the magnitude, speed and frequency of social change. In Proceedings of the 2018 International Conference of The Computational Social Science Society of the Americas: CSSSA 2018, Santa Fe, NM, USA, 25–28 October 2018. Available online: https://www.researchgate.net/publication/331073577_The_Impact_of_Indirect_Minority_Influence_on_Diversity_of_Opinion_and_the_Magnitude_Speed_and_Frequency_of_Social_Change_A_Brief_Report (accessed on 5 February 2024).
9. Jung, J.; Bramson, A.; Crano, W.D.; Page, S.E.; Miller, J.H. Cultural drift, indirect minority influence, network structure, and their impacts on cultural change and diversity. *Am. Psychol.* **2021**, *76*, 1039–1053. [CrossRef] [PubMed]
10. Ausloos, M.; Dawid, H.; Merlone, U. Spatial interactions in agent-based modeling. In *Complexity and Geographical Economics: Topics and Tools*; Commendatore, P., Kayam, S., Kubin, I., Eds.; Springer International Publishing Switzerland: Cham, Switzerland, 2015; pp. 353–377. [CrossRef]
11. Wood, W.; Lundgren, S.; Ouellette, J.A.; Busceme, S.; Blackstone, T. Minority influence: A meta-analytic review of social influence processes. *Psychol. Bull.* **1994**, *115*, 323–345. [CrossRef] [PubMed]
12. Paichler, G. Norms and attitude change I: Polarization and styles of behaviour. *Eur. J. Soc. Psychol.* **1976**, *6*, 405–427. [CrossRef]
13. Moscovici, S.; Personnaz, B. Studies in social influence. *J. Exp. Soc. Psychol.* **1980**, *16*, 270–282. [CrossRef]
14. Moscovici, S.; Lage, E.; Naffrechoux, M. Influence of a consistent minority on the responses of a majority in a color perception task. *Sociometry* **1969**, *32*, 365–380. [CrossRef]
15. Mugny, G.; Papastamou, S. When rigidity does not fail: Individualization and psychologization as resistances to the diffusion of minority innovations. *Eur. J. Soc. Psychol.* **1980**, *10*, 43–61. [CrossRef]
16. Crano, W.D.; Seyranian, V. Majority and minority influence. *Soc. Personal. Psychol. Compass* **2007**, *1*, 572–589. [CrossRef]
17. Maass, A.; Clark, R.D. Hidden impact of minorities: Fifteen years of minority influence research. *Psychol. Bull.* **1984**, *95*, 428–450. [CrossRef]
18. Butera, F.; Mugny, G.; Legrenzi, P.; Perez, J.A. Majority and minority influence, task representation and inductive reasoning. *Brit. J. Soc. Psychol.* **1996**, *35*, 123–136. [CrossRef]
19. Mugny, G. Negotiations, image of the other and the process of minority influence. *Eur. J. Soc. Psychol.* **1975**, *5*, 209–228. [CrossRef]
20. Zollman, K.J.S. Social network structure and the achievement of consensus. *Politics Philos. Econ.* **2012**, *11*, 26–44. [CrossRef]
21. Zollman, K.J.S. The epistemic benefit of transient diversity. *Erkenntnis* **2010**, *72*, 17–35. [CrossRef]
22. Gibbard, A. Manipulation of voting schemes: A general result. *Econometrica* **1973**, *41*, 587–601. [CrossRef]
23. Gibbard, A. Manipulation of schemes that mix voting with chance. *Econometrica* **1977**, *45*, 665–681. [CrossRef]
24. Dong, J.-Y.; Lu, X.-Y.; Li, H.-C.; Wan, S.-P.; Yang, S.-Q. Consistency and consensus enhancing in group decision making with interval-valued intuitionistic multiplicative preference relations based on bounded confidence. *Inform. Sci.* **2024**, *652*, 119727. [CrossRef]
25. Zou, W.-C.; Wan, S.-P.; Dong, J.-Y. Trust evolution based minimum adjustment consensus framework with dynamic limited compromise behavior for probabilistic linguistic large scale group decision-making. *Inform. Sci.* **2024**, *652*, 119724. [CrossRef]
26. Heft, A.; Buehling, K. Measuring the diffusion of conspiracy theories in digital information ecologies. *Convergence* **2022**, *28*, 940–961. [CrossRef]
27. Dentith, M.R.X. *The Philosophy of Conspiracy Theories*; Palgrave Macmillan: London, UK, 2014; Chapter 3. [CrossRef]
28. Furini, M. Identifying the features of ProVax and NoVax groups from social media conversations. *Comput. Hum. Behav.* **2021**, *120*, 106751. [CrossRef]
29. Rutjens, B.T.; Većkalov, B. Conspiracy beliefs and science rejection. *Curr. Opin. Psychol.* **2022**, *46*, 101392. [CrossRef] [PubMed]
30. Albarracín, D. Processes of persuasion and social influence in conspiracy beliefs. *Curr. Opin. Psychol.* **2022**, *48*, 101463. [CrossRef]
31. Galam, S. Modelling rumors: The no plane Pentagon French hoax case. *Phys. Stat. A Mech. Appl.* **2003**, *320*, 571–580. [CrossRef]
32. Galam, S. *Sociophysics: A Physicist's Modeling of Psycho-Political Phenomena*; Springer Science+Business Media, LLC: New York, NY, USA, 2012. [CrossRef]
33. Galam, S.; Moscovici, S. Towards a theory of collective phenomena: Consensus and attitude changes in groups. *Eur. J. Soc. Psychol.* **1991**, *21*, 49–74. [CrossRef]
34. Berger, L.R. A necessary and sufficient condition for reaching a consensus using DeGroot's method. *J. Am. Stat. Assoc.* **1981**, *76*, 415–418. [CrossRef]
35. DeGroot, M.H. Reaching a consensus. *J. Am. Stat. Assoc.* **1974**, *69*, 118–121. [CrossRef]
36. Potters, J.; van Winden, F. Lobbying and asymmetric information. *Public Choice* **1992**, *74*, 269–292. [CrossRef]
37. Nyczka, P.; Sznajd-Weron, K. Anticonformity or independence?–Insights from statistical physics. *J. Stat. Phys.* **2013**, *151*, 174–202. [CrossRef]
38. Rouchier, J.; Thoyer, S. Modelling a European decision making process with heterogeneous public opinion and lobbying: The case of the authorization procedure for placing genetically modified organisms on the market. In *Multi-Agent-Based Simulation III*; Hales, D., Edmonds, B., Norling, E., Rouchier, J., Eds.; Springer: Berlin/Heidelberg, Germany, 2003; pp. 149–166. [CrossRef]
39. Jackson, J.C.; Rand, D.; Lewis, K.; Norton, M.I.; Gray, K. Agent-based modeling: A guide for social psychologists. *Soc. Psychol. Personal. Sci.* **2017**, *8*, 387–395. [CrossRef]
40. Smith, E.R.; Conrey, F.R. Agent-based modeling: A new approach for theory building in social psychology. *Personal. Soc. Psychol. Rev.* **2007**, *11*, 87–104. [CrossRef] [PubMed]

41. Rodgers, J.L. The epistemology of mathematical and statistical modeling: A Quiet methodological revolution. *Am. Psychol.* **2010**, *65*, 1–12. [CrossRef] [PubMed]
42. Galam, S.; Gefen (Feigenblat), Y.; Shapir, Y. Sociophysics: A new approach of sociological collective behaviour. I. Mean-behaviour description of a strike. *J. Math. Sociol.* **1982**, *9*, 1–13. [CrossRef]
43. Stauffer, D. A biased review of sociophysics. *J. Stat. Phys.* **2012**, *151*, 9–20. [CrossRef]
44. Merlone, U.; Radi, D. Reaching consensus on rumors. *Phys. Stat. A Mech. Appl.* **2014**, *406*, 260–271. [CrossRef]
45. Ellero, A.; Fasano, G.; Sorato, A. Stochastic model of agent interaction with opinion leaders. *Phys. Rev. E* **2013**, *87*, 042806. [CrossRef]
46. Hare, A.P.; Bales, R.F. Seating position and small group interaction. *Sociometry* **1963**, *26*, 480–486. [CrossRef]
47. Sommer, R. Leadership and group geography. *Sociometry* **1961**, *24*, 99–110. [CrossRef]
48. Strodtbeck, F.L.; Hook, L.H. The social dimensions of a twelve man jury table. *Sociometry* **1961**, *24*, 397–415. [CrossRef]
49. Oestereich, A.L.; Pires, M.A.; Queirȯs, S.M.D.; Crokidakis, N. Phase transition in the Galam's majority-rule model with information-mediated independence. *Physics* **2023**, *5*, 911–922. [CrossRef]
50. Newmark, A.; Nownes, A.J. All of the above: Lobbying allied, undecided, and opposing lawmakers in committee and on the floor. *Polit. Res. Quart.* **2022**, *76*, 578–592. [CrossRef]
51. Kemeny, J.G.; Snell, J.L. *Finite Markov Chains*; Springer: New York, NY, USA, 1960. Available online: https://archive.org/details/finitemarkovchai0000unse (accessed on 5 February 2024).
52. The On-Line Encyclopedia of Integer Sequences (OEIS). The OEIS Foundation Inc., 2023. Available online: https://oeis.org/A000041 (accessed on 5 February 2024).
53. Feller, W. *An Introduction to Probability Theory and Its Applications. Volume 1*; John Wiley & Sons Inc.: New York, NY, USA, 1968. Available online: https://archive.org/details/introductiontopr0001unse (accessed on 5 February 2024).
54. Bouvier, J.B.; Xu, K.; Ornik, M. Quantitative resilience of linear driftless systems. In *2021 Proceedings of the Conference on Control and its Applications*; Matinez Diaz, S., Sanfelice, R., Eds.; Society for Industrial and Applied Mathematics: Philadelphia, PA, USA, 2021; pp. 32–39. [CrossRef]
55. Cinelli, M.; De Francisci Morales, G.; Galeazzi, A.; Quattrociocchi, W.; Starnini, M. The echo chamber effect on social media. *Proc. Natl. Acad. Sci. USA* **2021**, *118*, e2023301118. [CrossRef]
56. Macy, M.; Deri, S.; Ruch, A.; Tong, N. Opinion cascades and the unpredictability of partisan polarization. *Sci. Adv.* **2019**, *5*, eaax0754. [CrossRef] [PubMed]
57. Kanheman, D.; Sibony, O.S.; Sunstein, C.R. *Noise: A Flaw in Human Judgment*; William Collins/HarperCollinsPublishers: London, UK, 2021. Available online: https://ia804606.us.archive.org/11/items/ar_20211024/BOOKS.YOSSR.COM-Noise-A-Flaw-in-Human-Judgment.pdf (accessed on 5 February 2024).
58. Lax, D.A.; Sebenius, J.K. *3D Negotiation: Powerful Tools to Change the Game in Your Most Important Deals*; Harvard Business School Press: Boston, MA, USA, 2006.
59. Zollman, K.J.S. Network epistemology: Communication in epistemic communities. *Philos. Compass* **2013**, *8*, 15–27. [CrossRef]
60. Thorngate, W. "In general" vs. "it depends": Some comments of the Gergen-Schlenker debate. *Personal. Soc. Psychol. Bull.* **1976**, *2*, 404–410. [CrossRef]
61. Merlone, U.; Radi, D.; Romano, A. Opinion dynamics on networks. In *Complex Networks and Nonlinear Dynamics: Social and Economic Interactions*; Commendatore, P., Matilla-García, M., Varela, L.M., Cánovas, J.S., Eds.; Springer International Publishing Switzerland: Cham, Switzerland, 2016; pp. 49–63. [CrossRef]

physics

MDPI

Article

Statistical Mechanics of Social Hierarchies: A Mathematical Model for the Evolution of Human Societal Structures

Nestor Caticha [1], Rafael S. Calsaverini [2] and Renato Vicente [2,*]

[1] Instituto de Fisica, Universidade de Sao Paulo, São Paulo CEP 05508-090, SP, Brazil; ncaticha@usp.br
[2] Department of Applied Mathematics, Instituto de Matematica e Estatistica, Universidade de Sao Paulo, São Paulo CEP 05508-090, SP, Brazil; rafael.calsaverini@gmail.com
* Correspondence: rvicente@usp.br

Abstract: Social structure may have changed from hierarchical to egalitarian and back along the evolutionary line of humans. Within the tradition of sociophysics, we construct a mathematical model of a society of agents subject to competing cognitive and social navigation constraints and predict, using statistical mechanics methods, that its degree of hierarchy decreases with encephalization and increases with group size, hence suggesting human societies were driven from hierarchical to egalitarian structures by the encephalization during the last few million years and back to hierarchical due to fast demographic changes during the Neolithic. In addition, applied to a different problem, the theory leads to the following predictions for modern pre-literary humans: (i) an intermediate hierarchy degree in mild climates. In harsher climates, societies will be (ii) more egalitarian if organized in small groups (of less than 100 persons) but (iii) more hierarchical if in larger (of more than 1000 persons) groups. The predicted bifurcation, characteristic of a phase transition, is also seen in the empirical cross-cultural record (248 cultures in the Ethnographic Atlas).

Keywords: maximum entropy (MaxEnt); sociophysics; agent-based models; phase transition in societies

Citation: Caticha, N.; Calsaverini, R.S.; Vicente, R. Statistical Mechanics of Social Hierarchies: A Mathematical Model for the Evolution of Human Societal Structures. *Physics* **2024**, 6, 629–644. https://doi.org/10.3390/physics6020041

Received: 31 October 2023
Revised: 11 January 2024
Accepted: 23 January 2024
Published: 19 April 2024

1. Introduction

Behavioral phylogenetics makes it plausible that the common ancestor of *Homo* and *Pan* genera had a hierarchical social structure [1–5]. Paleolithic humans with a foraging lifestyle, however, had a largely egalitarian society, and yet hierarchical structures became again common in the Neolithic period. Today, nonliterate human societies fill the ethological spectrum [6] from egalitarian to authoritarian and despotic. This nonmonotonic journey U-shaped trajectory along the egalitarian–hierarchical spectrum during human evolution was stressed by Bruce M. Knauft [1]. Several anthropological explanations [1] have been constructed for Knauft's U.

A number of recent papers [7–9] have focused on the evolutionary mechanisms that could resolve the apparent Darwinian paradox of the early-stage viability of hierarchical social structures. Since leadership positions provide preferential access to resources, it is not clear why followers would comply in the absence of coercive institutions. One potential answer is group-level selection [8] aimed at reducing scalar stress while enhancing in-group cooperation. This stream of study reveals some of the driving forces behind the connection between group size and the prevalence of hierarchy. Our work is intended to complement existing evolutionary arguments. In the context of physics, our main emphasis is on formulating an equation of state for the system, grounded in micromotives, as opposed to focusing on the processes that govern the temporal evolution of the system's state.

Our approach to the study of social organization is based on an agent-based model and uses tools of information theory and statistical mechanics—a practice rooted in the tradition of sociophysics [10]. Sociophysics typically engages with mathematical tools to elucidate the interplay between variables in social systems. A path towards the mathematical analysis of cross-cultural data is to proceed first by formulating a hypothesis, which might take the form where two given variables are related, either by virtue of their correlation or by some causal influence. Regression analysis from the data could then be used in the attempt to falsify the hypothesis. While an important tool, this method fails to suggest new relations that might appear from a more theory-based approach to model building. Instead, we model the society by considering agents that interact through the exchange of information. In particular, we are interested in the perception of each agent of the social network of its society, taking into account cognitive constraints and social navigation demands. Then, from general principles of information theory, we obtain the probability of the agent's perceived social network conditioned on the context. By calculating expected values of network node degrees, we obtain order parameters, which indicate whether the network is in an egalitarian-symmetric or hierarchical-broken symmetry state. The context depends on a combination of parameters, which grows with cognitive capacity of the agents and decreases with group size, modulated by environmental pressure. Since social perceptions mediate motivations and hence possible behaviors, we can make predictions about the effect of these variables in the probable forms of social organization and give a scenario for the Knauft's U-shape.

We built the model to address the U-shaped curve of egalitarianism. In itself, that is not surprising. Any experimental result can be explained by a model built explicitly for that purpose. But the same model permits making predictions about another empirical situation, dealing with the influence of ecology on the expected hierarchy of modern human groups. The theory suggests a form of looking at the available ethnographic data [11] and allows a new interpretation of observed patterns involving social structure, community size and environment in terms of a competition between cognitive constraints and social navigation demands, which are related to phase transitions. Furthermore, it allows for changes beyond the U-behavior to a zig-zag-like, such as that described in Ref. [12].

2. Model and Definitions

The central concept in this article is the cognitive representation of the network of social interactions by each agent, the inferred or perceived network and not the actual social network. Each one of the n political agents of a group has a perceived social network, represented by a graph of n vertices. For every agent i, the network is $\boldsymbol{S}^i = \{s^i_{jk}\}$, where $s^i_{jk} = 1$, if agent i knows the social relation between agents j and k, independently of whether the relation is positive, neutral or negative. If agent i has no information about this relation, the bond is absent, $s^i_{jk} = 0$. This matrix is symmetric by its very design. For further details we suggest the following references [13–16] for models of the real social network with cognitive limitations and social interactions [17,18] for an agent-based model of how social stratification can emerge from efficiency demands of collective information processing. Reference [19] presents an evolutionary model for the emergence of inequity aversion as an adaptation selected from a pool of hierarchical groups. In addition, Refs. [7–9] treat other evolutionary approaches. The joint effect of demography and ecology on the emergence of hierarchy has been studied in Ref. [20]. Takao Terano and collaborators [15,16] used agent-based modeling and cognition-limited representations to analyze the emergence of money in terms of symmetry breaking in a doubly structural network representing a fixed social network and a dynamical network of exchangeability of commodities. We consider our paper as one that draws from and complements these earlier investigations.

Holding a particular belief about the network represents a cost to an agent. Our central hypothesis is that the cost can be represented by a function $C_0(\boldsymbol{S}^i)$ arising from cognitive and social constraints. The cost is built by noting that:

(1) The cognitive contribution to the cost increases with the number of memorized social relations N^i_{cog}. Then, the simplest form is

$$N^i_{\text{cog}} = \frac{1}{2}\sum_{j,k}^{n} s^i_{jk}. \tag{1}$$

(2) It is preferable to have first-hand information about the nature of associations of third parties, compared to the cost of relying on indirect information derived from heuristics such as the 'friend of a friend is a friend' and its variations. Two agents j and k can be connected by a bond or a path of bonds, through intervening agents, so that their social relation can be estimated. Call l^i_{jk} the length of the shortest path of bonds joining j and k in the perceived network $\boldsymbol{S}^i$. It is natural to assume that this lack of direct knowledge implies a social cost, which increases with the length of the shortest path between the agents.

Since $\boldsymbol{S}^i$ is the adjacency matrix, $\boldsymbol{M}^i(\lambda)$ is defined by the integer powers λ of $\boldsymbol{S}^i$ (see, e.g., Ref. [21]):

$$\boldsymbol{M}^i(\lambda) = \mathbf{S}^{i^{\lambda}}, \tag{2}$$

permit determining the shortest path

$$l^i_{jk} = \min \lambda, \quad \text{such that } \boldsymbol{M}^i(\lambda)_{jk} > 0. \tag{3}$$

Define $\bar{L^i}$, the mean over all pairs (j,k) (counted only once)

$$\bar{L^i} = \frac{2}{N(N-1)}\sum_{j,k} l^i_{jk}. \tag{4}$$

Both N^i_{cog} and $\bar{L^i}$ are functions of $\boldsymbol{S}^i$, the variables of the problem.

We define the joint cognitive-social cost of the representation as the sum of the cognitive term and the social cost. The purpose of this exercise is not to uncover a 'true' form for the function that describes the cost to an agent. Our focus is on understanding the fundamental outcomes of the basic competition between costs rather than a precise identification of a functional form.

To be able to sum such different things, we introduce a parameter α:

$$C_0(\boldsymbol{S}^i) = N^i_{\text{cog}} + \alpha\bar{L^i}. \tag{5}$$

The interpretation of the meaning of α comes from the following argument. For fixed values of N^i_{cog} and $\bar{L^i}$, as α increases, the relative contribution of the cognitive cost N^i_{cog} to the overall cost decreases. It is thus a proxy of the cognitive capacity of an agent. It can be assumed that α can be associated with the mass ratio between the neocortex and rest of the brain [22–24] and/or the medial frontal cortex including the anterior cingulate cortex [25–27]. The cost is infinite if there is no path between two agents in a perceived social network, so only connected graphs occur.

As this stands, an agent-perceived network is independent of the network of another agent. We now introduce interactions.

(3) Many different forms of communication lead to social learning [28]. We follow ideas from opinion dynamics [29] and suppose that, through gossiping, agents share information or learn from social partners about coalitions of pairs of other agents, leading to correlated perceived networks. In the simulations, we introduce the possibility of an agent m copying from an interlocutor i its link about any other pair (j,k).

Finally, we make a hypothesis that enters in the interpretation of the results but not in the construction of the model.

(4) We suppose that perceptions about social coalitions engender motivations that direct behaviors. We call this the Perceptions Motivate Behaviors (PMB) hypothesis. Hence, there is a correlation between the perceived social network and the real coalitions emerging. We believe that this particular step of our argument should be amenable

to experimental verification. It receives support from the finding that in traditional societies, perceived positions in a social hierarchy are important to determine the commanded authority [30]. Within the PMB hypothesis, when the perceived social network is symmetric, social agents are not inclined to accept inequalities. Attempts to dominate a group, either economically, politically or sexually, would be met with resistance. However, if the symmetry is broken, social agents would not be so reluctant to accept a hierarchical political organization, at least under high rates of information exchange (or gossiping), when the central agent is with high probability the same for most of the agents.

3. The Probability of a Perceived Social Network

The standard methodology of information theory and statistical mechanics, which includes Bayesian and entropic methods, leads to the attribution of a probability $P(\{\mathbf{S}^i\}|I)$ to a network $\mathbf{S}^i$, conditioned on available information I.

Consider $\mathbb{E}(C_0)$ the expected value of $C_0(\mathbf{S}^i)$ under $P(\mathbf{S}^i|I)$

$$\mathbb{E}(C_0) = \sum_{\mathbf{S}^i} C_0(\mathbf{S}^i)P(\mathbf{S}^i|I), \tag{6}$$

then the sum is over all 2^n possible configurations of $\mathbf{S}^i$. Suppose that either the information available is that $\mathbb{E}(C_0)$ has a specific value $\mathcal{C}$ or that it has a significantly constant value during a relatively long time or alternatively, it is supposed that the value of $\mathbb{E}(C_0)$ to be $\mathcal{C}$. In the last case, as $\mathcal{C}$ is varied, different behaviors will be revealed.

The method of inference by maximum entropy introduces a parameter β. This parameter (or Lagrange multiplier) sets the range of fluctuations of the cost above its minimum value. Semantically, it is analogous to a pressure under which the band lives. Large values above the minimum cost are unlikely for large β but very likely if β is small. This pressure may arise from the ecology or it might model normative or informational peer pressure, from the perception of external threats, or as peer pressure arising from moral standards [31,32]. The name 'pressure' describes β in the generalized manner used in thermodynamics [33], in the sense that it is an entropic intensive parameter conjugated to the extensive cost in $\mathcal{C}$. In a simple fluid, the temperature is conjugated to the energy, but here, $\mathcal{C}$ is not an energy nor β an inverse temperature.

The procedure calls for the maximization of the information entropy subject to the known constraints [34]. The entropy can be interpreted as a measure of the ignorance about $\mathbf{S}^i$. Many probability distributions are compatible with the information constraint in Equation (6). Therefore, we take the one distribution that is most ignorant while still compatible with the information and the normalization requirement:

$$P(\mathbf{S}^i|I) = \texttt{argmax}_P\Big\{ -\sum_{\mathbf{S}^i} P(\mathbf{S}^i)\log P(\mathbf{S}^i) \tag{7}$$

$$-\beta_0\big(\sum_{\mathbf{S}^i} P(\mathbf{S}^i) - 1\big) - \beta\big(\sum_{\mathbf{S}^i} P(\mathbf{S}^i)C_0(\mathbf{S}^i) - \mathcal{C}\big)\Big\}. \tag{8}$$

The result is the Boltzmann probability distribution

$$P(\mathbf{S}^i|I) = \frac{1}{Z}e^{-\beta C_0(\mathbf{S}^i)}, \tag{9}$$

where β, which controls the scale of fluctuations, is the intensive Lagrange multiplier conjugated to the extensive $\mathcal{C}$. The information content in the intensive β is equivalent to that in the extensive $\mathcal{C}$ due to Equation (6) since they are conjugated variables. The partition function Z depends on the number of agents, α and β, $Z = Z(n, \alpha, \beta)$.

The macroscopic state of the system is characterized by the values of appropriate order parameters, expected values of functions of interest, which we now define. We probe here whether represented agents are considered symmetrically or if distinctions are made. It is useful to consider the degree d^i_j of vertex j in a graph $\mathbf{S}^i$, the number of agents k with $s^i_{jk} = 1$:

$$d^i_j = \sum_k s^i_{jk}. \tag{10}$$

In addition, we define the maximum degree and the average degree, respectively,

$$d^i_{\max} = \mathtt{max}_j d^i_j, \quad \text{and} \quad d^i_{\text{avg}} = \frac{1}{n}\sum_{j=1}^{n} d^i_j, \tag{11}$$

for agent i. Natural order parameters are the expectation values $\mathbb{E}(d_{\max})$ and $\mathbb{E}(d_{\text{avg}})$ with respect to $P(\mathbf{S}^i|I)$.

We obtain estimates of the order parameters employing simple Metropolis Monte Carlo (MC) methods. We first considered isolated agents and did a MC simulation of the Boltzmann distribution (9). Calculations of the minimum distance were performed using Dijkstra's algorithm. Qualitatively equivalent results can be obtained by defining a distance to be infinite if larger than a fixed finite length. Then, we simulate the exchange of information between the agents, which couples the different perceived networks.

At this point a methodological disclaimer is necessary. Our approach can be described as an application of maximum entropy inference, akin to the canonical statistical analysis of physical systems at equilibrium. The appearance of a Boltzmann distribution as a result of our method should not be interpreted as an excessive reliance on physical analogies. Rather, it demonstrates the extensive applicability of this inference technique.

4. Results

The expected values $\mathbb{E}(d_{\max})$ of the maximum degree and $\mathbb{E}(d_{\text{avg}})$ of the average degree, with respect to the probability distribution $P(\mathbf{S}^i|I)$, are used to construct the order parameter H that signals hierarchical organization:

$$H = 2 - 2\frac{\mathbb{E}(d_{\text{avg}})}{\mathbb{E}(d_{\max})}. \tag{12}$$

This is a helpful definition since for large $H \lesssim 2$, the symmetry of the perceived networks is broken and for small $H \gtrsim 0$, representations are symmetric. The factor 2 is included for convenience when comparing with the empirical record. A central result from the simulations is that, given β, to a good approximation, H does not depend on α and n independently but on the value of the combination $z = 2\alpha/(n(n-1))$, the specific cognitive capacity per dyadic relation.

Our results can be described as follows, see Figure 1. For a given value of the pressure β, around 10, for sufficiently large specific cognitive capacity z (>1.2), H is close to zero. The typical perceived network of an agent is a fully connected graph (Figure 1, lower, on the right), meaning that cognitive costs are low enough and every possible pair of social interactions in the group can be stored. In the full graph, every agent is equal to any other. The representation of the network of social interactions is symmetric, and there is no agent that is seen as a central hub. The dynamics of social changes, a stochastic MC dynamics, is frozen.

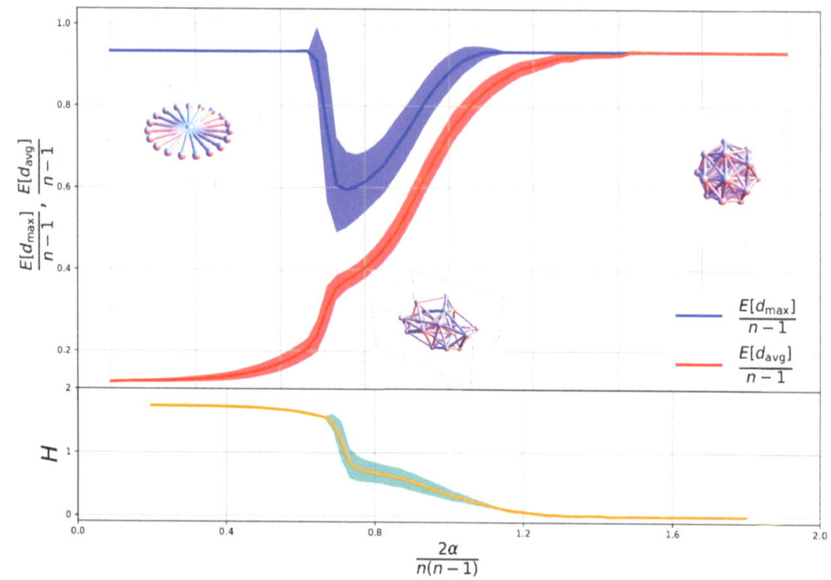

Figure 1. Upper: Monte Carlo estimates of the expected values of the maximum degree, averaged over the population, and the average degree of the social networks representation as a function of the specific cognitive capacity per dyadic relation, z, for $\beta = 10$. **Lower**: The order parameter H (12). Three regimes can be seen: for very large z, H goes to zero (symmetric phase) and almost all agents are equal. For very small z, H goes to 2, the broken symmetry phase. A particular agent occupies the central position of the network. An intermediate z transition region shows intermediate values of H. Error bars indicate the size of fluctuations obtained from the simulations. The insets are typical realizations of the inferred network of social interactions by an agent at that z position. Bonds are only drawn if the bond variable is present.

Either a smaller cognitive capacity or a larger size of the local social group, leading to intermediate z values, changes the above full graph picture (Figure 1, lower, in the center). Now, H is close to 1. Some bonds have to be absent in order to respect cognitive limits. The representations are still symmetric in a statistical sense, but fluctuations do occur, and a few agents will have a higher number of known social relations than others. The degrees of all vertices are not the same. Now, the dynamics of the perceived social network is not frozen, and different agents may temporarily occupy a more centralized position. To the perceiving agent, the members of its group are statistically equal, but its knowledge of their social connections is not uniform. Unequal roles start to appear in the inferred network, leading to the possibility of a social organization where there are 'big men' [35,36], and their distinctions might be described as those of first among equals, or rather among statistical equals in this theory.

For low z, resulting from an even larger number of agents or smaller cognitive capacity, a new regime for the organization of the inferred network emerges (Figure 1, bottom left), signaled by a larger value of $H \approx 2$. The symmetric role of the agents in the representation is broken. One of the represented agents has a special position, likely to be connected to most of the other agents, which in turn are connected almost uniquely to it. The dynamics again is frozen, and an agent crystallizes in a unique central position in the star topology of the inferred social network. Still, different agents have been sampled independently, and the central agents of the perceived networks are with high probability not the same.

Figure 2 shows the phase diagram, obtained by plotting the value of H in the (z, β) space. Figure 1 was obtained by a horizontal cut of Figure 2 at $\beta^{-1} = 0.1$.

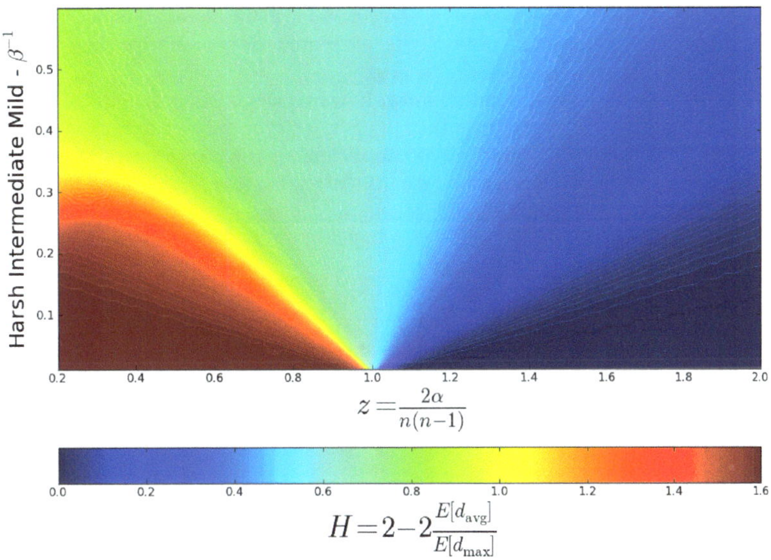

Figure 2. Hierarchical−egalitarian phase transition: phase diagram in the plane of specific cognitive capacity per dyadic relation, z, and the inverse ecological pressure, β^{-1}. The color code represents a measure of the social hierarchy measure or symmetry breaking parameter H. The red region is where the symmetry is broken (H near 2) and the maximum degree $\mathbb{E}(d_{\max})$ is much larger than the mean $\mathbb{E}(d_{\text{avg}})$. The blue region is the unbroken symmetry phase, $H \approx 0$.

5. Correlation of Networks through Gossip

Gossip, though not a standard technical term in statistical mechanics, is employed here to denote the interaction between agents that arises through the exchange of information about interpersonal relationships. The impact of gossip is to create correlations among the perceived social networks of various agents. The three regimes—full graph, intermediate and star—remain unaffected.

We now run the simulation for the n agents together. A parameter g ($0 < g < 1$) measures the intensity of information exchange through gossip.

Choose an agent i and a pair (j,k), independently of anything else, uniformly at random. A no gossip dynamic step occurs with probability $(1-g)$: a MC Metropolis update is performed on the bond of the pair (j,k). Let $\bar{C}_0$ be the joint cognitive-social cost with the bond s^i_{jk} replaced by $\bar{s}^i_{jk}$. With probability $\min(\exp(-\beta(\bar{C}_0 - C_0)), 1)$ let the change of s^i_{jk} by $\bar{s}^i_{jk}$ be accepted. Otherwise, s^i_{jk} is kept fixed. With probability g, another agent l is chosen, also independently and uniformly at random and its corresponding edge s^l_{jk} is copied to $\bar{s}^i_{jk}$.

The step performed with probability $(1-g)$ simulates the update of the social network representation by independent observations, learning new relations and forgetting about previously known relations. The gossip step, done with probability g, simulates the exchange of information where agent l tells and agent i learns or forgets something about the relation of agents j and k. Gossip can be introduced by more elaborate schemes, but this is sufficient for our modeling purposes.

After all i have been considered, a MC step has been completed. The α range $4.5 \le \alpha \le 90$ was divided uniformly into 100 intervals; the β range $0 < \beta \le 20$ was divided into 200 intervals. The values of n varied from 7 to 15. We run a MC simulation for each fixed n and for each pair of α and β. 10^6–2×10^6 MC steps were run for thermalization, and then data about the order parameters were collected for around 4×10^6 MCS. It is usual in studying phase transitions to look at the large n limit of the population, since singularities are only possible in the thermodynamic limit. However, since we are interested in the finite n case, we search for quantitative changes but not for critical exponents nor universality classes.

Figure 3 shows how frequent is the most frequent central agent as a function of the level of gossip g. Let $f_{ij} = P(c = j|i)$ be the probability that for the social network representation of agent i, the central element is j. The spreading of the probability distribution can be measured by the ratio of the average over the population of its entropy to the maximum possible value $\log n$:

$$\bar{S} = \frac{1}{\log n}\left(\frac{1}{n}\sum_{i}^{n}(-\sum_{j} f_{ij}\log f_{ij})\right). \tag{13}$$

The results indicate that large correlation occurs when gossip dynamics dominates independent dynamics, starting around $g \approx 0.5$.

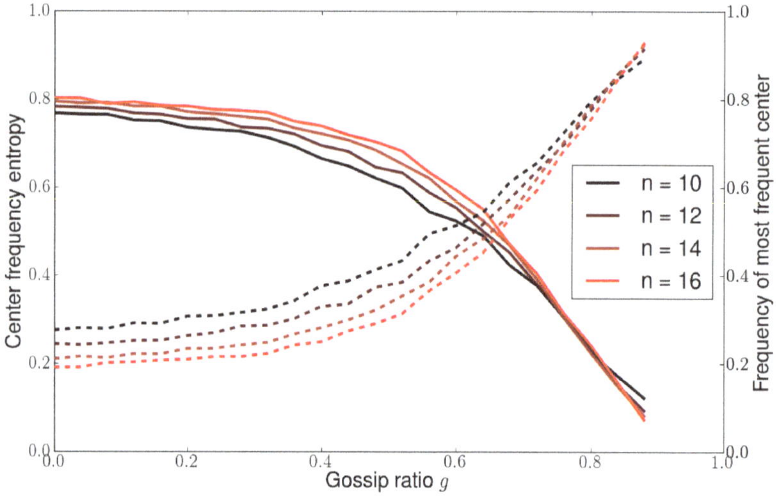

Figure 3. The effect of gossip inside the hierarchical phase. The entropy of the distribution of central agents (continuous curves) decreases with the increase of gossip, meaning that there is an onset of order where a particular agent preferentially occupies the centers of the stars. Every agent has a star network representation with a central agent. The agents share the same central agent with a probability that grows with g (dashed lines) showing another way of seeing the same ordering.

Figure 4 shows the acceptance rate for changes under the Metropolis dynamics in the network representations for three different βs. For high pressure, e.g., $\beta = 10$, and in the extreme cases of z, there is almost no tolerance to changes. In the large-z regime, upstarts that try to be different are not accepted. This is in accordance with the counter dominant behavior theory [2,5,37]. Perception of equality is hard to change, and this turns inequality unacceptable according to the PMB hypothesis.

In the low-z regime, the central agent is stable. Perceived network, highly correlated due to gossip, are very stable at both extremes of the scaling variable. The intermediate fluid phase decreases with increasing pressure β. Agents are only statistically equal and large dynamical fluctuations make the duration of the time intervals during which there is a change in the representational network much smaller than in the extreme egalitarian or hierarchical phases.

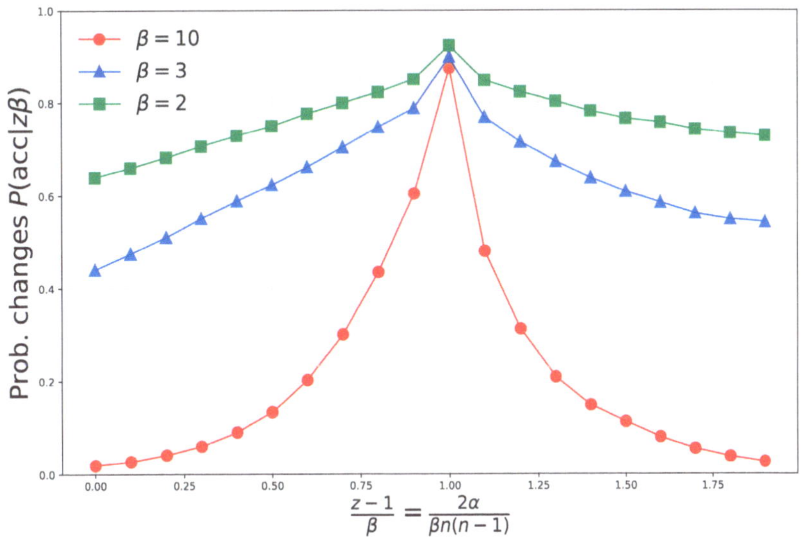

Figure 4. The probability of acceptance of changes $P(\text{acc}|z\beta)$ from the Monte Carlo simulations measures the tolerance to changes in the social network representations. As a function of z, for high pressure or harsher environments ($\beta = 10$), changes that would permit upstarts to be different from other agents are not tolerated for large z. This is analogous to counter dominance behavior theory [37]. Changes of the central agent of the star topology are unlikely to be accepted for small z. Milder pressures changes are accepted with high probability. Tolerance to changes is measured by the Monte Carlo acceptance probability.

6. Knauft's U-Shape

The broken symmetry phase with the gossip-induced correlations, together with the PMB hypothesis, allows the possibility of appearance of hierarchical social structures.

Since the relevant scaling between α and n is through z, for either larger band size for fixed cognitive resources or small enough cognitive resources for a given band size, a fully symmetric representation of the network of social relations becomes impossible and an egalitarian organization of the group becomes unlikely.

A stylized scenario for the non-monotonic change along the hierarchy spectrum (Figure 5) follows from a historical reconstruction of the dynamics of conditions external to the model. A smaller cognitive capacity (or encephalization) in the evolutionary past of the agents, and hence a smaller z, is suggestive of hierarchical organization. Along the human lineage after it separates from that leading to chimpanzees and bonobos, there was an increase in the cerebral volume and in the neocortex. The reason for this is not under scrutiny. Whether social navigation, need for cooperative hunting or toolmaking were the sole or joint driving forces, the increase is factual. As cognitive capacity increases (larger z), the social representation can become a full graph and motivate an egalitarian organization.

On a faster time scale, a growth of the number of members of bands characterizes the Neolithic, bringing back the conditions (small z) for the symmetry breaking of the representation of the social network under fixed cognitive conditions. Again, the reasons for this growth are totally outside the scope of the present mathematical model, but we can only speculate that technological advances such as food production and storage might be involved [2,5,37].

The possibility of a further change of z on a still shorter time scale, restoring a symmetric representation of society, can occur if the means for processing more information are enhanced, even in the absence of cognitive changes through evolution, but rather through advances such as printing technology or even computer networking. Thus, the model not only suggests a mechanism for the U-shape trajectory but also to conceive further reorganizations of the social network representation and its consequences on social organization due to external conditions changing the balance between computational or cognitive resources and the size of the group.

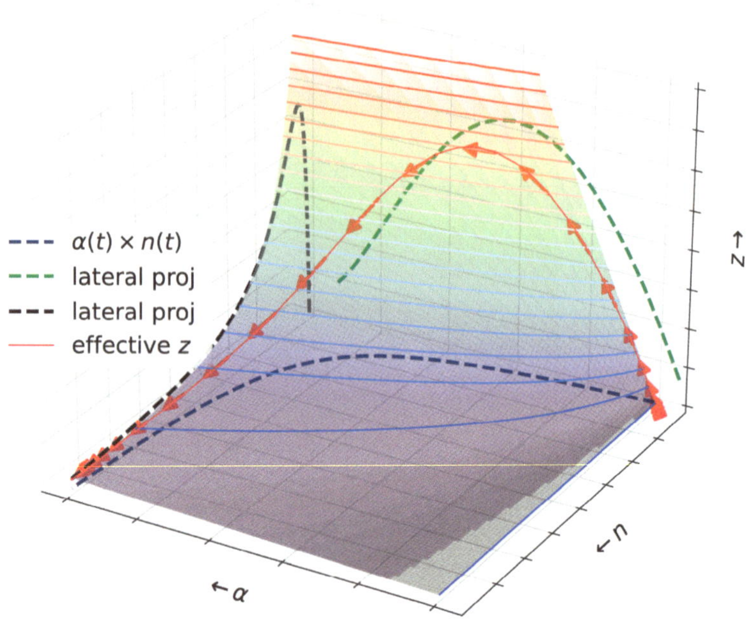

Figure 5. Schematic (inverted-) U-shape trajectory for the specific cognitive capacity, $z(\alpha, n) \propto 2\alpha/(n(n-1))$, as a function of time (arrows). The higher the value of z, the more symmetrical or egalitarian the society will be. This is just a representation of externally caused changes in the cognitive capacity $\alpha(t)$ and the mean size, $n(t)$, of social groups as a function of time, showing a rapid increase of $\alpha(t)$ followed by an increase of $n(t)$. The dashed lines are the projections onto the respective planes. The contours are drawn for constant z values.

7. Social Organization: Theoretical Predictions and the Empirical Record

We could end our presentation with the suggestion that the reason for Kanuft's U is the competition of navigational and cognitive costs. However, the model can be called to deal with a totally different problem. Can a signature of this competition between cognitive and social navigation constraints be seen today for modern humans? A clear theoretical prediction about the dependence of social stratification on ecological pressure β and group size ($z = z(n)$, α fixed) can be compared to data from the ethnographic record. The prediction is divided into two parts. First, for very mild climates, that is, small ecological pressure β, intermediate social structures are expected. As climates of increasing harshness are considered, different social organizations will occur. Second, this difference depends on group size. Cultures organized in small groups will be more egalitarian, and those in larger groups more hierarchical.

7.1. Theory: Conditional Probabilities and Order Parameters

The phase diagram in the β–z plane can be thought of as a phase diagram in the β–n plane if the encephalization, or cognitive capacities, of the agents is fixed (see Figure 6). To be able to confront with the empirical record, we divided the ranges of β and z into three regions each: harsh, intermediate and mild climates and 'small', 'medium' and 'large' groups, respectively. The phase diagram is thus divided into nine regions. The regions are chosen essentially so that all points in the β–z space in the harsh-large region are of the same color (blue). The same is done for the region of harsh–small (all red) and for mild–small and mild–large. The qualitative results are robust to changes in the detailed choices.

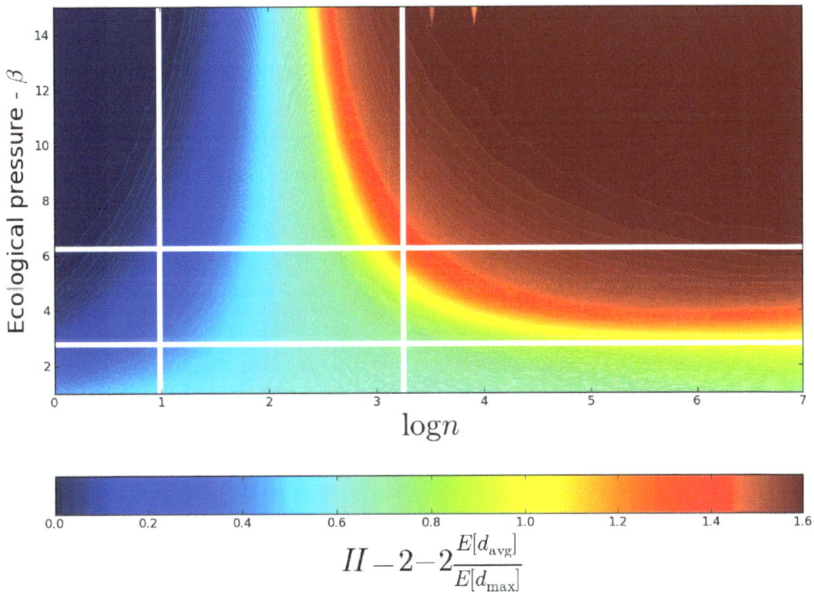

Figure 6. Phase diagram in the space of size of band versus ecological pressure, β. The color represents a measure of the social equality or symmetry H (12). The dark red region (large n) is where the symmetry is broken and the maximum degree is much larger than the mean. Dark blue (small n) is the symmetrical or egalitarian representation region. Separating those regions is a more fluid-like intermediate phase. The white lines show a choice of what is meant by large, intermediate and small both for β and n. The segmentation into nine regions is forced if the theory is to be compared to the empirical record.

A useful choice for the two values needed to separate the three regions are $\beta_H^{-1} = 0.15$, separating the harsh from the intermediate region and $\beta_M^{-1} = 0.35$ separating the intermediate form the mild region. For $\beta^{-1} > \beta_M^{-1}$, climate c is mild. For $\beta_H^{-1} < \beta^{-1} < \beta_M^{-1}$, climate c is intermediate, and for $\beta^{-1} < \beta_H^{-1}$, c is harsh.

For $z = \frac{2\alpha}{n(n-1)}$, the borders are set at $z_2 = 0.6$ and $z_1 = 1.4$ For $2 > z > z_1$, $s = 1$ small. For $z_2 > z > z_1$, $s = 2$ intermediate. For $0 < z < z_2$, $s = 3$ large. Then, we consider the mean order parameter $\bar{H}(s,c)$ for each region:

$$\bar{H}(s,c) = 2 - 2\frac{\int_{\beta\in c}\int_{z\in s} Ddzd\beta}{\int_{\beta\in c}\int_{z\in s} dzd\beta}, \tag{14}$$

where $D = \mathbb{E}(d_{\text{avg}})/\mathbb{E}(d_{\text{max}})$.

7.2. Ethnographical Data

Data were obtained from the Ethnographic Atlas (EA) [11] using the Standard Cross Cultural code [38].

The relevant variables for our study are s, h and c, which stand for size category, hierarchy category and climate category. All variables can take integer values 1, 2 or 3. They are obtained by grouping the EA variables into three groups as shown in Table 1. There are 248 cultures in Ref. [11] with information on these three variables.

Table 1. Grouping of the Ethnographic Atlas (EA) variables [11,38] into categories. See text for details.

↓ **Category/Value** →	**1**	**2**	**3**
s: size (v31 [a])	small	medium	large
h: social stratification (v66 [a])	egalitarian	simple structure	complex
c: climate (v95 [a])	harsh	intermediate	mild

[a] The name of the variable in the EA.

7.3. Data: Conditional Probabilities and Order Parameters

The categorical values are obtained by grouping the relevant variables of the EA according to Tables given in the Supplementary Materials as shown in Table 1, into three categories.

We extract the numbers of cultures $N(s,h,c)$ with a given set of values (s,h,c) and the marginal numbers $N(s,c)$ of cultures with a given pair of values of (s,c) independently of h. These are related by $N(s,c) = \sum_{h=1,2,3} N(s,h,c)$. The conditional probabilities,

$$P(h|sc) = \frac{N(s,h,c)}{N(s,c)}, \tag{15}$$

of a culture having a given class stratification, given its climate and group size.

Then, we calculate the average hierarchy of the cultures with the same values of n and c, that is, that belong to the same size and climate categories. We calculate the empirical average hierarchies conditional on size and climate:

$$\bar{H} = \mathbb{E}(h-1|sc) = \sum_{h=1,2,3} (h-1)P(h|sc), \tag{16}$$

which satisfies $0 \leq \bar{H} \leq 2$.

Fluctuations around the average

$$\sigma^2_{EA} = \mathbb{E}((h-\bar{h})^2|nc) = \sum_{h=1,2,3} (h-\bar{h})^2 P(h|nc), \tag{17}$$

can be calculated to define error bars (see Supplementary Materials).

7.4. Theory–Data Comparison

The empirical and theoretical predictions' order parameters are shown in Figure 7, left, and Figure 7, right, respectively. one can The typical bifurcation of a phase transition as different values of climate harshness are considered, occur in both the empirical data and the theoretical results. The qualitative agreement between theory and empirical record supports that our methodology is capable of suggesting new ways of looking at the available ethnographic records, which can now come under scrutiny by the community of quantitative ethnography dealing with cross-cultural studies.

That inequality rises with group size and that there are ecological factors involved has been previously considered [5,39–42], but not how the rise is modulated by ecological pressure nor the hypothesis that this is due to the competition of cognitive and social navigation needs and therefore the influence of climatic pressure on hierarchy can be reversed by demographics. It is not a coincidence that the transition between egalitarian and hierarchical regimes can be presented in the same language as phase transitions and critical phenomena in physical systems.

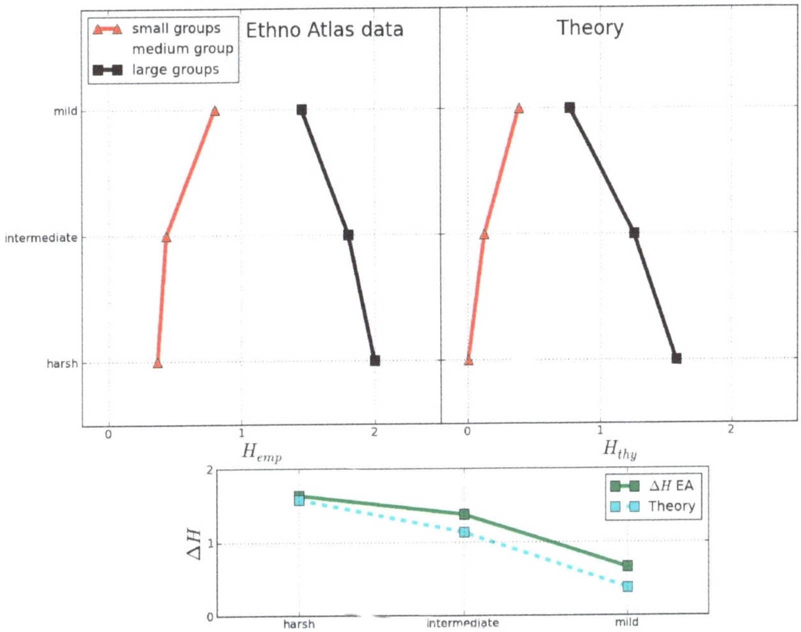

Figure 7. The bifurcation signature of the phase transition for Ethnographical Atlas (EA) data [11,38] (**upper right**) and theoretical prediction (14) (**upper left**). For mild climates, the expected hierarchies change with group size is quite small. For harsh climates, the expected hierarchy is larger for large groups and smaller for smaller groups. (**Lower**:) the difference in expected hierarchy ΔH between large and small groups decreases for milder climates. Harsh climates are tundra (northern areas), northern coniferous forest, high plateau steppe, desert (including arctic). Mild climates are temperate forest, temperate grasslands, Mediterranean, oases and certain restricted river valleys.

8. Discussion

Our approach, based on probability, entropic inference and information theory, is a methodological approach typical in the mathematical-physics modeling of systems that incorporates conditioning factors, however, they can be used when the conditioning comes from demographic, ecological, social and cognitive information.

Our main hypothesis is that a social-cognitive cost is relevant to characterize probabilistically the perceived social networks. The introduction of the conjugated parameter β, with the same informational content of the average cost, is an unavoidable theoretical consequence. It controls the size of fluctuations above the minimum possible value of the cost, prompting its interpretation as a generalized pressure. Gossip, a metaphor for information exchange, correlates the perceived networks. The cognitive capacity and the size of the group combine into a variable z, the specific cognitive capacity, and the perceived social state can be described in a space of just two dimensions (z, β). External to the model, the dynamics of encephalization and band size determine the historical evolution of z, leading to a scenario for non-monotonic hierarchical change [1]. Further changes in z could occur, e.g., due to technological advances, which translate into more effective information processing and better social navigation. In addition, an effective reduction of ecological pressure, following enhanced productivity, can occur. Then, a more egalitarian perception of the social networks will follow. The PMB hypothesis predicts that motivations and behaviors will change, but the theory does not go into the area of predicting how behaviors change, nor what institutions will emerge in order to permit such behaviors, nor the time scales of these changes. Our approach to the transition from hierarchical to egalitarian and back dispenses the issue of whether the hierarchical type of behavior laid dormant (Lars Rodseth in Ref. [1]) and remained present throughout the Pleistocene or whether the resurgence was due to convergent evolution (e.g., Ref. [43]). It can be turned on or off by the joint effects of cognitive resources, social demands, ecology and demography. These are similar to ordered–disordered transitions that result from pressure or temperature changes. The possibility of being solid ice is not dormant in water when it is heated up. At least that is not an appropriate metaphor, but it has been used in this context in the humanities.

One can speculate that the time spent in the large z egalitarian phase promoted conditions for the fixation of altruistic genes and the emergence of the 'do unto others' ideas since all are equal under the representation networks. One would not expect the fixation of altruistic behavior, which arises from punishment and collaboration [44–46] in other than the symmetric phase, but this should be amenable to model construction and analytic studies.

9. Conclusions

This simple model and the particular function we have used to represent the cognitive-social cost are far from complete. We do not claim specific numerical validation by confrontation with empirical data in any other way than just a qualitative one. More sophisticated forms of coalitions, other than dyadic pairing, should lead to increased richness of the phase diagram, without disrupting the rough overall picture. We have also avoided considering gender issues. Rampant sexual inequalities can exist in an egalitarian organization of males. Nevertheless, if competition between cognitive constraints and social navigation needs indeed occurs, then phase transitions from egalitarian to hierarchical perception follow from general arguments. It has been argued [47] that 'in the history of the human species, there is no more significant transition than the emergence and institutionalization of inequality'. We expect that these methods, which unify the theoretical analysis of the empirical findings behind the scenario for the U-shape dynamics and the conditions that influence the transformation of perception of social organization, will stimulate the use of information theory methods in the analysis of empirical research in cross-cultural studies.

Supplementary Materials: The following supporting information can be downloaded at: https://www.mdpi.com/article/10.3390/physics6020041/s1, Figure S1: Monte Carlo estimates of the maximum degree and the average degree of the social network representation as a function of $z = 2\alpha/n(n-1)$ for $\beta = 6$. This shows that the scaling of α with $n(n-1)/2$ is a good approximations that gets better for larger values of n; Figure S2: Conditional Probability distributions $P(h|s,c)$ obtained from the Ethnographic Atlas data set; Figure S3: The standard deviation of the distribution of expected hierarchies, or range of values of hierarchy as a function of ecological pressure for small, medium and large groups. Left: From the Ethnographic Atlas data [38]. Right: Theoretical prediction; Table S1: The number of data points (ethnographic groups) in the EA relevant for this study; Table S2: The results for the empirical stratification $\bar{H}$; Table S3: The results for the theoretical prediction H_T; Table S4: V31. Mean Size of Local Communities. Table S5: V66. Class Stratification; Table S6: V95. Climate: Primary Environment (Coded by Frank Moore from Phillips' Comparative Atlas); Table S7: EA Data: Cultures, Size, Class Stratification, Climate.

Author Contributions: Conceptualization and methodology, N.C. and R.V.; investigation, N.C. and R.S.C.; writing—original draft, N.C.; writing—review and editing, R.V.; visualization, R.S.C.; funding acquisition, N.C. and R.V. All authors have read and agreed to the published version of the manuscript.

Funding: This work was partially supported by Fapesp (2008/10830-2) and CNAIPS-USP, Brazil.

Data Availability Statement: Data is available from Supplementary Materials and https://growthecon.com/assets/Standard_Cross-Cultural_Sample-Codebook.pdf (accessed on 30 October 2023).

Acknowledgments: We thank Bruce Knauft for comments on the first stage of this project many years ago. While many of his comments led to changes in the presentation, many remain unaddressed and left for future work.

Conflicts of Interest: The authors declare no conflicts of interest.

References

1. Knauft, B.M.; Alber, T.S.; Betzig, L.; Boehm, C.; Knox Dentan, R.; Kiefer, T.M.; Otterbein, K.F.; Apddock, J.; Rodseth, L. Violence and sociality in human evolution [and comments and replies]. *Curr. Anthropol.* **1991**, *32*, 391–428. Available online: https://www.jstor.org/stable/2743815 (accessed on 20 January 2024). [CrossRef]
2. Boehm, C. *Hierarchy in the Forest: The Evolution of Egalitarian Behavior*; Harvard University Press: Cambridge, MA, USA, 1999. Available online: https://www.jstor.org/stable/j.ctvjf9xr4 (accessed on 20 January 2024).
3. Boehm, C. Ancestral hierarchy and conflict. *Science* **2012**, *336*, 844–847. [CrossRef]
4. Dubreuil, B. Paleolithic public goods games: Why human culture and cooperation did not evolve in one step. *Biol. Philos.* **2010**, *25*, 53–73. [CrossRef]
5. Dubreuil, B. *Human Evolution and the Origins of Hierarchies. The State of Nature*; Cambridge University Press: New York, NY, USA, 2010. [CrossRef]
6. Vehrencamp, S.L. A model for the evolution of despotic versus egalitarian societies. *Anim. Behav.* **1983**, *31*, 667–682. [CrossRef]
7. Gavrilets, S.; Auerbach, J.; van Vugt, M. Convergence to consensus in heterogeneous groups and the emergence of informal leadership. *Sci. Rep.* **2016**, *6*, 29704. [CrossRef] [PubMed]
8. Perret, C.; Hart, E.; Powers, S.T. From disorganized equality to efficient hierarchy: How group size drives the evolution of hierarchy in human societies. *Proc. R. Soc. B Biol. Sci* **2020**, *287*, 20200693. [CrossRef] [PubMed]
9. Lin, K.Y.; Schank, J.C. Small group size promotes more egalitarian societies as modeled by the hawk-dove game. *PLoS ONE* **2022**, *17*, e0279545. [CrossRef] [PubMed]
10. Galam, S. *Sociophysics: A Physicist's Modeling of Psycho-Political Phenomena*; Understanding Complex Systems; Springer Science+Business Media, LLC: New York, NY, USA, 2012. [CrossRef]
11. Murdock, G.P. *Ethnographic Atlas*; University of Pittsburgh Press: Pittsburgh, PA, USA, 1967. Available online: https://digital.library.pitt.edu/islandora/object/pitt:31735057895306 (accessed on 20 January 2024).
12. Turchin, P. *Ultrasociety: How 10,000 Years of War Made Humans the Greatest Cooperators on Earth*; Beresta Books LLC: Chaplin, CT, USA, 2015. Available online: https://www.pdfdrive.com/ultrasociety-how-10000-years-of-war-made-humans-the-greatest-cooperators-on-earth-e196524525.html (accessed on 20 January 2024).
13. Killworth, P.D.; Bernard, H.R. A model of human group dynamics. *Soc. Sci. Res.* **1976**, *5*, 173–334. [CrossRef]
14. Bernard, H.R.; Killworth, P.D. Why are there no social physics? *J. Steward Anthropol. Soc.* **1979**, *11*, 33–38. Available online: https://hrussellbernard.com/publications/ (accessed on 20 January 2024).
15. Yamadera, S. Examining the myth of money with agent-based modelling. In *Social Simulation: Technologies, Advances and New Discoveries*; Edmonds, B., Troitzsch, K.G., Hernández Iglesias, C., Eds.; Information Science Reference/IGI Global: Hershey, PA, USA, 2008; pp. 252–263. [CrossRef]
16. Kunigami, M., Kobayashi, M., Yamadera, S., Terano, T. On emergence of money in self-organizing doubly structural network model. In *Agent-Based Approaches in Economic and Social Complex Systems V. Post-Proceedings of the AESCS International Workshop 2007*; Terano, T., Kita, H., Takahashi, S., Deguchi, H., Eds.; Springer: Tokyo, Japan, 2009; pp. 231–241. [CrossRef]
17. Dávid-Barrett, T.; Dunbar, R.I.M. Processing power limits social group size: Computational evidence for the cognitive costs of sociality. *Proc. R. Soc. B Biol. Sci.* **2013**, *280*, 20131151. [CrossRef]
18. Dávid-Barrett, T.; Dunbar, R.I.M. Social elites can emerge naturally when interaction in networks is restricted. *Behav. Ecol.* **2014**, *25*, 58–68. [CrossRef]
19. Gavrilets, S. On the evolutionary origins of the egalitarian syndrome. *Proc. Natl. Acad. Sci. USA* **2012**, *108*, 14069–14074. [CrossRef]
20. Powers, S.T.; Lehmann, L. An evolutionary model explaining the Neolithic transition from egalitarianism to leadership and despotism. *Proc. R. Soc. B Biol. Sci.* **2014**, *281*, 20141349. [CrossRef] [PubMed]
21. Barabási, A.-L.; Pósfai, M. *Network Science*; Cambridge University Press: Cambridge, UK, 2016. Available online: http://networksciencebook.com (accessed on 20 January 2024).
22. Sawaguchi, T.; Kudo, H. Neocortical development and social structure in primates. *Primates* **1990**, *31*, 283–289. [CrossRef]
23. Dunbar, R.I.M. Neocortex size as a constraint on group size in primates. *J. Hum. Evol.* **1992**, *22*, 469–493. [CrossRef]
24. Shultz, S.; Dunbar, R.I.M. Species differences in executive function correlate with hippocampus volume and neocortex ratio across nonhuman primates. *J. Compar. Psych.* **2010**, *124*, 252–260. [CrossRef] [PubMed]
25. Amodio, D.M.; Frith, C.D. Meeting of minds: The medial frontal cortex and social cognition. *Nat. Rev. Neurosci.* **2006**, *7*, 268–277. [CrossRef] [PubMed]
26. Yoshida, K.; Saito, N.; Iriki, A.; Isoda, M. Social error monitoring in macaque frontal cortex. *Nat. Neurosci.* **2012**, *15*, 1307–1312. [CrossRef]
27. Allman, J.M.; Hakeem, A.; Erwin, J.M.; Nimchinsky, E.; Hof, P. Anterior cingulate cortex: The evolution of an interface between emotion and cognition. *Ann. N. Y. Acad. Sci.* **2001**, *935*, 107–117. [CrossRef]
28. Whiten, A. Primate culture and social learning. *Cogn. Sci.* **2000**, *24*, 477–508. [CrossRef]
29. Castellano, C.; Fortunato, S.; Loreto, V. Statistical physics of social dynamics. *Rev. Mod. Phys.* **2009**, *81*, 591–646. [CrossRef]
30. Earle, T. *How Chiefs Come to Power: The Political Economy in Prehistory*; Stanford University Press: Stanford, CA, USA, 1997. [CrossRef]

31. Caticha, N.; Vicente, R. Agent-based social psychology: From neurocognitive processes to social data. *Adv. Complex Syst.* **2011**, *14*, 711–731. [CrossRef]
32. Vicente, R.; Susemihl, A.; Jericó, J.P.; Caticha, N. Moral foundations in an interacting neural networks society: A statistical mechanics analysis. *Physica A Stat. Mech. Appl.* **2014**, *400*, 124–138. [CrossRef]
33. Callen, H.B. *Thermodynamics and an Introduction Thermostatistics*; John Wiley & Sons, Inc.: Hoboken, NJ, USA, 1985. Available online: https://archive.org/details/thermodynamicsan0000call/ (accessed on 20 January 2024).
34. Jaynes, E.T. *Probability Theory: The Logic of Science*; Cambridge University Press: New York, NY, USA, 2003. [CrossRef]
35. Sahlins, M.D. Poor man, rich man, big-man, chief: Political types in Melanesia and Polynesia. *Comp. Stud. Soc. Hist.* **1963**, *5*, 285–303. [CrossRef]
36. Lindstrom, L. 'Big man': A short terminological history. *Am. Anthropol.* **1981**, *83*, 900–905. [CrossRef]
37. Boehm, C.; Barclay, H.B.; Knox Dentan, R.; Dupre, M.-C.; Hill, J.D.; Kent, S.; Knauft, B.M.; Otterbein, K.F.; Rayner, S. Egalitarian society and reverse dominance hierarchy [and comments and reply]. *Curr. Anthropol.* **1993**, *34*, 227–254. Available online: https://www.jstor.org/stable/2743665 (accessed on 20 January 2024). [CrossRef]
38. White, D.R.; Burton, M.; Divale, M.; Gray, P.; Kprotayev, A.; Khalturina, D. Standard Cross Cultural Sample: Codebook. 2011. Available online: https://growthecon.com/assets/Standard_Cross-Cultural_Sample-Codebook.pdf (accessed on 20 January 2024).
39. Carneiro, R.L. On the relationship between size of population and complexity of social organization. *Southwest. J. Anthropol.* **1967**, *23*, 234–243. [CrossRef]
40. Bingham, P.M. Human uniqueness: A general theory. *Quart. Rev. Biol.* **1999**, *74*, 133–169. [CrossRef]
41. Fletcher, R. *The Limits of Settlement Growth: A Theoretical Outline*; Cambridge University Press: Cambridge, UK, 2007.
42. Summers, K. The evolutionary ecology of despotism. *Evol. Hum. Behav.* **2005**, *26*, 106–135. [CrossRef]
43. Smail, D.L. *On Deep History and the Brain*; University of California Press: Oakland, CA, USA, 2008. Available online: https://www.jstor.org/stable/10.1525/j.ctt1pp7mz (accessed on 20 January 2024).
44. Boyd, R.; Richerson, P.J. The evolution of reciprocity in sizable groups. *J. Theor. Biol.* **1988**, *132*, 337–356. [CrossRef]
45. Boyd, R.; Gintis, H.; Bowles, S. Coordinated punishment of defectors sustains cooperation and can proliferate when rare. *Science* **2010**, *328*, 617–620. [CrossRef] [PubMed]
46. Schonmann, R.H.; Vicente, R.; Caticha, N. Altruism can proliferate through population viscosity despite high random gene flow. *PLoS ONE* **2013**, *8*, e72043. [CrossRef] [PubMed]
47. Feinman, G.M. The emergence of inequality. In *Foundations of Social Inequality*; Price, T.D., Feinman, G.M., Eds.; Springer Science+Business Media: New York, NY, USA, 1995; pp. 255–279. [CrossRef]

physics

MDPI

Article

Aging in Some Opinion Formation Models: A Comparative Study

Jaume Llabrés [1], Sara Oliver-Bonafoux [1], Celia Anteneodo [2,3,*] and Raúl Toral [1]

1 Instituto de Física Interdisciplinar y Sistemas Complejos IFISC (CSIC-UIB), Campus Universitat de les Illes Balears, 07122 Palma de Mallorca, Spain; jaumellabres@ifisc.uib-csic.es (J.L.); saraoliver@ifisc.uib-csic.es (S.O.-B.); raul@ifisc.uib-csic.es (R.T.)
2 Department of Physics, Pontifical Catholic University of Rio de Janeiro, PUC-Rio, Rua Marquês de São Vicente, 225, Rio de Janeiro 22451-900, Brazil
3 National Institute of Science and Technology for Complex Systems, INCT-SC, Rio de Janeiro 22290-180, Brazil
* Correspondence: celia.fis@puc-rio.br

Abstract: Changes of mind can become less likely the longer an agent has adopted a given opinion state. This resilience or inertia to change has been called "aging". We perform a comparative study of the effects of aging on the critical behavior of two standard opinion models with pairwise interactions. One of them is the voter model, which is a two-state model with a dynamic that proceeds via social contagion; another is the so-called kinetic exchange model, which allows a third (neutral) state, and its formed opinion depends on the previous opinions of both interacting agents. Furthermore, in the noisy version of both models, random opinion changes are also allowed, regardless of the interactions. Due to aging, the probability of changing diminishes with the age, and to take this into account, we consider algebraic and exponential kernels. We investigate the situation where aging acts only on pairwise interactions. Analytical predictions for the critical curves of the order parameters are obtained for the opinion dynamics on a complete graph, in good agreement with agent-based simulations. For both models considered, the consensus is optimized via an intermediate value of the parameter that rules the rate of decrease of the aging factor.

Keywords: aging; opinion formation model; voter model; kinetic exchange model; social dynamics

Citation: Llabrés, J.; Oliver-Bonafoux, S.; Anteneodo, C.; Toral, R. Aging in Some Opinion Formation Models: A Comparative Study. *Physics* **2024**, *6*, 515–528. https://doi.org/10.3390/physics6020034

Received: 21 December 2023
Revised: 6 February 2024
Accepted: 16 February 2024
Published: 8 April 2024

1. Introduction

As is known, the ideas and concepts coming from the statistical and nonlinear physics have been applied successfully to understand social phenomena [1,2]. The main goal is to explain "macroscopic" emergent behavior in terms of the "microscopic" individual's components in the same way that one derives the laws of macroscopic physical systems, e.g., the equation of state, from microscopic information such as the forces amongst particles. One lesson we have learnt from the modern development of statistical physics is that the emergent behavior, manifested through macroscopic phase transitions, is largely dependent on some details of the microscopic interactions and that there are some universal features that only depend on a few ingredients, such as dimensionality, symmetries, etc. It is in this spirit that the modeling of social systems has relied on the development of very simple agent-based models that take as starting points some stylized facts about the interactions amongst agents. Consider, for example, the subject of the evolution of cultures. In a seminal paper, Robert Axelrod asked: "If people tend to become more alike in their beliefs, attitudes, and behavior when they interact, why do not all such differences eventually disappear?" [3]. He then introduced a model whose basic assumption is that interactions amongst individuals lead to an augmentation in the similarity between them. A detailed analysis [4] shows that, depending on the initial diversity, there is a phase transition between an ordered monocultural state and a disordered multicultural state, demonstrating the apparent paradox of the emergence of global polarization through the mechanism of local convergence. It was then established that the dimensionality of the space of interactions was a relevant variable of the model [5]. Similar ideas appear in other areas. In the context of opinion formation, for example, one expects, on the one hand, that social influences reduce differences between individuals via social contagion, but, on the other, interactions can also produce differentiation via repulsive forces, associated with anticonformist "contrarian" agents that tend to deviate from the behavior adopted by the neighbors in the network of connections [6,7]. Another example is Galam's model for minority opinion spread [8,9] that, although based on local majority rule, can lead to the hostile minority views having an advantage.

Other diverse mechanisms can compete with the tendency toward uniformity promoted by social contagion. One is the "noisy" behavior, ruled by a "social temperature" parameter, associated with imperfect imitations due to random factors or independent choices, such that random changes of opinion interpreted as idiosyncratic behavior can occur. Standard models that study the effect of independence include the noisy voter model and the noisy kinetic exchange model. In the former, agents adopt opinions ±1, copying the opinion of a random contact. This model has been studied under a variety of different, but formally equivalent setups [10–17]. In the latter, the states ±1 and 0 can be adopted via a rule that takes into account the current opinion of both interacting contacts [18,19]. In both cases, the rule of social influence acts with a probability of $(1-a)$ while otherwise, with a probability of a, a random change can occur. That is, these models rely on the assumption that opinion dynamics are mainly governed by mechanisms of imitation or social contagion that mold opinion formation through the interaction between individuals, but random factors are always present. In both cases, noise opposes the system's ability to reach a self-organized state with a winning opinion. Furthermore, both present a critical value a_c below which order can be achieved, the main difference being that $a_c \to 0$ in the thermodynamic limit, when the number of agents tends to infinity, for the noisy voter, while a_c remains finite in the kinetic exchange model in the same limit.

Another ingredient that has been proven to be relevant in the modeling of social systems is that of "aging". The effects of aging have been widely explored in several areas of research. In the field of computational biology [20–22], aging is understood as the increase of the mortality of a species as its population becomes older. In chemistry, aging appears when the properties of a material change over time without the influence of external forces but due to, for example, thermal degradation [23] or photo-oxidation [24]. In the context of social systems, aging is the concept that the larger resistance an agent offers to changing its state of opinion, the longer it has been holding the current state. Initially termed as "inertia" [25], it was shown that the slowing down of the microscopic dynamics induced via aging actually decreased the time needed to reach the macroscopically ordered state state of consensus. In this respect, and in alignment with previous studies [25–31], we consider that aging acts on the mechanism associated with social contagion.

In this paper, we compare the effects of aging on the two above-mentioned paradigmatic representatives of the class of noisy opinion models, the voter model and the kinetic exchange model, highlighting the similarities and differences amongst them. To this end, we consider an all-to-all interaction scheme in which every agent is connected to every other agent and develop an adiabatic approximation able to provide us with the phase diagram, including the location of the critical points separating regions of consensus from disordered regions. Afterward, we compare the location of the critical points for both models.

The rest of the paper is organized as follows: In Section 2, we introduce a general setup to study models of opinion dynamics under the presence of idiosyncratic, random changes of opinion, as well as the interaction between the agents modulated by an aging mechanism. In Section 3, we derive a mean-field-type approach for the noisy voter model under an adiabatic approximation, leaving the more technical details for Appendix A.1. In Section 3, we also compare the results of numerical simulations for two different functional forms for the aging probability, algebraic and exponential, focusing mainly on the magnetization, measuring the amount of order, or amount of consensus, in the system. In Section 4, after briefly reviewing the main results for the noisy kinetic exchange model, we compare the results for the magnetization and the dependence of the critical value of the noise intensity as a function of a parameter measuring the rate of decrease of the aging probability, for the voter and the kinetic exchange model. Finally, in Section 5, we end with some conclusions and an outlook for future work.

2. Models of Noisy Opinion Dynamics with Aging

In this study, we analyze two models that have been commonly used as prototypical examples in the study of the dynamics of opinion formation in social systems: the noisy voter model and the noisy kinetic exchange model. Both models consider a set of N agents endowed with an internal state variable s_i, $i = 1, \cdots, N$, representing the possible positions of agent i concerning a given topic. For the voter model, the internal variable can take two possible values $s_i \in \{-1, +1\}$, representing a position against or in favor of the topic. The kinetic exchange model introduces a third neutral state such that $s_i \in \{-1, 0, +1\}$. This internal variable evolves due to two different generic mechanisms that act stochastically: idiosyncratic changes and social influence. The difference between them is that the former randomly changes the state variable s_i independently of the states of other agents, while the latter changes s_i following a rule that depends on the state variables of another agent s_j. In the voter model, the social rule is that $s_i \to S(s_i, s_j) = s_j$, modeling the mechanism of imitation. In the kinetic exchange model, the social rule is $s_i \to S(s_i, s_j) = \text{sgn}[s_i + s_j]$, where $\text{sgn}[s]$ is the sign function. The rule is such that neutral agents can not modify the opinion of another agent, and agents with a well-defined opinion (either positive or negative) can convince a neutral agent or turn to the neutral state an agent with the opposite opinion. Let us mention that other rules for three-state models have been considered in the literature before [32–35]. In Ref. [32], it is shown that the inclusion of a third neutral state which individuals necessarily have to pass through increases the capability to reach consensus in the extreme values.

Previously, both models here considered have been studied under the influence of aging in the social mechanism [36,37]. In this paper, we generalize some of the previous studies and compare the effects that aging induces in both models.

To be precise, let us spell out in detail the rules of evolution for these models. Initially, we assign a random value to each state variable s_i and set all internal times τ_i to zero. Then,

1. At each iteration, an agent i is randomly selected.
2. With probability $(1 - a)$, the social rule is chosen: with probability $q(\tau_i)$, modeling the persistence or reaction of the individual i to change as a function of its age τ_i, a neighbor j is randomly selected and the opinion of agent i is modified according to the rule $s_i \to S(s_i, s_j)$.
3. Otherwise, with probability a, the idiosyncratic rule is chosen: the state s_i is replaced by a randomly selected value among all possible opinion states.
4. Irrespective of the update mechanism actually used by agent i (random change or pairwise interaction, with aging or not), its age τ_i is updated in the following way: if the state s_i has changed, then age is reset, i.e., $\tau_i \to 0$; otherwise, age is incremented in one unit, i.e., $\tau_i \to \tau_i + 1$.

We restrict ourselves in this paper to the all-to-all connectivity, where each agent is linked to every other agent. Hence, the set of neighbors of an agent i is the whole set of agents (excluding itself). Time is measured in Monte Carlo steps, such that one unit of time corresponds to N agent selections for updating.

Although other forms are possible, in the present study, we adopt the following two general functional forms for the probabilities (we use $q(\tau) \equiv q_\tau$ for brevity in the notation),

$$\text{algebraic decay}: \qquad q_\tau = \frac{q_\infty \tau + q_0 \tau^*}{\tau + \tau^*}, \tag{1}$$

$$\text{exponential decay}: \qquad q_\tau = q_\infty + (q_0 - q_\infty) e^{-\tau/\tau^*}, \tag{2}$$

where q_0 and q_∞, satisfying $0 \leq q_\infty < q_0 \leq 1$, denote, respectively, the initial ($\tau = 0$) and asymptotic ($\tau \to \infty$) values of q_τ, and τ^* is a parameter dictating the rate of decrease of the aging-probability, i.e., a larger value of τ^* implies a slower decay. Both forms have been considered before in the context of the study of the way that the voter model (without idiosyncratic changes) approaches the asymptotic consensus state [38]. A previous study of the noisy voter model [29] was limited to the algebraic case with $q_0 = 1/2$, $\tau^* = 2$. The aging-less case is recovered, formally taking the limit $\tau^* \to \infty$, implying $q_\tau = q_0$ for all values of the age τ. Our theoretical treatment is rather general, but we adopt the values $q_0 = 1$ and $q_\infty = 0$ in the numerical simulations.

3. Noisy Voter Model with Aging

In this Section, we present the results obtained for the noisy voter model with aging. We build over the treatment of Ref. [29] but present a more general one, valid for arbitrary forms of the aging-update probability.

In this model, each agent i is characterized by its binary state variable, $s_i \in \{-1, +1\}$, and its age or residence time in state s_i, τ_i. We denote with $x_\tau^\pm(t)$ the fractions of agents in states ± 1 and with age τ, such that $x(t) = \sum_{\tau=0}^{\infty} x_\tau^+(t)$ is the total fraction of agents in state $+1$ at time t, and $1 - x(t) = \sum_{\tau=0}^{\infty} x_\tau^-(t)$ is the total fraction of agents in state -1 at time t. As detailed in Appendix A.1, it is possible to derive a closed evolution equation for $x(t)$. The procedure starts by writing down rate equations for the densities $x_\tau^\pm(t)$. One then performs an adiabatic approximation that assumes that the evolution of $x_\tau^\pm(t)$ is attached to that of $x(t)$ which evolves at a longer time scale. Under this approximation, the equation for the dynamical evolution of the fraction $x(t)$ of agents in state $+1$ at time t is

$$\begin{aligned} \frac{dx}{dt} &= G^{\mathrm{v}}(x), \\ G^{\mathrm{v}}(x) &= \frac{a}{2}(1-2x) + (1-a)x(1-x)[\Phi^{\mathrm{v}}(x) - \Phi^{\mathrm{v}}(1-x)], \end{aligned} \tag{3}$$

where the function $\Phi^{\mathrm{v}}(x)$ is defined as

$$\Phi^{\mathrm{v}}(x) = \frac{\sum_{\tau=0}^{\infty} q_\tau F_\tau^{\mathrm{v}}(x)}{\sum_{\tau=0}^{\infty} F_\tau^{\mathrm{v}}(x)}, \tag{4}$$

with

$$F_0^{\mathrm{v}}(x) = 1, \quad F_\tau^{\mathrm{v}}(x) = \prod_{k=0}^{\tau-1} \gamma_2(q_k x, a), \quad \tau \geq 1, \tag{5}$$

and

$$\gamma_n(z, a) = \frac{a}{n} + (1-a)(1-z). \tag{6}$$

The steady state-solutions, x_{st}, are obtained by setting the time derivative of the density $x(t)$ to equal zero, or $G^{\mathrm{v}}(x_{\mathrm{st}}) = 0$, according to Equation (3). Note that, due to its very nature, the adiabatic approximation does provide the correct values of these steady-state solutions. This is because the exact calculation of the steady-state solutions requires that the rates of change of all densities are set to zero. This leads to the condition $G^{\mathrm{v}}(x_{\mathrm{st}}) = 0$, which is also a result of the adiabatic approximation. What the approximate dynamical Equation (3) provides is the stability of the different steady-state solutions as determined by the sign of the derivative $dG^{\mathrm{v}}(x)/dx$ evaluated at x_{st}.

In the aging-less case, $q_\tau = q_0$, $\forall \tau$, it is $\Phi^{\mathrm{v}}(x) = q_0$ and the only fixed point of Equation (3) is $x_{\mathrm{st}} = 1/2$, which is stable. This indicates that the disordered phase is the only asymptotic solution. It is quite clear from the structure of Equation (3) that $x_{\mathrm{st}} = 1/2$ is, trivially, also a steady-state solution for the aging situation, for an arbitrary function $\Phi^{\mathrm{v}}(x)$. The stability of this trivial solution and the existence of other steady-state solutions depend solely on the function $G^{\mathrm{v}}(x)$.

For the algebraic functional form of the aging-update probabilities given by Equation (1), it is possible, as explained in Appendix B, to obtain an analytical form for the function $\Phi^{\text{v}}(x)$ in terms of hypergeometric functions. However, even in this case, the non-trivial steady state values and their stability have to be obtained numerically. For the exponential form of the aging-update probabilities given by Equation (2), it does not seem to be possible to express $\Phi^{\text{v}}(x)$ in terms of other known functions, but we outline in Appendix B.2 a convenient numerical algorithm for its evaluation.

Both for the algebraic and the exponential forms of the aging-update probabilities, the numerical analysis concludes that, in the case that $q_0 = 1$, $q_\infty = 0$, the solution $x_{\text{st}} = 1/2$ is stable for $a > a_{\text{c}}(\tau^*)$ and unstable for $a < a_{\text{c}}(\tau^*)$ and that a new pair of symmetric stable solutions $x_{\text{st}}, 1 - x_{\text{st}}$ emerge for $a \leq a_{\text{c}}(\tau^*)$. These two solutions share the same stability, as it can easily be proven that $dG^{\text{v}}(x)/dx|_{x=x_{\text{st}}} = dG^{\text{v}}(x)/dx|_{x=1-x_{\text{st}}}$. This change of stability of the symmetric solution $x_{\text{st}} = 1/2$ is understood as a phase transition between a disordered phase at $a > a_{\text{c}}(\tau^*)$ where both opinions ± 1 coexist in equal proportion to an ordered, consensus phase for $a < a_{\text{c}}(\tau^*)$ where one of the opinions is majoritarian. The existence of this phase transition is one of the main results presented in Ref. [29] for a particular form of the aging probability (corresponding to the algebraic case with $q_0 = 1/2$, $\tau^* = 2$). The same transition was understood as a mapping of the problem with aging to another one with suitable nonlinear rates in the social mechanism [36]. We now present a general study of both the algebraic and exponential dependence of the aging probability and different values on the parameter τ^*.

In Figure 1, we plot the magnetization, $m_{\text{st}} = |2x_{\text{st}} - 1|$, as a function of the noise intensity, a, both for the algebraic and the exponential functional forms of the aging update probability and for several values of the parameter τ^*. In the same plot we show the results of numerical simulations of the stochastic rules for the agent-based model detailed in Section 2. The simulations agree remarkably well with the analytical results, validating the adiabatic approximation introduced in the analysis. There are some deviations with respect to the analytical calculation for values of the probability parameter a which are close to the critical value a_{c}. This is attributed to the necessary consideration of a finite number of agents $N = 10^4$ in the numerical simulations, while the theoretical treatment considers the thermodynamic limit $N \to \infty$.

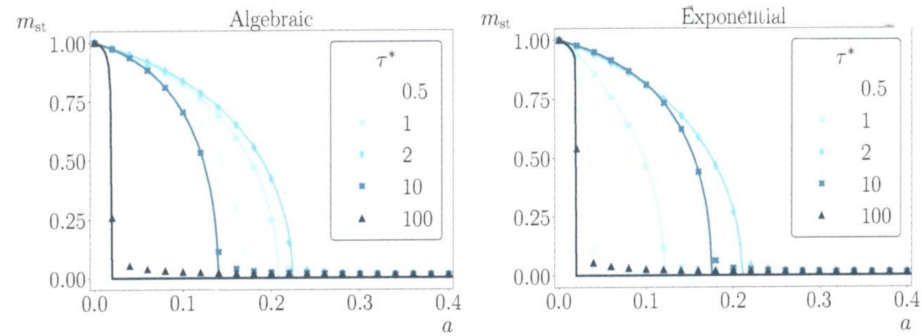

Figure 1. Noisy voter model with aging. Magnetization, $m_{\text{st}} = |2x_{\text{st}} - 1|$, as a function of the noise intensity, a, for different values of the τ^* parameter, for the algebraic (**left**) and exponential (**right**) kernels. Solid lines correspond to the theoretical results, while symbols represent the outcomes of numerical simulations with $N = 10^4$ agents, averaged over 5×10^6 Monte Carlo steps (MCS) after a transient of 5×10^6 MCS. Notice the non-monotonic dependency of the critical point with the parameter τ^*. See text for details.

For a given value of τ^*, the critical value a_{c} can be found by calculating the point at which the symmetric solution $x_{\text{st}} = 1/2$ changes from being stable to being unstable, i.e., by solving the equation $dG^{\mathrm{v}}(x)/dx|_{x=1/2} = 0$. A relatively simple calculation shows that this condition is equivalent to

$$(1 - a_c)\left.\frac{d\Phi^{\mathrm{v}}(x)}{dx}\right|_{x=1/2} = 2a_c. \tag{7}$$

Note that a_c also appears in the function $\Phi^{\mathrm{v}}(x)$. The dependence of the critical value a_c on the parameter τ^*, obtained by numerically solving the previous equation, is shown in Figure 2. The larger the value of a_c, the larger the region in parameter space in which a consensus-like phase is present. Remarkably, both for the algebraic and the exponential cases, the critical value tends toward zero for $\tau^* \to 0$ and $\tau^* \to \infty$, indicating that in those limits, the only asymptotic solution is the disordered one. The limit $\tau^* \to \infty$ leads to $q_\tau = 1$, $\forall\tau$, which corresponds to the standard version (without aging) of the noisy-voter model. It is known that this model displays a finite-size noise-induced phase transition between disorder and consensus at a critical value of the parameter $a_c = 2/N$. In the thermodynamic limit, it is $a_c \to 0$, in accordance with our result. The reason for the disappearance of the consensus in the limit $\tau^* \to 0$ stems from the fact that the social mechanism is activated with a probability $q(\tau)$ that tends toward 0 for all τ, except $\tau = 0$, for which $q(0) = q_0 > 0$. Hence, the only effective updating mechanism acting at all times is the set of random updates that necessarily lead to disorder. In between these two limits, there is an optimal value τ_c^* for which a_c is the maximum. For the algebraic case, it is $\tau_c^* = 2.01$, and the corresponding value of the noise intensity is $a_c = 0.224$, while for the exponential case, we find that $\tau_c^* = 3.77$ and $a_c = 0.242$. The critical lines $a_c(\tau^*)$ of the algebraic and exponential cases cross at the point $\tau_0^* = 2.33$, such that for $\tau^* < \tau_0^*$, the algebraic aging shows a larger consensus region (larger value of a_c) and the opposite for $\tau^* > \tau_0^*$.

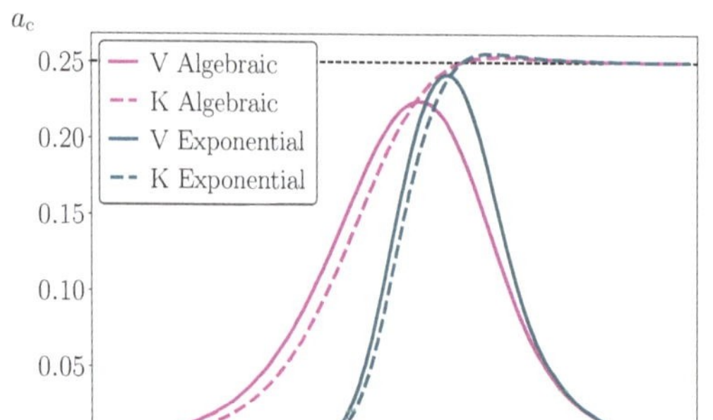

Figure 2. Theoretical curves of the critical noise value a_c versus the τ^* parameter in the algebraic and exponential cases for the noisy voter (V) and kinetic exchange (K) models. The horizontal lines correspond to the critical value in the aging-less case in the noisy voter (solid line) and in the kinetic exchange (dashed line) models.

4. Comparison with the Noisy Kinetic Exchange Model with Aging

We now compare the main similarities and differences between the noisy voter model and the kinetic exchange model, both under the presence of aging.

The kinetic exchange model introduces a third, neutral value of the state variable, $s_i \in \{-1, 0, +1\}$. At variance with the noisy voter model, a macroscopic characterization uses both densities $x^{\pm}(t)$ of agents in states ± 1 at time t. Via normalization, the density of the agents in state 0 is given by $x^0(t) = 1 - x^+(t) - x^-(t)$.

Along the lines used for the voter model and developed in more detail in Ref. [37], it is possible to write down a closed system of equations for the densities $x^{\pm}(t)$, namely,

$$\begin{aligned} \frac{dx^+}{dt} &= G^{\mathrm{K}}(x^+, x^-), \\ \frac{dx^-}{dt} &= G^{\mathrm{K}}(x^-, x^+), \\ G^{\mathrm{K}}(z,w) &= (1-a)[z(1-z-w)\Phi^{\mathrm{K}}(z+w) - z\,w\,\Phi^{\mathrm{K}}(w)] + \frac{a}{3}(1-3z). \end{aligned} \tag{8}$$

where

$$\Phi^{\mathrm{K}}(x) \equiv \frac{\sum_{\tau=0}^{\infty} q_\tau F_\tau^{\mathrm{K}}(x)}{\sum_{\tau=0}^{\infty} F_\tau^{\mathrm{K}}(x)}. \tag{9}$$

$$F_0^{\mathrm{K}}(x) = 1, \quad F_\tau^{\mathrm{K}}(x) = \prod_{k=0}^{\tau-1} \gamma_3(q_k\, x, a), \quad \tau \geq 1, \tag{10}$$

where $\gamma_n(z,a)$ is given by Equation (6). Although the dynamical system for the kinetic model, Equation (8), is quite different from that of the voter model, Equation (3), a quite similar structure is found for the functions $\Phi^{\mathrm{V}}(x)$ (4) and $\Phi^{\mathrm{K}}(x)$ (9). As explained in Appendix B, both functions are expressible in terms of hypergeometric functions for the algebraic dependence of the aging probabilities and can be computed using a very efficient numerical algorithm in the case of the exponential dependence.

Let us summarize now the main results of the analysis of the dynamical system Equation (8). In the aging-less case, $\Phi^{\mathrm{K}}(x) = q_0$, the dynamical equations always admit the steady-state solution $x_{\mathrm{st}}^+ = x_{\mathrm{st}}^- = 1/3$. This solution becomes unstable for values of the noise intensity a less than the critical value $a_{\mathrm{c}} = q_0/(3+q_0)$, where a pair of non-symmetric solutions $x_{\mathrm{st}}^+ \neq x_{\mathrm{st}}^-$ emerge. The behavior of the magnetization $m_{\mathrm{st}} = |x_{\mathrm{st}}^+ - x_{\mathrm{st}}^-|$ is given by

$$m_{\mathrm{st}} = \begin{cases} \dfrac{\sqrt{(q_0 - (3+q_0)a)(3q_0 - (1+3q_0)a)}}{\sqrt{3}\, q_0\,(1-a)}, & a \leq a_{\mathrm{c}}, \\ 0, & a \geq a_{\mathrm{c}}, \end{cases} \tag{11}$$

a result already obtained by Nuno Crokidakis [18] in the case that $q_0 = 1$.

When aging is included, the basic structure of the solution remains [37]: there is always a symmetric phase $x_{\mathrm{st}}^+ = x_{\mathrm{st}}^-$ steady-state solution which is stable for $a > a_{\mathrm{c}}(\tau^*)$. For $a < a_{\mathrm{c}}(\tau^*)$, this solution becomes unstable and a pair of non-symmetric stable solutions emerge. It is not possible to give explicit expressions for the solutions $x_{\mathrm{st}}^{\pm}$ for an arbitrary function $\Phi^{\mathrm{K}}(x)$, and they have to be determined numerically. The stability of these solutions is also determined numerically from the sign of the eigenvalues of the Jacobian matrix

$$\left.\begin{pmatrix} \dfrac{\partial \dot{x}^+}{\partial x^+} & \dfrac{\partial \dot{x}^-}{\partial x^+} \\ \dfrac{\partial \dot{x}^+}{\partial x^-} & \dfrac{\partial \dot{x}^-}{\partial x^-} \end{pmatrix}\right|_{x^{\pm} = x_{\mathrm{st}}^{\pm}} \tag{12}$$

evaluated at the fixed points. Here, the dot over a letter denotes the time derivative.

In Figure 3, we show the results of the magnetization m_{st} as a function of the noise intensity a for two values of the τ^* parameter for the noisy voter ($m_{\mathrm{st}} = |2x_{\mathrm{st}} - 1|$) and for the kinetic exchange ($m_{\mathrm{st}} = |x_{\mathrm{st}}^+ - x_{\mathrm{st}}^-|$) models with aging. In both cases, we find high agreement between theoretical results and numerical simulations using the stochastic rules of the processes.

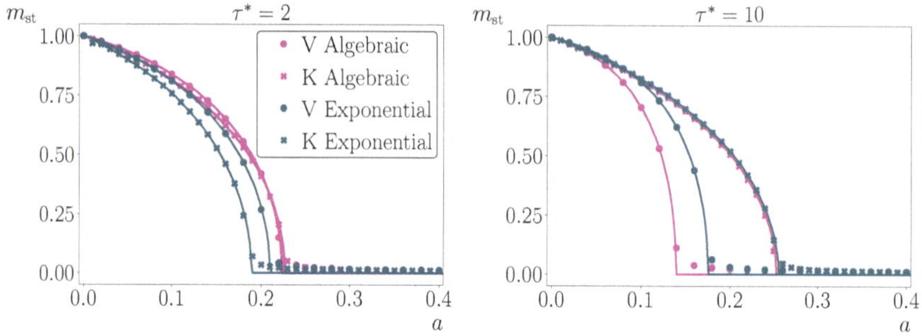

Figure 3. Comparison of the noisy voter model (V) and the noisy kinetic exchange model (K), both under the influence of aging. Magnetization m_{st} as a function of the noise intensity a in the algebraic and in the exponential cases for two values of the τ^* parameter: $\tau^* = 2$ (**left**) and $\tau^* = 10$ (**right**). Results of the mean-field theory are plotted with solid lines, while those of numerical simulations are displayed with symbols. The simulations have been performed with 10^4 agents and averaged over 5×10^6 MCS after 5×10^6 thermalization MCS.

As can be seen in Figure 2, for sufficiently small values of τ^* (strong or moderate aging, e.g., $\tau^* = 2$ in Figure 3), the most significant determining factor for the existence of a given consensus region is not the type of model considered (either voter or kinetic exchange), but the type of aging kernel. In particular, algebraic aging, which decays most slowly with τ, is the one that most enhances this effect. On the other hand, for large values of τ^* (weak aging, e.g., $\tau^* = 10$ in Figure 3), the kinetic exchange model gives rise to considerably larger consensus regions than the voter model, regardless of the type of aging considered (algebraic or exponential), which is consistent with the feature that in the aging-less limit ($\tau^* \to \infty$), the kinetic model has a non-null critical point in the thermodynamic limit, in contrast with the voter model.

The most interesting result for the kinetic model is the non-monotonic behavior of the magnetization: the critical value $a_c(\tau^*)$ exceeds the value corresponding to the aging-less case $a_c = 1/4$, and it then approaches asymptotically to $1/4$. As can be seen in Figure 2, this phenomenon occurs for both profiles of aging. Moreover, as in the noisy voter model with aging, in the kinetic model there is also a value τ_0^* such that for $\tau^* < \tau_0^*$, the algebraic aging gives rise to a larger consensus region than the exponential aging, while for $\tau^* > \tau_0^*$, the situation is reversed. For the noisy kinetic exchange model, it is $\tau_0^* = 5.40$.

5. Final Remarks

We have considered two models of opinion formation, namely the noisy versions of the 2-state voter and 3-state kinetic exchange models, and introduced the effect of aging through the probability of change q_τ acting on the interaction between agents. Moreover, we considered two different families of kernels for q_τ: either algebraically or exponentially decaying with τ.

A mean-field description has been given for the generic form of $q(\tau)$, unifying the previous results for the two models [29,37] and extending part of them to embrace a wider family of models with n opinion states. This description relies on obtaining the evolution equations for the density of the agents in each opinion state s and with the given age τ. Then, the steady state solutions of this system of (infinite) equations and the analysis of its stability (via an adiabatic approximation valid when separation of timescales holds) allowed for the obtaining of the plots of the order parameter m_{st} vs. the noise probability a, in agreement with agent-based simulations in a complete graph.

We have particularly considered the algebraic form $q_\tau = (q_\infty \tau + q_0 \tau^*)/(\tau + \tau^*)$ and the exponential form $q_\tau = q_\infty + (q_0 - q_\infty)\exp(-\tau/\tau^*)$, with $0 \leq q_\infty < q_0 \leq 1$. In numerical examples, we focused on the case that $(q_0, q_\infty) = (1, 0)$. Our results indicate that qualitatively similar tendencies emerge for both kernels in both models as follows.

For sufficiently large τ^* (weak aging), there is good agreement with the aging-less situation, as expected. For all values of τ^*, the magnetization, m_{st}, as a function of the noise intensity, a, displays a continuous transition from order to disorder at a critical value a_c. In all the analyzed cases, we observed an optimal value of τ^* to obtain a consensus (maximizing this critical value a_c). It is also noteworthy that, while both models produce different results under weak aging, when aging is strong or moderate, the critical value a_c becomes nearly insensitive to the particular social interaction rule (s_i, s_j) and number of opinion states n.

An interesting continuation would be to go beyond all-to-all interactions and to study aging effects on the opinion dynamics in random networks. Moreover, other forms of the aging kernel might also bring new features.

Author Contributions: All authors contributed equally to this work. All authors have read and agreed to the published version of the manuscript.

Funding: Partial financial support has been received from the Agencia Estatal de Investigación (AEI, MCI, Spain) MICIU/AEI/10.13039/501100011033 and Fondo Europeo de Desarrollo Regional (FEDER, UE) under Project APASOS (PID2021-122256NB-C21) and the María de Maeztu Program for units of Excellence in R&D, grant CEX2021-001164-M. C.A. also acknowledges the partial support received from Conselho Nacional de Desenvolvimento Científico e Tecnológico (CNPq)-Brazil (311435/2020-3) and Fundação de Amparo à Pesquisa do Estado de Rio de Janeiro (FAPERJ)-Brazil (CNE E-26/201.109/2021).

Data Availability Statement: Data are contained within the article.

Conflicts of Interest: The authors declare no conflicts of interest.

Appendix A. Rate Equations

In the general case of n possible state values s, the rate equations for the density x^s_τ of the agents in state s and age τ are

$$\frac{dx^s_\tau}{dt} = \Omega_{ss}(\tau-1) - \sum_{s'} \Omega_{ss'}(\tau), \quad \tau \geq 1, \tag{A1}$$

where $\Omega_{ss'}(\tau)$ is the transition rate from state s to state s'. The first term accounts for those updates that did not result in a change of state and hence increase their internal time from $\tau-1$ to τ, while the second accounts for those that resulted in a change of state and therefore resulted in a change $\tau \to 0$ of the internal time. For $\tau = 0$, the rate equation includes all those changes that result in a reset of the internal time,

$$\frac{dx^s_0}{dt} = \sum_{\tau=0}^{\infty} \sum_{s' \neq s} \Omega_{s's}(\tau) - \sum_{s'} \Omega_{ss'}(0). \tag{A2}$$

Adding Equations (A1) and (A2) over all values of τ, one obtains the rate equations for the density of agents in each state s as $x^s = \sum_{\tau=0}^{\infty} x^s_\tau$. We now derive the rate expressions for the noisy voter. A similar treatment for the kinetic exchange model can be found in Ref. [37].

Appendix A.1. Noisy Voter Model with Aging

The state can take two possible values $s \in \{-1, +1\}$. For the all-to-all connectivity adopted in this paper, the set of transition rates $\Omega_{ss'}$ comprises

$$\begin{aligned}
\Omega_{--}(\tau) &= x^-_\tau \left(\frac{a}{2} + (1-a)(1 - x^+ q_\tau)\right), \\
\Omega_{-+}(\tau) &= x^-_\tau \left(\frac{a}{2} + (1-a) x^+ q_\tau\right), \\
\Omega_{+-}(\tau) &= x^+_\tau \left(\frac{a}{2} + (1-a) x^- q_\tau\right), \\
\Omega_{++}(\tau) &= x^+_\tau \left(\frac{a}{2} + (1-a)(1 - x^- q_\tau)\right).
\end{aligned} \tag{A3}$$

In this paper, we have used the notation x for x^+ and $1-x$ for x^-. We show now the derivation of the expression for the first of these rates, $\Omega_{--}(\tau)$: The probability that an agent with internal time τ and state value $s=-1$ remains -1 first requires the selection of an agent in that state (probability x_τ^-); then, if the social rule is chosen (probability $1-a$), the copying mechanism activates with a probability of q_τ and the selected neighbor must be in the state -1 (probability x^-), but if the copying mechanism is not activated (probability $1-q_\tau$), the state remains -1. If, on the other hand, the idiosyncratic rule is activated (probability a) and the new state is chosen to equal -1 with a probability of $1/2$, this leads to

$$\Omega_{--}(\tau) = x_\tau^- \left((1-a)(x^- q_\tau + (1-q_\tau)) + a\frac{1}{2} \right), \tag{A4}$$

which, after the replacement of $x^- = 1-x^+$, leads to the first of Equation (A3). Other transition rates in these equations are obtained similarly.

From these rates and Equations (A1) and (A2), one can derive the rate equations for x_τ^s as

$$\begin{aligned} \frac{dx_\tau^-}{dt} &= \Omega_{--}(\tau-1) - x_\tau^-, \\ \frac{dx_\tau^+}{dt} &= \Omega_{++}(\tau-1) - x_\tau^+, \end{aligned} \tag{A5}$$

for $\tau \geq 1$, and

$$\begin{aligned} \frac{dx_0^-}{dt} &= \frac{a}{2}x^+ + (1-a)x^- y^+ - x_0^-, \\ \frac{dx_0^+}{dt} &= \frac{a}{2}x^- + (1-a)x^+ y^- - x_0^+, \end{aligned} \tag{A6}$$

for $\tau = 0$, where we have defined

$$y^s \equiv \sum_{\tau=0}^{\infty} q_\tau x_\tau^s. \tag{A7}$$

To obtain a closed equation for the time evolution of the densities x^s, we use an adiabatic approximation whereby we assume that the time derivative Equation (A5) of the age-dependent density x_τ^s can be set to zero, allowing for the expression of $x_\tau^\pm$ in terms of $x_{\tau-1}^\pm$. The solution of this recursive relation is

$$\begin{aligned} x_\tau^- &= x_0^- F_\tau(x^+), \\ x_\tau^+ &= x_0^+ F_\tau(x^-), \end{aligned} \tag{A8}$$

with

$$F_0(x) = 1, \qquad F_\tau(x) \equiv \prod_{k=0}^{\tau-1} \gamma_2(q_k x, a), \quad \tau \geq 1, \tag{A9}$$

and the function γ_2 is defined in Equation (6). Adding Equation (A8) over all $\tau \geq 0$, we obtain

$$\begin{aligned} x^- &= x_0^- \sum_{\tau=0}^{\infty} F_\tau(x^+), \\ x^+ &= x_0^+ \sum_{\tau=0}^{\infty} F_\tau(x^-), \end{aligned} \tag{A10}$$

which, substituted in Equation (A8), leads to

$$x_{\tau}^{-} = x^{-}\frac{F_{\tau}(x^{+})}{\sum_{\tau} F_{\tau}(x^{+})},$$
$$x_{\tau}^{+} = x^{+}\frac{F_{\tau}(x^{-})}{\sum_{\tau} F_{\tau}(x^{-})}, \tag{A11}$$

which are now expressed in terms of the global variables $x^{\pm}$.

Adding Equations (A5) and (A6) over all values of τ, we obtain the corresponding equations for the density x^{s} of each state s.

$$\frac{dx^{-}}{dt} = \frac{a}{2}(x^{+} - x^{-}) + (1-a)(x^{-}y^{+} - x^{+}y^{-}),$$
$$\frac{dx^{+}}{dt} = \frac{a}{2}(x^{-} - x^{+}) + (1-a)(x^{+}y^{-} - x^{-}y^{+}). \tag{A12}$$

Moreover, note that one of the two equations can be eliminated, since $x^{+} + x^{-} = 1$. Then, in order to obtain a closed evolution equation for the global variable x^{+}, one needs to express the variables y^{s} appearing in Equation (A12) in terms of x^{+}. This can be done by using Equation (A11) into Equation (A7), yielding

$$y^{-} = (1 - x^{+})\Phi(x^{+}), \quad y^{+} = x^{+}\Phi(1 - x^{+}), \tag{A13}$$

where we have introduced the function

$$\Phi(x) \equiv \frac{\sum_{\tau=0}^{\infty} q_{\tau}F_{\tau}(x)}{\sum_{\tau=0}^{\infty} F_{\tau}(x)}. \tag{A14}$$

Let us note here, for consistency, that in the aging-less case, $q_{\tau} = 1$, it is $\Phi(x) = 1$ and, hence, $y^{s} = x^{s}$, for $s = -1, +1$.

The replacement of Equation (A13) in Equation (A12) leads to the rate equation for $x \equiv x^{+}$, Equation (3). Note that the function $\Phi(x)$ depends on the aging profile q_{τ} both through the explicit dependence on Equation (A14) and in the expression of the $F_{\tau}(x)$ of Equation (A9). As shown in Appendix B, the function $\Phi(x)$ can be related to hypergeometric functions in the case of an algebraic dependence on the aging probability.

Appendix B. Calculation of Sums Involving $F_{\tau}(x)$

Appendix B.1. Algebraic Aging

In the case of a rational function of the age of the general form

$$q_{\tau} = \frac{q_{\infty}\tau + q_{0}\tau^{*}}{\tau + \tau^{*}}, \tag{A15}$$

where $\tau^{*} > 0$ and $0 \le q_{\infty} < q_{0}$, the function $F_{\tau}(x)$ defined is given by

$$F_{\tau}(x) \equiv \prod_{k=0}^{\tau-1} \gamma_{n}(q_{k}\, x, a) = \gamma_{n}(q_{\infty}x, a)^{\tau}\frac{(\tau^{*}\xi_{n}(x,a))_{\tau}}{(\tau^{*})_{\tau}}, \quad \tau \ge 1, \tag{A16}$$

where

$$\xi_{n}(x,a) \equiv \frac{\gamma_{n}(q_{0}x,a)}{\gamma_{n}(q_{\infty}x,a)}, \tag{A17}$$

with the function $\gamma_{n}(z,a)$ defined in Equation (6), and $(z)_{\tau} \equiv \Gamma(z+\tau)/\Gamma(z)$ is the Pochhammer symbol and Γ is the gamma function.

In order to compute the function $\Phi(x) = \dfrac{\sum_{\tau=0}^{\infty} q_{\tau}F_{\tau}(x)}{\sum_{\tau=0}^{\infty} F_{\tau}(x)}$, one needs the following sums:

$$\sum_{\tau=0}^{\infty} F_\tau(x) = {}_2F_1(1, \tau^* \xi_n(x,a); \tau^*; \gamma_n(q_\infty x, a)), \tag{A18}$$

$$\sum_{\tau=0}^{\infty} q_\tau F_\tau(x) = q_0\, {}_2F_1(1, \tau^* \xi_n(x,a); 1+\tau^*; \gamma_n(q_\infty x, a)) + \frac{q_\infty}{1+\tau^*} \gamma_n(q_0 x, a)\, {}_2F_1(2, 1+\tau^* \xi_n(x,a); 2+\tau^*; \gamma_n(q_\infty x, a)). \tag{A19}$$

In this paper, we cover the algebraic case corresponding to $q_0 = 1$ and $q_\infty = 0$.

Appendix B.2. Exponential Aging

In the definition of $F_\tau(x)$, Equation (A16), where $\gamma_n(z,a)$ is given by Equation (6), the substitution of $q_\tau = e^{-\tau/\tau^*}$ yields

$$F_\tau(x) = \Big(\underbrace{1 - \frac{n-1}{n}a}_{\alpha}\Big)^\tau \Big(\underbrace{nx\frac{1-a}{n-(n-1)a}}_{b}; e^{-1/\tau^*}\Big)_\tau, \tag{A20}$$

where $(r;s)_0 = 1$ and $(r;s)_k = \prod_{i=0}^{k-1}(1 - rs^i)$ for $k \geq 1$ is the q-Pochhammer symbol, and where $\alpha \in (1/n, 1)$ and $b \in (0,1)$. Therefore, the function $\Phi(x)$ is given by

$$\Phi(x) = \frac{\sum_{\tau=0}^{\infty} q_\tau F_\tau(x)}{\sum_{\tau=0}^{\infty} F_\tau(x)}. \tag{A21}$$

A way to compute numerically the infinite series in the numerator and denominator of Equation (A21) is to cut them off by finite sums up to an upper index $\tau = M$. However, the convergence seems to be slow, and large values of M are needed for a precise calculation, especially for small values of τ^* or a. To overcome this difficulty, an efficient procedure has been described in Ref. [37]. It starts by introducing the function

$$\phi(z,b,s) = \sum_{\tau=0}^{\infty} (b;s)_\tau\, z^\tau, \tag{A22}$$

from which Equation (A21) reads

$$\Phi(x) = \frac{\phi(\alpha s, b, s)}{\phi(\alpha, b, s)}, \quad s = e^{-1/\tau^*} \in (0,1). \tag{A23}$$

Then, we use the iteration relation that follows from its definition, Equation (A22), namely,

$$\phi(\alpha, bs^k, s) = 1 + \alpha(1 - bs^k)\,\phi(\alpha, bs^{k+1}, s), \quad k = L, L-1, \ldots, 0 \tag{A24}$$

with the initial condition

$$\phi(\alpha, bs^{L+1}, s) = \frac{1}{1-\alpha}, \tag{A25}$$

that results from the approximation $bs^{L+1} \approx 0$ and the use of $(0;s)_k = 1$ in the definition (A22). We have taken L such that $bs^{L+1} < \epsilon = 10^{-12}$, or $L \sim \log(\epsilon/b)/\log s = \tau^* \log(b/\epsilon)$, although other smaller values of ϵ produced the same results.

References

1. Castellano, C.; Fortunato, S.; Loreto, V. Statistical physics of social dynamics. *Rev. Mod. Phys.* **2009**, *81*, 591–646. [CrossRef]
2. Conte, R.; Gilbert, N.; Bonelli, G.; Cioffi-Revilla, C.; Deffuant, G.; Kertesz, J.; Loreto, V.; Moat, S.; Nadal, J.P.; Sanchez, A.; et al. Manifesto of computational social science. *Eur. Phys. J. Spec. Top.* **2012**, *214*, 325–346. . [CrossRef]
3. Axelrod, R. The dissemination of culture: A model with local convergence and global polarization. *J. Confl. Resolut.* **1997**, *41*, 203–226. [CrossRef]

4. Castellano, C.; Marsili, M.; Vespignani, A. Nonequilibrium phase transition in a model for social Influence. *Phys. Rev. Lett.* **2000**, *85*, 3536–3539. [CrossRef] [PubMed]
5. Klemm, K.; Eguíluz, V.M.; Toral, R.; Miguel, M.S. Role of dimensionality in Axelrod's model for the dissemination of culture. *Phys. A Stat. Mech. Appl.* **2003**, *327*, 1–5. [CrossRef]
6. Galam, S. Contrarian deterministic effects on opinion dynamics: "The hung elections scenario". *Phys. A Stat. Mech. Appl.* **2004**, *333*, 453–460. [CrossRef]
7. Gimenez, M.C.; Reinaudi, L.; Galam, S.; Vazquez, F. Contrarian Majority Rule Model with External Oscillating Propaganda and Individual Inertias. *Entropy* **2023**, *25*, 1402. [CrossRef] [PubMed]
8. Galam, S. Minority opinion spreading in random geometry. *Eur. Phys. J. B* **2002**, *25*, 403–406. [CrossRef]
9. Forgerini, F.L.; Crokidakis, N.; Carvalho, M.A.V. Directed propaganda in the majority-rule model. *Int. J. Mod. Phys. C* **2024**. in print. [CrossRef]
10. Lebowitz, J.L.; Saleur, H. Percolation in strongly correlated systems. *Phys. A Stat. Mech. Appl.* **1986**, *138*, 194–205. [CrossRef]
11. Fichthorn, K.; Gulari, E.; Ziff, R.; Considine, D.; Redner, S.; Takayasu, H.; Fichthorn, K.; Gulari, E.; Ziff, R. Noise-induced bistability in a Monte Carlo surface-reaction model. *Phys. Rev. Lett.* **1989**, *63*, 1527. [CrossRef] [PubMed]
12. Considine, D.; Redner, S.; Takayasu, H. Comment on "Noise-induced bistability in a Monte Carlo surface-reaction model". *Phys. Rev. Lett.* **1989**, *63*, 2857. [CrossRef] [PubMed]
13. Kirman, A. Ants, rationality, and recruitment. *Quart. J. Econ.* **1993**, *108*, 137–156. .. [CrossRef]
14. Granovsky, B.L.; Madras, N. The noisy voter model. *Stoch. Process. Their Appl.* **1995**, *55*, 23–43. [CrossRef]
15. Alfarano, S.; Lux, T.; Wagner, F. Time variation of higher moments in a financial market with heterogeneous agents: An analytical approach. *J. Econ. Dyn. Control* **2008**, *32*, 101–136. [CrossRef]
16. Diakonova, M.; Eguíluz, V.M.; San Miguel, M. Noise in coevolving networks. *Phys. Rev. E* **2015**, *92*, 032803. [CrossRef] [PubMed]
17. Carro, A.; Toral, R.; San Miguel, M. The noisy voter model on complex networks. *Sci. Rep.* **2016**, *6*, 24775. [CrossRef] [PubMed]
18. Crokidakis, N. Phase transition in kinetic exchange opinion models with independence. *Phys. Lett. A* **2014**, *378*, 1683–1686. [CrossRef]
19. Vieira, A.R.; Crokidakis, N. Noise induced absorbing phase transition in a model of opinion formation. *Phys. Lett. A* **2016**, *380*, 2632–2636. [CrossRef]
20. Penna, T.J.P. A bit-string model for biological aging. *J. Stat. Phys.* **1995**, *78*, 1629–1633. [CrossRef]
21. Azbel, M.Y. Unitary mortality law and species-specific age. *Proc. R. Soc. Lond. B Biol. Sci.* **1996**, *263*, 1449–1454. [CrossRef]
22. Azbel, M.Y. Phenomenological theory of mortality. *Phys. Rep.* **1997**, *288*, 545–574. . [CrossRef]
23. Celina, M.; Gillen, K.; Wise, J.; Clough, R. Anomalous aging phenomena in a crosslinked polyolefin cable insulation. *Radiat. Phys. Chem.* **1996**, *48*, 613–626. [CrossRef]
24. Robinson, A.L.; Donahue, N.M.; Shrivastava, M.K.; Weitkamp, E.A.; Sage, A.M.; Grieshop, A.P.; Lane, T.E.; Pierce, J.R.; Pandis, S.N. Rethinking organic aerosols: Semivolatile emissions and photochemical aging. *Radiat. Phys. Chem.* **2007**, *315*, 1259–1262. [CrossRef] [PubMed]
25. Stark, H.U.; Tessone, C.J.; Schweitzer, F. Decelerating microdynamics can accelerate macrodynamics in the voter model. *Phys. Rev. Lett.* **2008**, *101*, 018701. [CrossRef] [PubMed]
26. Fernández-Gracia, J.; Eguíluz, V.M.; San Miguel, M. Update rules and interevent time distributions: Slow ordering versus no ordering in the voter model. *Phys. Rev. E* **2011**, *84*, 015103. [CrossRef] [PubMed]
27. Pérez, T.; Klemm, K.; Eguíluz, V.M. Competition in the presence of aging: Dominance, coexistence, and alternation between states. *Sci. Rep.* **2016**, *6*, 21128. [CrossRef] [PubMed]
28. Rozanova, L.; Boguñá, M. Dynamical properties of the herding voter model with and without noise. *Phys. Rev. E* **2017**, *96*, 012310. [CrossRef]
29. Artime, O.; Peralta, A.F.; Toral, R.; Ramasco, J.J.; San Miguel, M. Aging-induced continuous phase transition. *Phys. Rev. E* **2018**, *98*, 32104, [CrossRef]
30. Abella, D.; San Miguel, M.; Ramasco, J.J. Aging effects in Schelling segregation model. *Sci. Rep.* **2022**, *12*, 19376. [CrossRef]
31. Abella, D.; San Miguel, M.; Ramasco, J.J. Aging in binary-state models: The Threshold model for complex contagion. *Phys. Rev. E* **2023**, *107*, 024101. [CrossRef] [PubMed]
32. Svenkeson, A.; Swami, A. Reaching consensus by allowing moments of indecision. *Sci. Rep.* **2015**, *5*, 14839. [CrossRef] [PubMed]
33. Balenzuela, P.; Pinasco, J.P.; Semeshenko, V. The Undecided Have the Key: Interaction-Driven Opinion Dynamics in a Three State Model. *PLoS ONE* **2015**, *10*, e0139572. [CrossRef]
34. Vazquez, F.; Redner, S. Ultimate fate of constrained voters. *J. Phys. A Math. Gen.* **2004**, *37*, 8479–8494. [CrossRef]
35. Gekle, S.; Peliti, L.; Galam, S. Opinion dynamics in a three-choice system. *Eur. Phys. J. B* **2005**, *45*, 569–575. [CrossRef]
36. Artime, O.; Carro, A.; Peralta, A.F.; Ramasco, J.J.; San Miguel, M.; Toral, R. Herding and idiosyncratic choices: Nonlinearity and aging-induced transitions in the noisy voter model. *Compt. Rend. Phys.* **2019**, *20*, 262–274. [CrossRef]

37. Vieira, A.R.; Llabrés, J.; Toral, R.; Anteneodo, C. Noisy kinetic-exchange opinion model with aging. *arXiv* **2023**, arXiv:2311.13675. https://doi.org/10.48550/arXiv.2311.13675.
38. Peralta, A.F.; Khalil, N.; Toral, R. Ordering dynamics in the voter model with aging. *Phys. A Stat. Mech. Appl.* **2019**, *552*, 122475. [CrossRef]

physics

MDPI

Article

A Theory of Best Choice Selection through Objective Arguments Grounded in Linear Response Theory Concepts

Marcel Ausloos [1,2,3,4,*], Giulia Rotundo [5] and Roy Cerqueti [6,7]

1 School of Business, University of Leicester, Brookfield, Leicester LE2 1RQ, UK
2 Group for Research and Applications of Physics in Economy and Sociology (GRAPES), Rue de la Belle Jardinière 483, Angleur, B-4031 Liège, Belgium
3 Department of Statistics and Econometrics, The Bucharest University of Economic Studies, Caeia Dorobantilor 15-17, 010552 Bucharest, Romania
4 Universitatea Babe-Blolyai, Str. Mihail Kogălniceanu nr. 1, 400084 Cluj-Napoca, Romania
5 Department of Statistical Sciences, Sapienza University of Rome, Piazzale Aldo Moro 5, I-00185 Rome, Italy; giulia.rotundo@uniroma1.it
6 Department of Social and Economic Sciences, Sapienza University of Rome, Piazzale Aldo Moro 5, I-00185 Rome, Italy; roy.cerqueti@uniroma1.it
7 Groupe de Recherche Angevin en Économie et Management (GRANEM), University of Angers, SFR Confluences, F-49000 Angers, France
* Correspondence: ma683@le.ac.uk

Abstract: In this study, we propose how to use objective arguments grounded in statistical mechanics concepts in order to obtain a single number, obtained after aggregation, which would allow for the ranking of "agents", "opinions", etc., all defined in a very broad sense. We aim toward any process which should a priori demand or lead to some consensus in order to attain the presumably best choice among many possibilities. In order to specify the framework, we discuss previous attempts, recalling trivial means of scores—weighted or not—Condorcet paradox, TOPSIS (Technique for Order Preference by Similarity to Ideal Solution), etc. We demonstrate, through geometrical arguments on a toy example and with four criteria, that the pre-selected order of criteria in previous attempts makes a difference in the final result. However, it might be unjustified. Thus, we base our "best choice theory" on the linear response theory in statistical physics: we indicate that one should be calculating correlations functions between all possible choice evaluations, thereby avoiding an arbitrarily ordered set of criteria. We justify the point through an example with six possible criteria. Applications in many fields are suggested. Furthermore, two toy models, serving as practical examples and illustrative arguments are discussed.

Keywords: best objective choice; complex systems; linear response theory; opinion formation; selection by ranking; sociophysics

Citation: Ausloos, M.; Rotundo, G.; Cerqueti, R. A Theory of Best Choice Selection through Objective Arguments Grounded in Linear Response Theory Concepts. *Physics* **2024**, 6, 468–482. https://doi.org/10.3390/physics6020031

Received: 3 January 2024
Revised: 10 February 2024
Accepted: 15 February 2024
Published: 27 March 2024

1. Introduction

Nowadays, statistical physics has found wide open research topics outside of its classical aims in economics and sociology nowadays [1–5]. Thus, consider the interplay between sociology and physics: sociophysics.

Forget Hobbes, Quetelet, Comte, Verhulst, and others there should be no need to point out, as an introduction or justification for this paper, that Galam is a pioneer of modern sociophysics [6,7] (see also the relevant papers [8–15] of interest for consideration in this paper).

One of Galam's goals was to set a framework, provide techniques, and search for conclusions on the dynamics of opinion formation in various societies. Galam's studies lead to the finding of the conditions for consensus (kind of an equilibrium), chaotic states, and intermediary complex phases in multi dimension diagrams. However, it is not quite apparent that his findings pertain to what people would refer to or call "the best choice".

But, what is the best choice? It should be admitted that what one would call the best choice is a highly relative and subjective state or concept. The answer contains ingredients from both personal ("selfish") and global ("self-effacing") points of view. Moreover, the dependence on exogenous and endogenous conditions is quite significant, but such considerations are left for political discussions elsewhere. We consider that the final states in Galam's studies or models, and subsequent studies by many, result from a too heavily weighted stochastic set of constraints or hypotheses. Opinion dynamics surely result from individual "votes" (meaning choices) due to herding or because of contrarians [9,12,16]. Nevertheless, do the dynamics imply a good choice? Or worse, is the choice (meaning vote) the best choice? That seems to be a crucial point, not only at election times in democracies, but also in choices like (we limited the references to a few papers) in the media [17], including music genrefication networks [18]; in economics [19], including regional studies [20] and drawdown market price sizes at speculative times [21], in academia, including scholarly journal rankings [22], research networks clustering [23], world universities ranking [24,25] or samba schools ranking [26,27]; and in sport [28–31].

More explicitly, in the academic domain, what is the best choice when hiring or promoting a colleague, when appointing a vice-chancellor, or when selecting teams for research grants? In the sport domain, the ranking of football teams or of cyclist racing teams is obtained through apparently objective numbers, but the rules can often (or even always) be debated and challenged [32,33]. The same remark holds for the Nobel Prize, the Pulitzer Prize, the Goncourt Prize, the Oscar or Cesar Awards which are given through highly subjective, not objective, criteria; not discounting facing a choice between roads going from X to Y, for going on holidays or to a restaurant, etc.? What is the best equipment or car or cell phone to buy? What is the best food, from a health point of view? All these questions mix subjective and objective criteria.

This boils down to the fundamental and practical question: what are the criteria needed for reaching the best choice? In other words, how should one conclusively rank a set of "things", "people" or "teams" in a constrained set of criteria?

This leads us to remember that the "final choice" leads to a somewhat paradoxical situation, the Condorcet paradox [34–36], for example. Recall that the Condorcet method [34–36] is a voting system that will always select the candidate that voters prefer over each other candidate, when comparing between them one at a time. Further considerations on comparing preferred choices lead to Arrow's incompatibility theorem [37]. Moreover, the order of criteria might lead to ambiguities [38].

The discussion, and the subsequent answers, should pertain to a comparison of the evaluation methods according to criteria [39,40]. Most of these are based on previous achievements, even though their forecasting value, not mentioning consistency for future achievements or impact, are far from certain.

This drastically annoying deduction seems to stem from the plethora of "parameters", i.e., possible criteria. Practically, one turns toward aggregation processes [41–43], going from multi dimensions toward a single number. This makes life complicated enough when one turns toward the modelling. Thus, one wishes to have some indubitable argument; often, that means having rigorous mathematical theorems. However, this is often hard to implement, in particular for laymen (or lay women, or lay others). Therefore, mathematical arguments might be by-passed through physics concepts which allow metaphors and analogies, like in modern statistical mechanics or like in Galam's view of social thought dynamics, for example, on networks.

Therefore, after outlining elements for discussion from information theory, arguments from geometry, and surely arguments on complex systems and sociophysics, we propose a powerful argument grounded in linear response theory (LRT): in LRT, the coefficients (magnetic susceptibility, transport coefficients, etc.), measured in the laboratory and estimated theoretically, are discussed and defined through the correlations between the (fluctuations of the) dependent variables. Thus, we suggest calculating all correlation functions implying the relevant variables in the sociophysics research topics of interest, in particular when searching for opinion formation, as due to Galam. This idea seems to be missing in earlier considerations. We claim that this idea is leading to a more objective hierarchy of values in opinion formation choice, i.e., topics.

Within this set of considerations, a study framework can be conceived together with applications, thereby leading to the following structural content of this study.

Section 2 contains a brief review of "old" and "modern" techniques for selecting and ranking agents or events, like the rank-size "laws". We discuss such attempts, recalling trivial "means of scores", be they weighted or not, the Condorcet paradox, etc. We recall preference aggregation techniques, like the maximum likelihood rule (MLR) (Section 2.1), and the Technique for Order Preference by Similarity to Ideal Solution (TOPSIS) (Section 2.2).

However, the aggregation problems have two different aspects: rank aggregation or score aggregation. These are briefly distinguished in Section 2.3.

In particular, in Section 2.4, we start from geometrical aspects for ranking scores, or measures, and for later aggregating multi-valued measures. Furthermore, we demonstrate through such geometrical arguments that the order of criteria might make a difference on the final result, but this might be unjustified. It is easily observed that, if only three criteria are used, the procedure leads to an undebatable result. We show on a set of four criteria for a toy model that the order is drastically relevant and influences the outcome after aggregation.

We base our best choice theory in Section 3, upon the linear response theory. We indicate that one should be calculating correlations functions between all possible choice evaluations. Appendix A recalls the fundamentals of a linear response theory in statistical physics.

We conclude with some emphasis, if it is needed, on the finding that the best ranking is obtained from a method based on rigorous arguments for obtaining a final score and ranking. We are aware that subjective arguments often influence the final decision. We suggest applications in a few domains in Section 4. Two examples can be found in Appendix B.

2. Modern Ideas and Methods

In this Section, we consider ideas and methods for the ranking and selecting of the best outcome due to inequalities between events and agents, also called "variables", thereby assuming that a hierarchy takes places and leads to the best choice [44].

Before recalling modern ideas, let us remind the reader that the selection process leads to some so-called rank-size (RS) displays or tables. Indeed, RS analysis is the basic way of measuring disorder in a population: the largest "size" gets the first rank, and a hierarchy, from which an ordering through inequality is deduced [44]. This is a hierarchical description. *In fine*, this leads to considering whether empirical laws follow patterns, thereby suggesting models.

The RS law derives from the analytical form presented by the variables, often ranked in descending order as a function of the (discrete) index i, giving a "rank" to each of the g_i, $i = 1, \ldots, N$ variables. The cumulative law $\sum_i g_i$ leads to the cumulative concentration distribution curve (CCDC). The sum $\sum_i^N g_i$ over all the i elements can serve as a normalisation value. One immediately obtains the "normalised" CCDC . When the normalised CCDC goes over the 80% threshold, this defines the Pareto rank, r_P. The Pareto principle expects this rank r_P to be equal to $N/5$. Thus, the RS method is convenient, and better suitable for large populations, i.e., when the size or/and rank ranges can be large. This is rarely the case when only a few selection criteria are implied.

For completeness, nevertheless, let us mention that Marfels [45] distinguishes several types of concentration ratios according to weighting schemes and their structure, which can be discrete or cumulative [46]. Beside such ratios, inequality aspects are often discussed in order to touch upon socio-economic concerns [47].

Furthermore, one can recall that, when the goal is to find a compromise between the various rankings, the statistical median is thought to be the most appropriate solution [48,49].

However, in situations when decision making should be a way of compromising between conflicting decisions, the use of MLR, discussed in the following Section 2.1, is justified.

2.1. Maximum Likelihood Rule

Recall that a choice demands some ranking. This is usually done by combining ordered preference lists into a single consensus value, as in a reviewing procedure [49,50]. We outline the MLR, which is based directly on rankings and not on the scores.

In brief, methods of preference aggregation, such as the MLR, are based on the concept of pairwise preference notions [51], which are much used in economics and in opinion formation, alongside the simple majority voting rule, without a discussion of the scores. It is also known as the Kemeny [52] rule (or the Kemeny–Young method [48]). In order to go beyond the limit of the method, one introduces a variant, taking into account the behavioural argument of "blindness to small changes" (BSC) [53].

Mathematically, an MLR ordering is defined as one that minimises the total number of discrepancies among all the reviewers in their *pairwise preferences* between all options. It can be viewed as a voting scheme that determines not just a single chosen winner, but an entire ordered list. Therefore, the MLR ordering generally satisfies the most possible reviewers regarding their stated rankings of options. The ordering does not use whatever information about how much higher a candidate is ranked over another, but only a relative ordering (it is worth pointing out that MLR is a Condorcet method).

Practically, one can say that the method counts the *pairwise preferences* and applies the majority rule over each of them.

Examples abound on applying this rule and finding "solutions" [54]. One short illustration is given in Appendix B.

The second step of the method is counting the number of times where one candidate is ranked over another. In doing so, the MLR emphasises that the first-ranked choice is preferable compared to all other options in individual pairwise comparison. Similarly, the MLR second ranked choice would be be preferred to all other options (except the first ranked), and so on.

In order to complete this Section, we may consider that the MLR implies that candidates with the same score are considered to belong to the same "indifference set". If this occurs, one decides to list the "candidates" (specifying the "agents" or "events") one after the other, e.g., in alphabetical order, in such a set. One may consider other ways to manage equal scoring, but there is no need to foresee any special treatment here to take care of ex aequo positions in the ranking. Nevertheless, in most evaluation procedures, a linear order is requested for the final ranking; the most unwanted point is when two "agents" are ex aequo on the first rank. In general, in order to achieve such a goal, or resolve such a dilemma, extra information is a posteriori added for discriminating equally scored candidates. This might be unjustified but this "solution" is left for more suitable considerations and is beyond outside of the scope of the present study.

2.2. Technique for Order Preference by Similarity to Ideal Solution

The TOPSIS, was originally developed by Hwang and Yoon [55], with further pertinent developments, e.g., in Refs. [56–60], is a method of compensatory aggregation that compares a set of alternatives by identifying weights for each criterion, normalising scores for each criterion and calculating the geometric distance between each alternative and the ideal alternative, which is the best score in each criterion.

TOPSIS works attractively in various areas; see, e.g, [61].

2.3. Score Rather than Rank Aggregation

However, some fundamental emphases between processes in the preferential ordering problems must be made through distinguishing rank aggregation from score aggregation. Often, the ranks are not known, but are deduced from scores. However, as pointed out, one does not always know how the scores are obtained (recall the number of stars in the

Michelin guide or the Gault and Millau scores for restaurants, or the ranking order on Trip Advisor). In such cases, scores are incompatible and sometimes incomprehensible; ranking only makes some sense, allowing for ex aequos.

However, if scores are known from a reliable set of measures and for a given set of criteria, some objective construction can follow, leading to some meaningful score aggregation. A scoring function can be defined as $f^{(X_j)}(s_i)$, with $s_1, s_2, s_3, \ldots, s_m$, being the scores on each i-th criterion, for a given "agent" X_j, with $j = 1, \ldots, n$.

The aggregation results, from the set of n "events", "agents" and the m scoring criteria, sorted, for example, in decreasing order, lead to finding the top-k "candidates" according to the scoring function, imposing $f([s_i]) < f([s_i'])$. The objective is straightforward: to compute the top-k "agents" with the minimum cost. While the target is apparent, the methods can be quite different, depending on the choice of f. The very first problem stems from the sorting out algorithm. Let us point to the Fagin algorithm [62], to the medrank algorithm [63] and to the threshold algorithm [64]. For some completeness, another method is the Borda count [65]: for each ranking, one assigns a score D equal to the number of objects it defeats. The total weight of D is the number of points it accumulates from all rankings.

However, these rank or score ranking methods can be challenged because they are missing a key ingredient, i.e., the selection order of the criteria.

2.4. On the Sequence of Criteria: A Geometrical Perspective

In brief, many (if not all) earlier studies ignore a crucial factor: the sequence of criteria. It can be pointed out and straightforwardly illustrated that the ascertainment of criteria has an effect on ranking values; this is common knowledge to anyone having participated in surveys [66,67] and/or selection processes.

Suppose that there are three criteria giving a numerical value (a, b and c, respectively) for the agent or event. One can define a coordinate system with three axes stemming from some origin O in equivalent directions (such that the angle between each axis is, therefore, $2\pi/3$) and plot the values on each "criterion axis". Next, one way to aggregate the three values is to consider them as three sides of a triangle, and calculate the triangle area, which becomes the "score".

In this case, one may recall that it would be necessary, in order to find the area of a triangle with three sides, that one uses the Heron's formula: indeed, the area S of a triangle with three known sides, a, b and c, is calculated from

$$S = \sqrt{s(s-a)(s-b)(s-c)}\,, \tag{1}$$

where s is the semi-perimeter of the triangle, i.e., $s = (a+b+c)/2$.

On the other hand, knowing two sides b and c around an angle A, the formula to calculate the area of a triangle is given by $(1/2)\,b\,c \cdot \sin A$. Obviously,

$$S = b\,c/2 \quad \text{if} \quad A = \pi/2, \tag{2}$$

while

$$S = (\sqrt{3}/4)\,b\,c \quad \text{if} \quad A = 2\pi/3 \quad (\text{or } \pi/3). \tag{3}$$

In the latter case, this means that whatever the order of criteria, the total area, i.e., the sum of the three triangles areas, is invariant and equal to $S = (\sqrt{3}/4)\,(b\,c + a\,b + a\,c)$.

However, the matter is different when there are more than three criteria or sustaining axes. Consider the case of four criteria. Let the axes form a coordinate system with four axes in symmetric directions. The permutation of axes leads to six different polygons. A toy example, with $a = 1$, $b = 2$, $c = 3$ and $d = 4$, can be seen in Figure 1. The four inner triangles in each polygon are rectangular triangles for which each area is easily obtained (the total area is given in Figure 1 in arbitrary units); it can be well noticed that, according to symmetry, only three different areas are relevant. Therefore, this leads to three different sizes if the area is considered to be the aggregated number for ranking the agents or events.

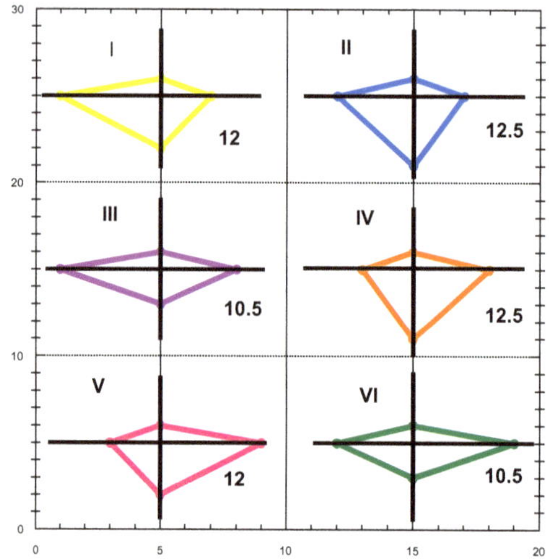

Figure 1. Demonstration that, starting from four criteria, the shape of the polygon based on the variable values leads to different areas. The toy side length values are $a = 1$, $b = 2$, $c = 3$ and $d = 4$ in arbitrary units), leading to six possible polygons, but three different sizes as indicated. See text for details.

When the number of criteria increases, from which many more polygons can be drawn, the area of such polygons can always be decomposed into a number of triangles, for which each area can be calculated using Equation (1), remembering that, using the length of two sides (a and b) of the known angle (C) between them, in any triangle, one can easily calculate the third side length of the triangle through the Al-Kashi formula, i.e.,

$$c^2 = a^2 + b^2 - 2\,a\,b \cdot \cos A\,. \tag{4}$$

3. The Problem and Its Solution

Thus, the way to proceed is as follows. First, take a survey with as many possible criteria that are needed for a population, whatever its size. One may suppose, without loss of generality, that the choice is based on Likert scales, whatever the useful range $[0,5]$, $[0,7]$, $\ldots$, $[0,20]$.

Notice that it could be of interest to consider the mean and standard deviation of the data distribution on each axis, such that one can obtain a possible "universal comparison" of final rankings and hierarchies. If the statistical characteristics make sense, the measured values could be normalised or, at least, constrained to vary in an identical range. One could assume that all the values involved in the decision process belong to $[0,1]$. The value 0 is achieved in the worst case for the considered parameter (bad realisation) and, accordingly, 1 is taken in the "best choice" case. Nevertheless, such a scaling is not a fundamental request. Although of interest, this normalisation aspect is not discussed further here and is left for further investigation.

Then, one reports the values on equidistant axes. These are disposed in a regular star-shaped network with a central node, a common starting point and identical angles between consecutive axes; the angle value actually depends on the number of axes (criteria). The non-common nodes of the consecutive segments are connected. The resulting polygon has a number of sides identical to the number of evaluation parameters; the polygon is not necessarily regular and can be considered as being made of triangles. If N is the number of axes, the largest possible number of different polygons is $(N-1)!/2$. The number of different triangles is $N!$.

In this paper, let us focus on hexagons, since evaluation processes are quite often associated with six different parameters.

In Figure 2, we present an example for two hexagons out of sixty possible cases; let a, b, c, d, e, f (e.g., $\in [0,1]$) be the values of the six parameters. The surface of the resulting hexagon $\mathcal{S}$ can be straightforwardly computed. The evaluation score coincides with the value of $\mathcal{S}$, so that a high $\mathcal{S}$ means a high score. However, notice that $\mathcal{S}$ belongs to the range associated to the limiting cases $a = b = c = d = e = f = 0$ (worst result of the evaluation process, with $\mathcal{S} = 0$) and $a = b = c = d = e = f = 1$ (best result of the evaluation exercise, with $\mathcal{S} = 3\sqrt{3}/2$).

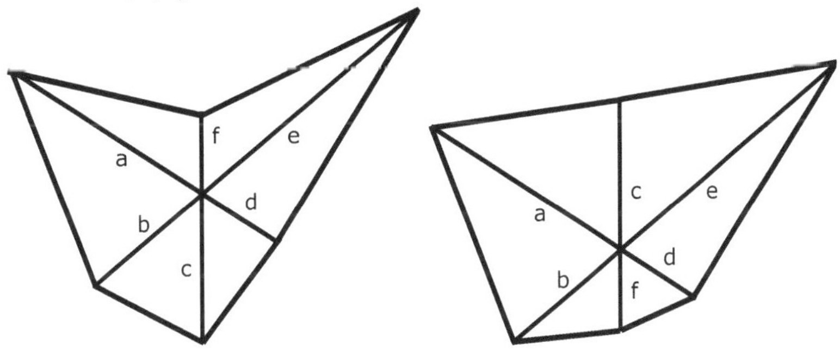

Figure 2. A toy case of six criteria for evaluation is presented with the parameter values a, b, c, d, e, and f with swapped axes c and f (**left** versus **right**). Notice that a different disposition in the relative order of the axes leads to a different area for the hexagon. See text for more details.

Certainly, the surface of the hexagons shown in Figure 2 depends on how the segments are disposed: the values of the six measures are always a, b, c, d, e and f, but the placement of c and f have been reverted in both displays.

Let $\mathcal{S}_1$ and $\mathcal{S}_2$ be the surface of the hexagons in Figure 2, left, and Figure 2, right, respectively. One has

$$\mathcal{S}_1 = \frac{\sin(\pi/3)}{2}(ab + bc + cd + de + ef + fa) = \frac{\sqrt{3}}{4}[(ab + bc + cd + de + ef + fa)],$$

$$\mathcal{S}_2 = \frac{\sin(\pi/3)}{2}(ab + bf + fd + de + ec + ca) = \frac{\sqrt{3}}{4}[ab + bf + fd + de + ec + ca];$$

In the case $b + d - e - a \neq 0$, one has that $\mathcal{S}_1 = \mathcal{S}_2$, if and only if $c = f$.

Such a remark leads to the questionability of the evaluation exercise through polygons. Indeed, the resulting polygons, differently shaped according to the disposition of the axes, have different surfaces. This means that the evaluation is strongly dependent on how selection parameters are included in the graphical representation of the problem, as we again stress.

To overcome such an inconsistency, and in order to obtain a more indubitable scoring, we propose taking the average $\bar{\mathcal{S}}(a, b, c, d, e, f)$ of all the possible polygon surfaces, thus over all the possible dispositions of the axes, here corresponding to six evaluation criteria:

$$\bar{\mathcal{S}}(a, b, c, d, e, f) = \frac{\sin(\pi/3)}{5!} \sum_{\mathbf{H}(a,b,c,d,e,f)} (x_1x_2 + x_2x_3 + x_3x_4 + x_4x_5 + x_5x_6 + x_6x_1), \quad (5)$$

where

$$\mathbf{H}(a, b, c, d, e, f) \equiv (x_1, \ldots, x_6) \in \{a, b, c, d, e, f\} \quad (6)$$

with $x_i \neq x_j$ for each $i \neq j$.

By construction, $0 \leq \bar{S}(a,b,c,d,e,f) \leq 3\sqrt{3}/2$. Such a mean surface measure achieves the intended job effectively, i.e., providing a consistent score of the evaluated object whose six parameters (criteria) take values a,b,c,d,e,f.

This is consistent with the measure concept, resulting in calculating the correlation between fluctuations in statistical physics, as can be found in linear response theory: see Appendix A. Examples of applications are given in Appendix B.

4. Conclusions

To summarize, the present study is devoted to resolve the contradictory situation when searching for an objective ranking procedure and subsequently observing some hierarchy through a set of criteria for a given population. This sort of scientometric practice is tied to considerations found in geometry and statistical physics, complementing aspects of research on opinion dynamics found in Galam's numerous studies.

Observe the technical and somewhat philosophical frames in the present ranking study. Decisions often concern rather small sub-systems, like the funding of a project, the evaluation of a research group, or that of an individual at his or her hiring or promotion. Statistical laws cannot be immediately applied for such cases in which, due to the (very finite) size of the system, fluctuations which are not necessarily due to stochastic causes play a crucial role. Indeed, recall that scientific work is part of a social system; its actors are human beings. Moreover, statistics are like snapshots and do not often allow for predicting the future sufficiently well. In practice, indeed, any newly introduced criterion or "measure" will be met by the capacity of humans to interact and make decisions.

Without going back to the Bible and raising the question about who is the just and who will be saved (certainly, according to a single judge), one may, within modern scientific reasoning on choice, admit that when several partners and judges are implied, the resulting choice (solution) ends up into a difficulty pointed out by Arrow [37]. He considered the preference aggregation problem, that is the problem of passing from a set of known individual preferences to a pattern of social decision making. Arrow's today oft-quoted theorem, while being variously worded, has nevertheless shown that the difficulties met in the building process of preference aggregation are highly general. The theorem implies that no rank order voting system can convert the ranked preferences of individuals into a community-wide (complete and transitive) ranking, besides also meeting a specific set of natural criteria.

Why indeed is it impossible to reach a choice? Commonly, one constructs several filters and a priori decides on the order of their applications, like in the case of a decision tree (DT). That leads to a discussion on ranking the filters rather than the candidates.

Indeed, in the DT scheme, due to the order of filtering criteria, the final choice is significantly biased. The order of filters is often adapted a priori in order to select the final choice, e.g., maintaining a candidate, or candidates, in competition with others during the selection process, for hypocritical, political or other reasons.

In opinionology, the selection process consists in projecting from a multidimensional space onto various planes, and finally finding the intersection of the distribution. Recall that in physics, it is like projecting on various "external field axes" or making a scalar product. There might not be any "solution". This may further mean that the set (of "candidates", "events", etc.) to be ordered is either not to be ranked or that the filters are not appropriate.

Thus, we propose, starting from Galam's and others' considerations on the dynamics of choices, a procedure for obtaining the "best choice" through an objective statistical physics method.

We would like make a remark at this stage. One may say that our method is not better or worse than another, but is merely another type of classification; for example, it will not be applied by the Hollywood Academy to award Oscar Prizes. Let us notice that we are not using the word "better". We emphasise that our method is more rigorous, contains less arbitrariness and is based (through an analogy) on major statistical physics concepts. Certainly, we might regret that our method might not be applied in Los Angeles. Actually, we admit that we are not aware of all the criteria used by the Academy in order to award Oscars. The same holds true in other branches of opinion formation. In a decision process, the final ranking might be due to hidden or specifically weighted criteria; idem for the final choice. Our method does not apply outside the objective world. We do not consider subjective criteria.

Surely, the situation might not be closed, since the choice of criteria is left for many discussions between agents, be they surveyees, surveyors or other stakeholders.

Nevertheless, our argument, after outlining elements of reflexion from information theory, geometry and surely complex systems, as found in sociophysics considerations, proposes a strong argument for calculating choice values. It is grounded in linear response theory. In the latter, as it is recalled in Appendix A, the coefficients, that are measured in laboratories and estimated theoretically, are defined through the correlations between the dependent variables; thus, we suggest carrying out the same in sociophysics, i.e., calculating all correlation functions implying the relevant variables, thereby avoiding an arbitrarily ordered set of criteria. In particular, this could be useful and meaningful in the modelling of the opinion formation processes due to Galam. This methodology is missing in previous sociophysics studies. Thus, on a LRT basis, which is rather easily implemented, this should lead to a more objective hierarchy of values ahead of selection processes.

Author Contributions: Conceptualization, methodology, software, validation, formal analysis, investigation, data curation, writing—original draft preparation, writing—review and editing, visualization, M.A., G.R. and R.C.; supervision, funding acquisition, M.A. All authors have read and agreed to the published version of the manuscript.

Funding: Work by M.A. was partially supported by the project "A better understanding of socio-economic systems using quantitative methods from physics" funded by European Union—NextgenerationEU and Romanian Government, under National Recovery and Resilience Plan for Romania, contract no.760034/23.05.2023, code PNRR-C9-I8-CF 255/29.11.2022, through the Romanian Ministry of Research, Innovation and Digitalization, within Component 9, "Investment I8".

Data Availability Statement: The data mentioned in this study are freely available on the internet as referred to or on request from the corresponding author.

Conflicts of Interest: The authors declare no conflict of interest. The funding sponsors had no role in the design of the study; in the collection, analyses, or interpretation of data; in the writing of the manuscript, and in the decision to publish the results.

Appendix A. Linear Response Theory

The LRT was independently invented by Green [68,69] and by Kubo [70]: it describes the coefficients relating the effect of a perturbation on a thermodynamic system in equilibrium. The LRT has given a general proof of the fluctuation–dissipation theorem, which states that the linear response of a given system to an external perturbation is expressed in terms of the correlations between the fluctuations of the relevant (with respect to the perturbation) variables characterizing the system in thermal equilibrium [71].

Kubo [70] considered the application of a magnetic field to an equilibrium system and demonstrated that the magnetic susceptibility can be defined through the average of the correlations between the magnetic moment density fluctuations. The LRT also well applies to the description of the electrical conductivity or the thermal conductivity [72,73], even if the system is characterised as being in a non-equilibrium state.

Consider a perturbation $B(t)$, e.g., a magnetic field, and search for the response $M(t)$ of the system, i.e., the magnetisation. Usually, one demands to obtain $M(t) = \chi B(t)$.

By analogy, in opinion formation and related rankings, one may consider that the "response" is going to be the final score. This is extracted through the correlations between the fluctuations of the values attributed to "candidates" (in a wide sense, usually called "agents'; however, they might not be humans) through the various criteria. The feature that criteria are introduced for later agent ranking is considered to induce "perturbations" in the original system. The external field is the exogenous decision of introducing criteria.

Subsequently, one may consider that the external (field) perturbation is applied at some time t_0 to some (for example) agents in a given population; the "population system" is thus moved away from its equilibrium and is characterised through a non-equilibrium ensemble average. Thus, this leads to the measured score of each agent as a result of interactions in response to the opinion field perturbation, as can be seen in most of Galam's models.

In practice, considering a weak interaction with some external field, one can obtain the resulting score by performing an expansion in powers of the perturbation [68–73]. The leading term in this expansion is independent of the field, but the next term describes the deviation from the equilibrium behaviour in terms of a linear dependence on the external perturbation through the correlations between the system fluctuations. The average score is the linear response function, i.e., the quantity that contains the (microscopic) information on the system and how it responds to the perturbing field.

Appendix B. Two Examples

Among the many possible examples among the various research fields outlined in Sections 1 and 4, we have selected two examples which are likely of interest to many readers and researchers in physics, particularly in sociophysics. The first example concerns the promotion of researchers. The second example pertains to the evaluation of football (soccer) players. The former stresses how to reach a final ranking from ranks obtained through various filters. Three methods, outlined in the main text, are compared. The latter example stresses that a meaningful ranking can be obtained from a final aggregation number based on measured values and analysed through correlations proposed in the main text.

Appendix B.1. Example 1: Researcher Promotion

Consider, as an illustration, that five researchers (A, B, C, D, E) are applying for a promotion. The relevant selection committee has decided to base its ranking of the value of the candidates on four criteria related to the impact factor (IF) of the journal where a paper has been published; from a practical point of view, let us note that the five candidates were required to submit their best ten papers , where the "best" being estimated due to the IF.

The committee calculates

- the IF mean of the best five papers: $\overline{B5}$;
- the IF mean of the worst five papers: $\overline{W5}$;
- the IF mean on the oldest five papers: $\overline{F5}$;
- the IF mean on the five most recent papers: $\overline{L5}$.

The timing of the publication is considered irrelevant at this stage. The resulting mean values and the corresponding candidate rank are given in Table A1 for each criterion.

Table A1. Profile of five candidates having published ten papers, listed in chronological order, with the candidate evaluation through various criteria, proposed by examiners. The best score value is underlined for each criterion. See text for details.

Candidate	$\overline{B5}$	Rank	$\overline{W5}$	Rank	$\overline{F5}$	Rank	$\overline{L5}$	Rank
A	<u>7.80</u>	1	3.20	2	<u>7.80</u>	1	3.20	4
B	7.00	4	2.20	3	2.20	4	7.00	2
C	7.60	2	<u>3.60</u>	1	7.60	2	3.60	3
D	6.40	5	1.00	5	5.60	3	1.80	5
E	7.60	2	2.20	3	2.20	4	<u>7.60</u>	1

Let the committee admit that (a) the ranks are more relevant than the scores and (b) the ranking of candidates is in descending order of the criteria scores. Then:

(i) It can immediately be seen that candidate A has the highest $\overline{B5}$ (=7.8) and $\overline{F5}$ (=7.8); author E has the maximum score on the last five ($\overline{L5}$) (=7.60), but author B does not reach any highest score, among these competitors or under any criterion. Let us also remark that, in this toy example, due to the nature of the evaluation, candidate D lays at the bottom on each ranking, no matter the specific criteria that are used. A complication arises in the need to rank the four others, since each of them is a winner for at least one committee member (or criterion): A and C even gain twice, but on different criteria certainly; B and E are twice ex aequos, but not near the top places. Thus, we have shown that different criteria lead to different rankings, but more so that the toy model implies that there is no obvious final choice as was, indeed, codified by Arrow's theorem. Moreover, there is no indubitable hierarchy along this simple aggregation process.

(ii) Next, consider the MLR method (Section 2.1) based on rankings, not on the scores, as given in Table A1. Recall that a MLR ordering is defined as one that minimises the total number of discrepancies among all the criteria in a pairwise preference scheme. The MLR ordering does not use any information about how much higher a candidate is ranked over another, but only a relative ordering. Practically, the method counts the pairwise preferences. The second step of the method consists in counting the number of times in which one candidate is ranked over another (from Table A1). In so doing, the MLR emphasises that the first-ranked choice wins against all other options in individual pairwise comparisons. Similarly, the MLR second-ranked choice would win against all other options (except the first-ranked), and so on. The present case outcome is reported in Table A2.

Table A2. Maximum likelihood rule (MLR) application: table reporting how many times, according to Table A1, the candidate in a row is ranked before the candidate in a column ex aequos. The parentheses indicate ex aequos, while the duplicated parentheses indicate two ex aequos.

Candidate	A	B	C	D	E
A	–	3	2	4	3
B	1	–	1	3	((0))
C	2	3	–	4	(2)
D	0	1	0	–	1
E	1	((2))	(1)	3	–

It can be observed from Table A2 that D has the lowest number of potential preferences for promotion, and hence is properly ranked as last. It can be remarked that the coalition of criteria formed only by $\overline{B5}$ and $\overline{L5}$ can be a decisive in the ranking of E before B. However, the rank relations between A and C are not so neat. Therefore, the selection committee would be faced, again, with the application of Arrow's theorem.

(iii) Finally, let us consider the newly proposed method. It boils down to calculating the surfaces of rectangular triangles and their subsequent averaging, i.e., here, for three types of polygons with four sides. Using Equation (2), one obtains the results displayed in Table A3. It looks that this ranking is justified. Thereafter, one may conclude that the proposed method is more justified and advantageous than classical ones.

Table A3. Ranking of five candidates according to the method proposed in this paper. The "Area" denotes the surface of each relevant 4-sided polygon, the "Total" denotes the sum of the area of such polygons, and the "Average" denotes the average area.

Candidate	Area (×2)			Total	Average	Rank
A	12	15	15	42	14	1
B	40	42	42	124	41.33	4
C	16	15	15	46	15.33	2
D	80	80	80	240	80	5
E	24	25	21	70	23.33	3

Appendix B.2. Example 2: Football (Soccer) Players

While writing this paper, we came across a related application we consider below. It concerns football players, ranking them on their "market value", "potential", salaries, age and many statistical "criteria" see [74] or, for teams, see [75]. Another example of "player value" can be found, for example in Ref. [76], where eight criteria are displayed. Notice that two axes—"yellow cards" and "red cards"—might both have a 0 value, reducing the display to a hexagonal pattern. Radar picture comparisons can be created [77].

In particular, e-players can form their own team, based on real players, considering six criteria, displayed on a regular hexagon and thereafter transformed into a colourful pattern, see, e.g., [78].

The six skill measures, obtained through several sub-criteria, are given in Table to the Table which those belong and abetter clarifyin there. Please consider Table A4.

Table A4. The six skill variables considered for measuring the "value" of a football player Eden Hazard, with their respective measure according to Ref. [78]. The data in the matrix corresponds to the area (divided by $[\sin(\pi/3)]/2$, see Equation (3)) of the fifteen possible triangles to be arranged in sixty different hexagons. Here, SHO stands for "shooting" and determines finishing skill and shot power, including penalty success; PAS stands for "passing" and denotes ability to successfully pass the ball with vision; DRI stands for "dribbling" and denotes ball control, agility and balance; DEF stands for "defending" and notes tackling and interceptions; PHY stands for "physicality" and notes strength, stamina and aggressiveness; PAC stands for "pace" and notes the speed and acceleration of the player.

Measure	Skill	SHO	PAS	DRI	DEF	PHY	PAC
77	SHO	-	6237	6545	2695	4774	6083
81	PAS	6237	-	6885	2835	5022	6399
85	DRI	6545	6885	-	2975	5270	6715
35	DEF	2695	2835	2975	-	2170	2765
62	PHY	4774	5022	5270	2170	-	4898
79	PAC	6083	6399	6715	2765	4898	-

Notice that it is not quite obvious how the "overall score" is obtained. Since only one hexagon is shown, it is certainly one for which the axes are a priori chosen, which does not conform to our recommendations about a tentative ranking (of football players) upon some unique value resulting from some aggregation process.

Thereafter, in order to avoid extra decimals, we measure the area of triangles and corresponding hexagons in $\sqrt{3}/4$ units; see Equation (3).

Consider one case in order to describe how the resulting aggregation player value is obtained, ahead of some selection process [78] according to which the "best overall" value of 82 of the player is obtained. However, the average of this playes skill values equals 69.83, as can be obtained from the values given in the Measure of Table A4. Furtrhermore, Table A4 reports the data corresponding to the area (divided by $[\sin(\pi/3)]/2$, see Equation (3)) of the fifteen possible triangles. For space saving, not all sixty relevant hexagons can be displayed, nor the entire list of their areas. For completeness, the statistical characteristics of the polygon areas' distribution are given in Table A5. The first line reports the characteristics for the distribution of triangle areas on the diagram displayed in Ref. [78]; this corresponds to at least six-axes diagram. The second line details the distribution of the 60-hexagon areas considered according to the present method. It is remarkable that the LRT result shows a considerably narrower distribution, thus more convincing statistics. The sign of the skewness is also in favour of the LRT method.

Table A5. Main statistical characteristics of the distributions of six-triangle areas forming the single hexagon [78] and of the areas each of sixty possible hexagons [78] along the six skill axes, following the linear response theory (LRT) method. See text for details. Here, "Min", "Max" and "Std Dev" stand for the minimum, maximum and standard deviation of the distributions, respectively.

Source	Size	Min	Max	Total Area	Mean	Std Dev	Skewness	Kurtosis
[78]	6	2170	6885	29,248	4874.7	1912.2	−0.4457	−1.4123
LRT	60	28,298	29,600	1,734,432	28,907	437.04	0.3076	−1.5345

Not considering here a comparison with other players, we should point out that such a comparison demands a final aggregation score. In order to have a universal rule, we suggest that the best approach is to measure the "player hexagon area(s)" with respect to the perfect player, i.e., the largest regular hexagon. For the case of Ref. [78] and the LRT case, one obtains 0.4875 and 0.4818, respectively.

As a conclusion of this Appendix, let us notice that we have first debated how to reach a hierarchical selection through three methods using numerical examples. This illustration is based on a toy case in which set of five candidates are applying for a promotion through a four-criteria selection process. Next, we presented the entire scheme, resulting into the aggregation score for a given individual examined through six values or criteria, according to two methods.

We have stressed the objective advantages of the LRT based method in both examples.

Therefore, one may conclude from both toy cases here that the usability and the advantage of the LRT methodology both are well supporting considerations in the sociophysics interdisciplinary field when searching for objective ranking, leading to justified choice. Let us emphasise that the main justification is the use of correlation functions, as in the LRT, i.e., a pertinent statistical physics theory basis.

References

1. Stauffer, D. Grand unification of exotic statistical physics. *Phys. A Stat. Mech. Appl.* **2000**, *285*, 121–126. [CrossRef]
2. Stauffer, D. Introduction to statistical physics outside physics. *Phys. A Stat. Mech. Appl.* **2004**, *336*, 1–5. [CrossRef]
3. Săvoiu, G.; Simăn, I.I. Sociophysics: A new science or a new domain for physicists in a modern university. In *Econophysics. Background and Applications in Economics, Finance, and Sociophysics*; Săvoiu, G., Ed.; Academic Press/Elsevier Inc.: Oxford, UK, 2013; Chapter 10. [CrossRef]
4. Kutner, R.; Ausloos, M.; Grech, D.; Di Matteo, T.; Schinckus, C.; Stanley, H.E. Econophysics and sociophysics: Their milestones & challenges. *Phys. A Stat. Mech. Appl.* **2019**, *516*, 240–253. [CrossRef]
5. Grech, D.; Miśkiewicz, J. (Eds.). *Simplicity of Complexity in Economic and Social Systems*; Springer Nature Switzerland AG: Cham, Switzerland, 2020. [CrossRef]
6. Galam, S.; Gefen (Feigenblat), Y.; Shapir, Y. Sociophysics: A new approach of sociological collective behaviour. I. Mean-behaviour description of a strike. *J. Math. Sociol.* **1982**, *9*, 1–13. [CrossRef]
7. Galam, S. Sociophysics: A review of Galam models. *Int. J. Mod. Phys. C* **2008**, *19*, 409–440. [CrossRef]
8. Galam, S. Rational group decision making. A random field Ising model at $T = 0$. *Phys. A Stat. Mech. Appl.* **1997**, *238*, 66–80. [CrossRef]
9. Galam, S. Contrarian deterministic effects on opinion dynamics: "The hung elections scenario". *Phys. A Stat. Mech. Appl.* **2004**, *333*, 453–460. [CrossRef]
10. Galam, S. Modeling the forming of public opinion: An approach from sociophysics. *Glob. Econ. Manag. Rev.* **2013**, *18*, 2–11. [CrossRef]
11. Galam, S. Is it necessary to lie to win a controversial public debate? An answer from sociophysics. In *Nonlinear Phenomena in Complex Systems: From Nano to Macro Scale*; Matrasulov, D., Stanley, H.E., Eds.; Springer Science+Business Media: Dordrecht, The Netherlands, 2014; pp. 37–45. [CrossRef]
12. Galam, S. The invisible hand and the rational agent are behind bubbles and crashes. *Chaos Solitons Fractals* **2016**, *88*, 209–217. [CrossRef]
13. Galam, S.; Cheon, T. Tipping points in opinion dynamics: A universal formula in five dimensions. *Front. Phys.* **2020**, *8*, 566580. [CrossRef]
14. Galam, S. Opinion dynamics and unifying principles: A global unifying frame. *Entropy* **2022**, *24*, 1201. [CrossRef] [PubMed]
15. Biondi, Y.; Giannoccolo, P.; Galam, S. Formation of share market prices under heterogeneous beliefs and common knowledge. *Phys. A Stat. Mech. Appl.* **2012**, *391*, 5532–5545. [CrossRef]

16. Dhesi, G.; Ausloos, M. Modelling and measuring the irrational behaviour of agents in financial markets: Discovering the psychological soliton. *Chaos Solitons Fractals* **2016**, *88*, 119–125. [CrossRef]
17. Lehmann, D.R. Television show preference: Application of a choice model. *J. Mark. Res.* **1971**, *8*, 47–55. [CrossRef]
18. Lambiotte, R.; Ausloos, M. On the genre-fication of music: A percolation approach. *Eur. Phys. J. B Condens. Mat. Compl. Syst.* **2006**, *50*, 183–188. [CrossRef]
19. Camagni, R.; Capello, R.; Caragliu, A. The rise of second-rank cities: What role for agglomeration economies? *Eur. Plan. Stud.* **2015**, *23*, 1069–1089. [CrossRef]
20. Novac, A.; Dumitrescu-Moroianu, N. Dynamic model of regional convergence. *Rom. Stat. Rev. Suppl.* **2020**, *6*, 49–60. Available online: https://www.revistadestatistica.ro/supliment/2020/06/model-dinamic-de-convergenta-regionala/ (accessed on 12 February 2024).
21. Rotundo, G.; Navarra, M. On the maximum drawdown during speculative bubbles. *Phys. Stat. Mech. Appl.* **2007**, *382*, 235–246. [CrossRef]
22. Golubic, R.; Rudes, M.; Kovacic, N.; Marusic, M.; Marusic, A. Calculating impact factor: How bibliographical classification of journal items affects the impact factor of large and small journals. *Sci. Engin. Ethics* **2008**, *14*, 41–49. [CrossRef]
23. Cerqueti, R.; Iovanella, A.; Mattera, R. Clustering networked funded european research activities through rank-size laws. *Ann. Oper. Res.* **2023**, in print. [CrossRef]
24. Liu, N.C.; Cheng, Y. The academic ranking of world universities. *High. Educ. Eur.* **2005**, *30*, 127–136. [CrossRef]
25. Florian, R.V. Irreproducibility of the results of the Shanghai academic ranking of world universities. *Scientometr.* **2007**, *72*, 25–32. [CrossRef]
26. López, A.M. Number of Points Obtained by Leading Samba Schools at Carnival Parades in Rio de Janeiro, Brazil between 1985 and 2022. *Statista*, 16 February 2024. Available online: https://www.statista.com/statistics/1241979/ranking-samba-schools-carnival-rio-de-janeiro-brazil/ (accesseed on 12 February 2024).
27. Taylor, J.M. The politics of aesthetic debate: The case of Brazilian carnival. *Ethnology* **1982**, *21*, 301–311. [CrossRef]
28. Mitchell, J.H.; Haskell, W.L.; Raven, P.B. Classification of sports. *J. Am. Coll. Cardiol.* **1994**, *24*, 864–866. [CrossRef] [PubMed]
29. Ausloos, M.; Gadomski, A.; Vitanov, N.K. Primacy and ranking of UEFA soccer teams from biasing organization rules. *Phys. Scr.* **2014**, *89*, 108002. [CrossRef]
30. Malcata, R.M.; Vandenbogaerde, T.J.; Hopkins, W.G. Using athletes' world rankings to assess countries' performance. *Int. J. Sport. Physiol. Perform.* **2014**, *9*, 133–138. [CrossRef] [PubMed]
31. Ficcadenti, V.; Cerqueti, R.; Varde'i, C.H. A rank-size approach to analyse soccer competitions and teams: The case of the Italian football league "Serie A". *Ann. Oper. Res.* **2023**, *325*, 85–113. [CrossRef]
32. Ausloos, M. Hierarchy selection: New team ranking indicators for cyclist multi-stage races. *Eur. J. Oper. Res.* **2023**, *314*, 807-816. [CrossRef]
33. Ausloos, M. Shannon entropy and Herfindahl-Hirschman Index as team's performance and competitive balance indicators in cyclist multi-stage races. *Entropy* **2023**, *25*, 955. [CrossRef]
34. Condorcet, Marquis de, de Caritat, J.-A.-N. *Essai sur l'Application de l'Analyse la Probabilité des Décisions Rendues à la Pluralité des Voix*; L'Imprimerie Royale: Paris, France, 1785. Available online: https://gallica.bnf.fr/ark:/12148/bpt6k417181 (accessed on 12 February 2024).
35. Young, H.P.; Levenglick, A.A. A consistent extension of Condorcet's election principle. *SIAM J. Appl. Math.* **1978**, *35*, 285–300. [CrossRef]
36. Young, H.P. Condorcet's theory of voting. *Am. Polit. Sci. Rev.* **1988**, *82*, 1231–1244. [CrossRef]
37. Arrow, K.J. A difficulty in the concept of social welfare. *J. Polit. Econ.* **1950**, *58*, 328–346. [CrossRef]
38. He, Y.; Deng, Y. Ordinal belief entropy. *Soft Comput.* **2023**, *27*, 6973 6981. [CrossRef]
39. Krawczyk, M.J.; Wołoszyn, M.; Gronek, P.; Kułakowski, K.; Mucha, J. The Heider balance and the looking-glass self: Modelling dynamics of social relations. *Sci. Rep.* **2019**, *9*, 11202. [CrossRef] [PubMed]
40. Krawczyk, M.J.; Kułakowski, K. Structural balance of opinions. *Entropy* **2021**, *23*, 1418. [CrossRef] [PubMed]
41. Columbu, G.L.; De Martino, A.; Giansanti, A. Nature and statistics of majority rankings in a dynamical model of preference aggregation. *Phys. A Stat. Mech. Appl.* **2008**, *387*, 1338–1344. [CrossRef]
42. Munda, G. Choosing aggregation rules for composite indicators. *Soc. Indic. Res.* **2012**, *109*, 337–354. [CrossRef]
43. Nehama, I. Approximately classic judgement aggregation. *Ann. Math. Artif. Intell.* **2013**, *68*, 91–134. [CrossRef]
44. Pumain, D. *Hierarchy in Natural and Social Sciences*; Springer Science+Business Media B.V.: Dordrecht, The Netherlands, 2006. [CrossRef]
45. Marfels, C. Absolute and relative measures of concentration reconsidered. *Kyklos* **1971**, *24*, 753–766. [CrossRef]
46. Bikker, J.A.; Haaf, K. Competition, concentration and their relationship: An empirical analysis of the banking industry. *J. Bank. Financ.* **2002**, *26*, 2191–2214. [CrossRef]
47. Cowell, F. *Measuring Inequality*; Oxford University Press: Oxford, UK, 2011. [CrossRef]
48. Young, H.P. Optimal voting rules. *J. Econ. Perspect.* **1995**, *9*, 51–64. [CrossRef]
49. García, J.A.; Rodriguez-Sánchez, R.; Fdez-Valdivia, J.; de Moya-Anegón, F. A web application for aggregating conflicting reviewers' preferences. *Scientometr.* **2014**, *99*, 523–539. [CrossRef]
50. Torres, R. Limiting dictatorial rules. *J. Math. Econ.* **2005**, *41*, 913–935. [CrossRef]

51. Le Cam, L. Maximum likelihood: An introduction. *Int. Stat. Rev.* **1990**, *58*, 153–171. [CrossRef]
52. Kemeny, J.G. Mathematics without numbers. *Daedalus* **1959**, *88*, 577–591. Availabe online: https://www.jstor.org/stable/20026529 (accessed on 12 February 2024).
53. Varela, L.M.; Rotundo, G. Complex network analysis and nonlinear dynamics. In *Complex Networks and Dynamics. Social and Economic Interactions*; Commendatore, P., Matilla-García, M., Varela, L.M., Cáovas, J .S., Eds.; Springer International Publishing AG: Cham, Switzerland, 2016; pp. 3–25. [CrossRef]
54. Lambiotte, R.; Ausloos, M. Coexistence of opposite opinions in a network with communities. *J. Stat. Mech. Theory Exp.* **2007**, *2007*, P08026. [CrossRef]
55. Hwang, C.-L.; Yoon, K. *Multiple Attribute Decision Making: Methods and Applications. A State-of-the-Art Survey*; Springer: Berlin/Hiedelberg, Germany, 1981. [CrossRef]
56. Yoon, K. A reconciliation among discrete compromise situations. *J. Oper. Res. Soc.* **1987**, *38*, 277–286. [CrossRef]
57. Hwang, C.-L.; Lai, Y.-J.; Liu, T.-Y. A new approach for multiple objective decision making. *Comput. Oper. Res.* **1993**, *20*, 889–899. [CrossRef]
58. Lai, Y.-J.; Liu, T.-Y.; Hwang, C.-L. TOPSIS for MODM. *Eur. J. Oper. Res.* **1994**, *76*, 486–500. [CrossRef]
59. Parida, P.K.; Mishra, D.; Behera, B. A new approach for selection of candidate by TOPSIS technique. *J. Ultra Sci. Phys. Sci. A* **2017**, *29*, 541–546. . [CrossRef]
60. Chakraborty, S.; Chatterjee, P.; Das, P.P. *Multi-Criteria Decision-Making Methods in Manufacturing Environments*; Apple Academic Press: New York, NY, USA, 2023; Chapter 6. [CrossRef]
61. Chakraborty, S. TOPSIS and Modified TOPSIS: A comparative analysis. *Decis. Analyt. J.* **2022**, *2*, 100021. [CrossRef]
62. Fagin, R. Generalized First-Order Spectra and Polynomial-Time Recognizable Sets. In *Complexity of Computation*; Karp, R.M., Ed.; American Mathematical Society: Providence, RI, USA, 1974; pp. 43–73. Available online: https://www.kenclarkson.org/RonFaginPaper/RonFagin.html (accessed on 12 February 2024).
63. Kohler, R. A segmentation system based on thresholding. *Comput. Graph. Image Process.* **1981**, *15*, 319–338. [CrossRef]
64. Morariu, N.; Iancu, E.; Vlad, S. A neural network model for time series forecasting. *Rom. J. Econ. Forecast.* **2009**, *4*, 213–223. Available online: https://ideas.repec.org/a/rjr/romjef/vy2009i4p213-223.html (accessed on 12 February 2024).
65. Ho, T.K.; Hull, J.J.; Srihari, S.N. Decision combination in multiple classifier systems. *IEEE Trans. Pattern Anal. Mach. Intell.* **1994**, *16*, 66–75. [CrossRef]
66. Alwin, D.F.; Krosnick, J.A. The measurement of values in surveys: A comparison of ratings and rankings. *Public Opin. Quart.* **1985**, *49*, 535–552. [CrossRef]
67. Sanchez, M.E. Effects of questionnaire design on the quality of survey data. *Public Opin. Quart.* **1992**, *56*, 206–217. [CrossRef]
68. Green, M.S. Markoff random processes and the statistical mechanics of time-dependent phenomena. *J. Chem. Phys.* **1952**, *20*, 1281–1295. [CrossRef]
69. Green, M.S. Markoff random processes and the statistical mechanics of time-dependent phenomena. II. Irreversible processes in fluids. *J. Chem. Phys.* **1954**, *22*, 398–413. [CrossRef]
70. Kubo, R. The fluctuation-dissipation theorem. *Rep. Prog. Phys.* **1966**, *29*, 255–284. [CrossRef]
71. Andrieux, D.; Gaspard, P. Fluctuation theorem and Onsager reciprocity relations. *J. Chem. Phys.* **2004**, *121*, 6167–6174. [CrossRef]
72. Mori, H. Transport, collective motion, and Brownian motion. *Prog. Theor. Phys.* **1965**, *33*, 423–455. [CrossRef]
73. Ausloos, M. Relation between the Mori-Green-Kubo formulae and their Boltzmann approximation for electronic transport coefficients. *J. Phys. A Math. Gen.* **1978**, *11*, 1621–1632. [CrossRef]
74. Players. Available online: https://sofifa.com/players (accessed on 12 February 2024).
75. Teams. Available online: https://sofifa.com/teams (accessed on 12 February 2024).
76. LALIGA. Robert Lewandowski. Available online: https://www.laliga.com/en-GB/player/robert-lewandowski (accessed on 12 February 2024).
77. LALIGA. Player Comparison. Available online: https://www.laliga.com/en-GB/comparator/players?player1=witsel (accessed on 12 February 2024).
78. Hazard, E. Available online: https://sofifa.com/player/183277/eden-hazard/230036/ (accessed on 12 February 2024).

physics

Article

Social Depolarization: Blume–Capel Model

Miron Kaufman [1,*], Sanda Kaufman [2] and Hung T. Diep [3]

1 Department of Physics, Cleveland State University, Cleveland, OH 44115, USA
2 Levin School of Urban Affairs, Cleveland State University, Cleveland, OH 44115, USA; s.kaufman@csuohio.edu
3 Laboratoire de Physique Théorique et Modélisation, Institut Sciences et Techniques, CY Cergy Paris Université, CNRS, UMR 8089, 2, Avenue Adolphe Chauvin, 95300 Pontoise, France; diep@cyu.fr
* Correspondence: m.kaufman@csuohio.edu

Abstract: This study belongs to an emerging area of research seeking ways to depolarize societies in the short run (around events such as elections) as well as in a sustainable fashion. We approach the depolarization process with a model of three homophilic groups (US Democrats, Republicans, and Independents interacting in the context of upcoming federal elections). We expand a previous polarization model, which assumed that each individual interacts with all other individuals in its group with mean-field interactions. We add a depolarization field, which is analogous to the Blume–Capel model's crystal field. There are currently numerous depolarization efforts around the world, some of which act in ways similar to this depolarization field. We find that for low values of the depolarization field, the system continues to be polarized. When the depolarization field is increased, the polarization decreases.

Keywords: political polarization; depolarization; anticipatory scenarios; agent-based models; opinion dynamics; statistical physics approaches for social dynamics

Citation: Kaufman, M.; Kaufman, S.; Diep, H.T. Social Depolarization: Blume–Capel Model. *Physics* **2024**, *6*, 138–147. https://doi.org/10.3390/physics6010010

Received: 16 October 2023
Revised: 27 November 2023
Accepted: 13 December 2023
Published: 22 January 2024

1. Introduction

For at least the past three decades, scholars have observed, defined [1], measured [2], and modeled social polarization, its drivers, its effects [3,4], and its trends around the world [5–8] (The references included here are illustrative of the numerous articles addressing the increasing polarization around the world.) Increasing polarization tendencies have been documented for the short run [9,10] and predicted in the long run [11] unless some event or intervention changes its course (for example, after long months of a deep split in the Israeli polity, accompanied by numerous weekly demonstrations, the sudden violent events of October 2023 played the role of a focusing event: differences were mostly set aside, and the entire society concentrated on mutual help and a unified response).

The problems generated by severe societal polarization are felt in many places and in many ways—and in particular, in a diminished ability to solve serious societal problems demanding consensus. Besides the numerous real consequences to which they can lead, decisions resulting from such fraught processes tend to be non-robust, further accentuating political divisions and acrimony, and causing the public to lose trust in the democratic process and possibly disengage.

For example, the two political parties in the US Congress—Democrats and Republicans—are deeply polarized. As a direct result, both the Senate and the House of Representatives have great difficulties in making necessary joint decisions about important policies, such as budget allocations. Each has extremely slim majorities: in the Senate, there are forty nine Republicans to forty eight Democrats and three Independents (caucusing with Democrats who in effect hold the majority); and in the House, there are 221 Republicans to 212 Democrats [12]. The resulting decision dynamics mostly lead to impasse. Many issues up for vote are very contentious, including, for example, nominations for necessary government positions, which remain unoccupied for extended time periods [13]. Most proposals are decided on party line, meaning that they pass by the slimmest of majorities, at one or two vote differences. As a result, some senators and representatives acquire more power than warranted by their constituencies or seniority, because their votes are commodities sought by both parties. They can exact favors for their states or districts, angering the public [14]. Yearly budget decisions are delayed to the last legal minute under threat of government closure [14]. The consequences of failure to vote for the budget can be far-ranging, extending beyond the US border. For example, in Fall 2023, the House of Representatives approved a bi-partisan 45-day stopgap spending bill to prevent government closure, which did not contain continued financial support for Ukraine. Moreover, despite relatively broad consensus on financially assisting Israel in its war which broke out on 7 October 2023, this aid could not be extended because the House of Representatives had first to vote for a Speaker—a party-line decision.

Polarization is especially notable and acute preceding elections in democratic countries, such as the United States [11,15]. For example, the latest yearly *Gallup* polls conducted over the past 20 years regarding 24 issues central to past and current debates found increasing gaps between Democrat and Republican views (some deeper than others) for all issues, amounting to severe political polarization [15]. These findings are consistent and buttressed by those of Refs. [10,16].

Several factors have been viewed as causes, symptoms, and/or drivers for the observed increasing polarization among Americans. They include the widespread loss of mutual trust [17] and loss of trust in government, media, and science [18,19]. Media and science have aligned themselves with politics [20], further diminishing public trust. Lack of trust may be behind the seemingly distinct, nonoverlapping, and contradictory sets of data, facts, news, and opinions held by members of the two groups. Using an agent-based model (quite different than the model we propose), in Ref. [21], the authors have studied the role of communication in polarization.

Acute homophily [22] also contributes to polarization. It impels individuals to communicate almost exclusively with members of their own group who share their point of view, reinforcing the information disjunction. For example, people tend to evaluate information by the party affiliation of the source, more or rather than by its content, arguments, and evidence [23,24]. This way of evaluating information quality leads to the wholesale rejection of any statements coming from the opposing group. It also leads to the partition of society into groups whose members have non-intersecting perceptions of reality. In turn, this prevents the groups from solving problems, which necessitates finding agreement regarding specific policies and laws even while continuing to disagree on values. In Ref. [3], the authors examined several causes of polarization and proposed an additional category: bad-faith actors [4] fueling and reinforcing differences for their own purposes.

At times, such as currently in the US, a perfect storm is generated by deep value differences among opposing groups, combined with homophily (little or no direct intergroup communication) in the context of short political cycles which exacerbate both the differences and the homophily. In the face of the societal damage driven by polarization, scholars have been turning their attention to various theoretical and practical depolarization approaches, exploring tools at various scales, from local to country-wide [25–29]. Although research on depolarization is not recent (e.g., [30,31]), in Ref. [5], it is observed that, with a few exceptions [32–34], there are no corresponding definitions, measures, and models for depolarization.

As polarization reaches pernicious levels in some places including the US [24], efforts to counter it are emerging in research [3,5,35,36], media [17], and communities (see [35] for several examples of depolarizing initiatives at the community and organization levels). In the latter, one depolarizing strategy entails actively breaking homophily [37,38] to help individuals realize that even when they have very different values, they may share some interests that can be satisfied through joint decisions. In this paper, we contribute to these efforts by using sociophysics modeling to explore depolarization possibilities in the case of the US ahead of the 2024 elections. We ask what, if any, types of events or actions might work to counter the observed acute polarization trend, and how durable their effects can be.

Sociophysics methods [39] including statistical mechanics are particularly suitable to the study of polarization and depolarization because they are parsimonious ways to handle the complexity of social systems phenomena without requiring the use of extensive databases. For several examples of this approach, and how sociophysics can be applied specifically to the study of polarization, e.g., [11,40–42]. Together with agent-based models [43,44], sociophysics tools help construct anticipatory scenarios of polarization under different assumptions [45], circumventing some of the serious challenges to prediction posed by complexity [46].

Recently, we have proposed [11,42] a statistical mechanics model for exploring the dynamics of polarization in the US political system, between Democrat- and Republican-affiliated groups in the population, with consideration of a third group, Independents. The latter is now relatively large (representing a historic highest percent at about 41% of the American voter population [47] and therefore critical in determining election outcomes: neither of the two formal parties can win without attracting a considerable number of Independent voters. The model's results are qualitatively similar to poll outcomes in time, e.g., [16]. Using this model, we generated and explored [11,42] scenarios of whether leadership, also discussed by Ref. [35], and/or external events—for example, a massive natural disaster or a serious external threat, labeled a "focusing event"—might bring the groups closer to each other at least for some time. It was found [11,42] that although leadership and focusing events can help reduce polarization, their impact is rather temporary, consistent with conclusions in Ref. [5], who found in numerous case studies around the world that polarization reductions rarely last beyond about 10 years.

The voting public in the US is currently split down the middle between the two major parties and has been for at least the last 8 years [48], impairing the ability of governments at all levels to make necessary decisions on key topics such as the economy, the environment, energy, health care, immigration, and foreign relations, including financial assistance to other countries and global organizations. We ask here how we can overcome the current impasse resulting from the confluence of deep value differences, short political cycles that exacerbate them, novel technology effects, and homophily, which impedes direct communication across party lines. Perhaps the answers lie in the confluence of a multiplicity of conditions, which together push extremes toward each other (not necessarily to the middle).

We propose to use sociophysics modeling again to examine several conditions that together might depolarize the public effectively (especially important ahead of the 2024 national elections) by nudging opinion extremes toward each other to overcome homophily and enable idea exchanges in the short run, and even for longer durations. To this end, we expand the three-group polarization model of Refs. [11,42] by adding to it a depolarization field D similar to the Blume–Capel [49–51] crystal field. By design, increasing the value of D reduces the polarization of the system, as assessed by a measure proposed in Ref. [11]. We find that the depolarization field D has an additional, desirable effect: for each group, the distribution of individuals' stances, ranging from extreme left/liberal to extreme right/conservative, is a Gaussian normal distribution. This is consistent with empirical distributions captured through polling, such as by the Pew Research Center [16] for example. The challenge remains to identify actions that together have an effect similar to the D field. A three-state model with agents belonging to a single group was considered in Ref. [52]. By contrast, our model considers three groups (two homophilic) and the stance of each individual is a continuous variable.

The balance of this article begins with the description of our model in Section 2, followed by numerical results in Section 3. We offer our final remarks and plans for future work in Section 4.

2. Method: A Dynamic Mean-Field Model

We approach the study of depolarization in the context of US political contests. They are waged among three major US political groups historically providing candidates for president, congress, as well as state and local positions: Democrats and Republicans (which are currently highly polarized [48]), and Independents. The latter tend to lean toward, and reinforce, the number of Democrat or Republican voters in specific elections. Especially now, when they constitute a relatively large group of voters compared to the other two [47], Independents are viewed as a deciding factor in elections. For this reason, both Democrats and Republicans attempt to attract them to their respective positions.

Our three-group model represents the interactions in time of voters affiliated with the two parties—Democrats and Republicans—and with the nonaffiliated Independents. The model allows the testing of various interventions, which might contribute to depolarization by bringing extreme positions closer to the center.

We assume that in each of the three political groups, each individual has preferences with respect to several key political issues [11,34]. These individual preferences range between extremely left-leaning and extremely right-leaning, but more or less aligned with the positions taken by their respective parties. Democratic party- and Republican party-affiliated individuals actively interact with each other mostly within their own group, enhancing internal cohesion. Nevertheless, they also keep an eye on the stances of the members of the other groups.

Within any of the Democrat and Republican groups, everyone interacts with everyone, on a complete Erdös–Renyi network. The Independents, not being formally organized into a bloc or identifiable, interact with other Independents in a much weaker fashion or not at all. In what follows, we label Democrats as group 1, Republicans as group 2, and Independents as group 3.

In each group, each individual has a stance s that reflects their preferences regarding single issues under current political debate—economics, health care, defense, immigration, climate change—or a package of such issues. The stance s varies between -1 and +1, where -1 corresponds to the Democrats' extreme progressive/left position, +1 corresponds to the republicans' extreme conservative/right position, and 0 corresponds to the middle-of-the-road position.

The Democrats and the Republicans are homophilic, meaning that individuals in a given group prefer to communicate with other individuals from the same group through intra-group couplings J, and little or not at all with anyone from the other group. Thus, the magnitude of the couplings J quantifies the cohesiveness of each group. The other groups' mean stances influence the stances of individuals in the group under consideration through inter-group couplings K. The groups' leaderships act on individuals' stances through the fields H and D. The temperature T represents the effects of the context on the individual stances. For example, at the moment, different states of the economy or politics might impinge on the degree of polarization and on the extent to which leaders or other factors can reduce it.

We use the Boltzmann probability weight to compute the probability distributions for individual stances in each group. The underlying assumption in this study is that statistical physics models describe social systems at a scale that includes a large number of individuals. The Boltzmann distribution function maximizes entropy (disorder) for a given energy function. Quantitative properties of large-scale phenomena exhibit regularities that are not sensitive to short-scale details.

We compute the average stance of each group using this Boltzmann probability distribution $\exp(-E/T)$, where E is the energy. The negative energy associated with an individual in group 1 is:

$$-E_1 = (J_1 s_1 + K_{12} s_2 + K_{13} s_3 + H_1) s - D_1 s^2 \,,$$

where s denotes the stance of that individual and s_1, s_2, and s_3 denote the mean stances of groups 1, 2, and 3, respectively. The fields H_1 and D_1 represent the action of the leaders on individuals of group 1. For $H_1 > 0$, the mean stance is pushed towards positive values, while for $H_1 < 0$, the mean stance is pushed to negative values.

When positive, the field D_1 favors depolarization through $D_1 s^2$, whose effect is to push stances s toward the center: $s \sim 0$; while when D_1 is negative, it favors polarization: $|s| \sim 1$. The crystal field D_1, which in our context controls depolarization, was used in Refs. [49–51] to study the thermodynamics of UO_2 and He^3–He^4 mixtures. In our implementation, the stance s is a continuous variable, whereas in the original model [49–51], the spin $s = -1, 0, 1$. For $D >> 0$, the mean stance approaches zero after a few time iterations, while for $D << 0$, the mean stance $|s|$ approaches unity after a few time iterations.

To write the mean-field theory for our model, we introduce the Langevin–Blume–Capel function:

$$LBC(h,d) = \frac{\int_{-1}^{1} s e^{hs-ds^2} ds}{\int_{-1}^{1} e^{hs-ds^2} ds}, \tag{1}$$

where h and d stand for H/T and D/T, respectively.

We employ a numerical adaptative integration with a tolerance of 0.001 to evaluate this function. The average stance s at time $t + 1$ is assumed to be determined by preferences of the group at an earlier time t. This lag represents the time it takes to change individuals' stances. The time t is expressed in units of the lag time. Thus, for each of the three groups, respectively:

$$\begin{aligned} s_{1,t+1} &= LBC(h_1 + j_1 s_{1,t} + k_{12} s_{2,t} + k_{13} s_{3,t}, d_1), \\ s_{2,t+1} &= LBC(h_2 + j_2 s_{2,t} + k_{21} s_{1,t} + k_{23} s_{3,t}, d_2), \\ s_{3,t+1} &= LBC(h_3 + j_3 s_{3,t} + k_{31} s_{1,t} + k_{32} s_{3,t}, d_3), \end{aligned} \tag{2}$$

where h_1 stands for H_1/T, k_{12} stands for K_{12}/T, d_1 stands for D_1/T, etc. The inter-group interaction parameters K_{12} and K_{21} are not necessarily equal, as members of one group may feel cooperative toward another group, who might not reciprocate. The model includes fifteen parameters: three *j*s, six *k*s, three *h*s, and three *d*s.

We use here the polarization measure defined in Ref. [11] as the distance at any point in time between the mean stances of groups 1 and 2:

$$P = (s_2 - s_1)/2, \tag{3}$$

This definition is consistent with Pew polls, e.g., [10,16] reporting distributions of stances among Republicans and Democrats.

It is defined so that $-1 \le P \le 1$. The unpolarized case $P = 0$ corresponds to equal stances $s_1 = s_2$. Polarization is extreme when $P = 1$, corresponding to the Republicans' stance $s_2 = 1$ (most conservative/right) and the Democrats' stance $s_1 = -1$ (most progressive/left) or when $P = -1$, corresponding to the Republicans' stance $s_2 = -1$ and the Democrats' stance $s_1 = 1$.

The distribution of individuals of group 1 among the stances s at time t is:

$$\begin{aligned} \rho_1(s,t) &= \frac{e^{(j_1 s_{1t} + k_{12} s_{2t} + k_{13} s_{3t} + h_1)s - d_1 s^2}}{\int_{-1}^{1} e^{(j_1 s_{1t} + k_{12} s_{2t} + k_{13} s_{3t} + h_1)s - d_1 s^2} ds}, \\ \rho_2(s,t) &= \frac{e^{(j_2 s_{2t} + k_{21} s_{1t} + k_{23} s_{3t} + h_2)s - d_2 s^2}}{\int_{-1}^{1} e^{(j_2 s_{2t} + k_{21} s_{1t} + k_{23} s_{3t} + h_2)s - d_2 s^2} ds}, \\ \rho_3(s,t) &= \frac{e^{(j_3 s_{3t} + k_{32} s_{2t} + k_{31} s_{1t} + h_3)s - d_3 s^2}}{\int_{-1}^{1} e^{(j_3 s_{3t} + k_{32} s_{2t} + k_{31} s_{1t} + h_3)s - d_3 s^2} ds}. \end{aligned} \tag{4}$$

$\rho_1(s,t)ds$ gives the fraction of all individuals in group 1 who have a stance in the interval $(s, s + ds)$ at time t. Similarly, ρ_2 and ρ_3 are the distribution functions for groups 2 and 3, respectively.

3. Numerical Results and Discussion

To assess the influence of the depolarization field D on the three groups' stances, we generate four scenarios for which we use the same J and K values (these values were selected based on a qualitative analysis to replicate 2017 poll results [10,16] on Democrats' and Republicans' increasingly polarized stances) as in our previous article [11]. The field values are consistent with following assumptions. Individuals in group 1 are more cohesive than individuals in group 2: $j_1 > j_2$. Individuals in group 3 have no cohesion $j_3 = 0$. They exert no influence on the other two groups: $k_{13} = k_{23} = 0$. Individuals in group 3 are contrarian to group 1, which is in power: $k_{31} < 0$. They are not influenced by group 2 individuals: $k_{32} = 0$. Thus, all results that follow are obtained: $j_1 = 5, j_2 = 3, j_3 = 0, k_{12} = -4$, $k_{21} =-5, k_{31} = -3, k_{13} = 0, k_{23} = 0, k_{32} = 0, h_1 = 0, h_2 = 0$, and $h_3 = 0$. We also fix $d_3 = 5$ for all cases discussed below.

In the scenario shown in Figure 1a, we set $d_1 = d_2 = 0$, meaning no depolarization action (this was the case for all scenarios we studied in Ref. [11], where d_3 was also 0). As a result, the system polarizes over time. Starting at $t = 0$ with s_1 and s_2 quite close to 0, in time the distance between the Democrats' and Republicans' mean stances increases. This can also be seen in Figure 1b, where the polarization measure increases from 0 to about 0.8. In Figure 1c (where $t = 5$), we show the stance distributions for the three groups. Since the value of d_3 is positive, the distribution of stances for group 3 (Independents) is Gaussian. In contrast, the Democrats' and the Republicans' distributions $\rho(s,t)$ depend monotonically on the stance since $d_2 = d_2 = 0$.

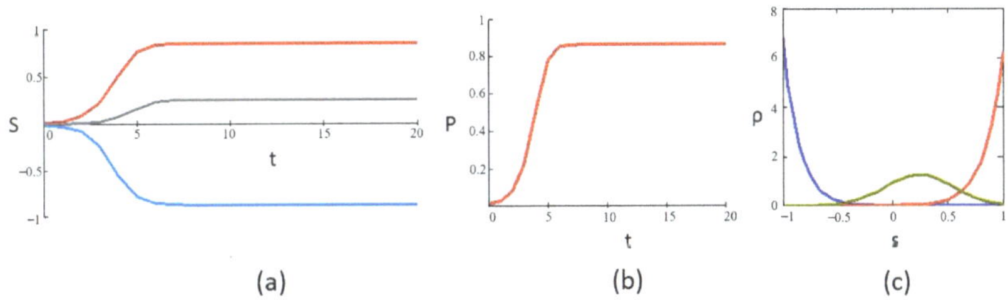

Figure 1. (**a**) Mean group stances versus time, (**b**) polarization time, and (**c**) stance distributions for groups 1 (blue line), 2 (red), and 3 (green) at $t = 5$ with the parameters $j_1 = 5, j_2 = 3, j_3 = 0, k_{12} = -4$, $k_{21} = -5, k_{31} = -3, k_{13} = 0, k_{23} = 0, k_{32} = 0, d_1 = 0, d_2 = 0, d_3 = 5, h_1 = 0, h_2 = 0$, and $h_3 = 0$. See text for details.

In the scenario of Figure 2a, we set the intervention $d_1 = d_2 = 3$. Consequently, the system polarization diminishes over time. Starting at $t = 0$ with $s_1 = -1$ and $s_2 = 1$, over time the distance between the mean stances diminishes. Figure 2b also reflects this effect: polarization P decreases from 1 to about 0.6. In Figure 2c, for $t = 5$, we show that since the d values are all positive, the distributions for the three groups are all Gaussian.

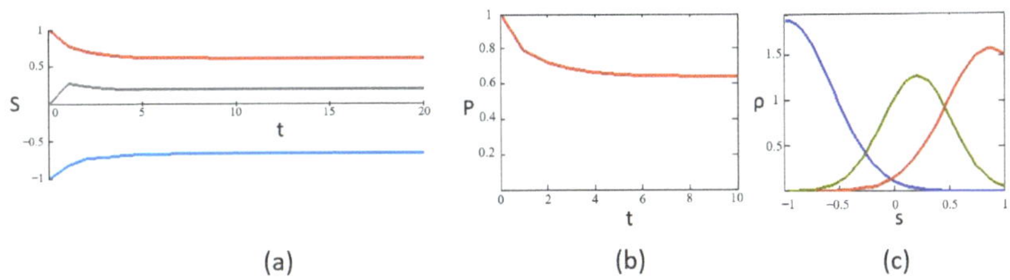

Figure 2. (**a**) Mean group stances versus time, (**b**) polarization versus time, and (**c**) stance distributions for groups 1 (blue line), 2 (red), and 3 (green) at $t = 5$ with the parameters $j_1 = 5, j_2 = 3, j_3 = 0, k_{12} = -4$, $k_{21} = -5, k_{31} = -3, k_{13} = 0, k_{23} = 0, k_{32} = 0, d_1 = 3, d_2 = 3, d_3 = 5, h_1 = 0, h_2 = 0$, and $h_3 = 0$. See text for details.

In the scenario of Figure 3a, we further increase the depolarization fields $d_1 = d_2 = 5$. The system depolarizes over time. Starting at $t = 0$ with $s_1 = -1$ and $s_2 = 1$, over time the distance between the mean stances diminishes. This is also shown in Figure 3b, where the polarization decreases from 1 at $t = 0$ to about 0 at $t = 20$. In Figure 3c, we show the stance distributions for the three groups. Since the ds are positive, the distributions are all Gaussian.

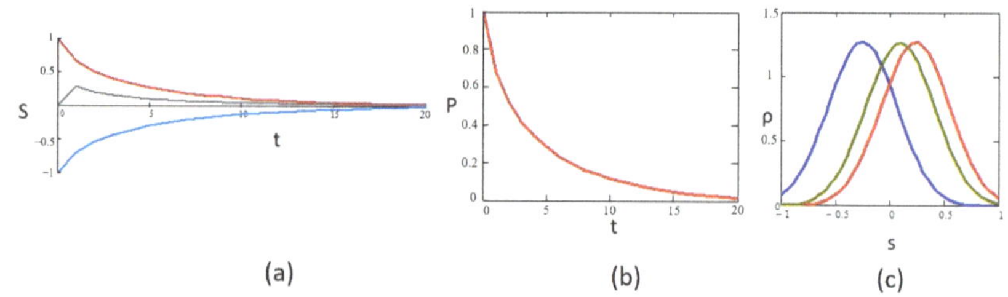

Figure 3. (**a**) Mean group stances versus time, (**b**) polarization versus time, and (**c**) stances distributions for groups 1 (blue line), 2 (red), and 3 (green) at $t = 5$ with the parameters $j_1 = 5$, $j_2 = 3$, $j_3 = 0$, $k_{12} = -4$, $k_{21} = -5$, $k_{31} = -3$, $k_{13} = 0$, $k_{23} = 0$, $k_{32} = 0$, $d_1 = 5$, $d_2 = 5$, $d_3 = 5$, $h_1 = 0$, $h_2 = 0$, and $h_3 = 0$. See text for details.

Finally, we consider a fourth scenario, which in Ref. [11] exhibited time oscillations of the three stances: $j_1 = 5$, $j_2 = 3$, $j_3 = 0$, $k_{12} = 4$, $k_{21} = -5$, $k_{31} = -3$, $k_{13} = 0$, $k_{23} = 0$, $k_{32} = 0$, $h_1 = 0$, $h_2 = 0$, and $h_3 = 0$. Again, we add depolarization fields: $d_1 = d_2 = 3$ and $d_3 = 5$. Now in Figure 4a, the three groups' stances exhibit damped oscillations over time. The system depolarizes in time and the polarization P also exhibits damped oscillations (Figure 4b). The distributions of stances for the three groups, shown in Figure 4c ($t = 5$), are Gaussian because the depolarization field values are positive for each group.

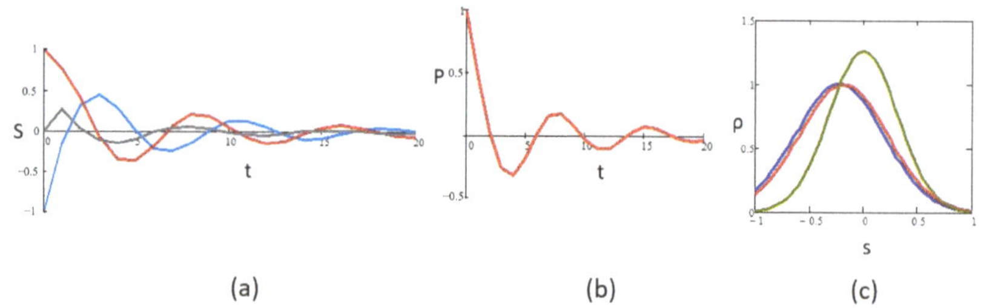

Figure 4. (**a**) Mean group stances versus time, (**b**) polarization versus time, and (**c**) stances distributions for groups 1 (blue line), 2 (red), and 3 (green) at $t = 5$ with the parameters $j_1 = 5$, $j_2 = 3$, $j_3 = 0$, $k_{12} = 4$, $k_{21} = -5$, $k_{31} = -3$, $k_{13} = 0$, $k_{23} = 0$, $k_{32} = 0$, $d_1 = 3$, $d_2 = 3$, $d_3 = 5$, $h_1 = 0$, $h_2 = 0$, and $h_3 = 0$.

The increase in value of the depolarization fields of groups 1 and 2 reduces polarization over time. In the oscillatory case, which occurs when k_{12} and k_{21} have different signs, the depolarization field dampens those oscillations.

We have shown through these scenarios that when a positive depolarization field is added to the three-group system, the distribution of stances becomes Gaussian. It means that it is maximized at a stance within the interval $(-1, 1)$, signifying a non-extreme stance. When the groups' stances are non-extreme, there exists the possibility of breaking out of the extreme polarization.

4. Concluding Remarks

Using the case of the US political system where three polarized groups—Democrats, Republicans, and Independents—interact, we explored scenarios of depolarization under the effect of intervention. We found that when a depolarization field D is added, the polarization decreases over time. If the D field is sufficiently strong, the polarization decreases to zero.

In practice, the field D can result from the actions of the groups' leaderships and/or from events that can refocus all groups on concerted actions in the face of some external threats. Examples of both have occurred in the past in the United States, such as during World War II, the Vietnam War, after the 9/11 attack on the World Trade Center in New York City, and elsewhere. The field D can also be the result of numerous local grassroots initiatives (also proposed in Ref. [17] as an antidote to polarization) currently occurring in the United States. They may add up to country-wide depolarization. Several examples can be found in Ref. [36]. These initiatives are akin to massively parallel intervention—a multiplicity of independent, locally driven actions—proposed in Refs. [3,53]—which together, when reaching a critical mass, can have a depolarizing effect among the political groups, at least in the short run. Sustaining depolarization in the longer run remains a challenge. As Ref. [5] observed in cases around the world, in general, repolarization tends to resume after about 10 years. Since in our model, the extent of depolarization depended on the strength of the intervention, we plan to explore further what factors might extend the duration of depolarized states.

This model includes, besides the single site field D, a field H. While D promotes compromise stance $s \sim 0$, the H field fosters extreme positions: $|s| \sim 1$. This could be the result of social influencers and/or media activities. We plan to further explore the group dynamics in the 15-dimensional parameter space, including the role of the H field. For example, we intend to elucidate the influence of the D field on the chaotic dynamics that we observed in the three-group model with competing K interactions. We also plan to study the model with intra-group short-range interactions by means of Monte-Carlo simulations. We will extend our study to account for different levels of randomness by varying the temperature.

Author Contributions: Conceptualization, M.K., S.K. and H.T.D.; methodology, M.K.; software, M.K.; validation, M.K.; formal analysis, M.K. and S.K.; data curation, S.K.; writing—original draft preparation, M.K. and S.K.; writing—review and editing, S.K. and H.T.D. All authors have read and agreed to the published version of the manuscript.

Funding: This research received no external funding.

Data Availability Statement: The data that support the findings of this study are available from the corresponding author upon reasonable request.

Conflicts of Interest: The authors declare no conflicts of interest.

References

1. Baldassarri, D.; Gelman, A. Partisans without constraint: Political polarization and trends in American public opinion. *Am. J. Sociol.* **2008**, *114*, 408–446. [CrossRef]
2. Bramson, A.; Grim, P.; Singer, D.J.; Berger, W.J.; Sack, G.; Fisher, S.; Holman, B. Understanding polarization: Meanings, measures, and model evaluation. *Philos. Sci.* **2017**, *84*, 115–159. [CrossRef]
3. Burgess, G.; Burgess, H.; Kaufman, S. Applying conflict resolution insights to the hyper-polarized, society-wide conflicts threatening liberal democracies. *Confl. Resolut. Quart.* **2022**, *39*, 355–369. [CrossRef]
4. Burgess, G.; Burgess, H. Challenging "Bad-Faith" actors who seek to amplify and exploit our conflicts. In *Beyond Intractability/Moving Beyong Intractability. Conflict Frontiers Seminar*; March 2021; Available online: https://www.beyondintractability.org/frontiers/bad-faith-actors (accessed on 10 December 2023).
5. McCoy, J.; Press, B.; Somer, M.; Tuncel, O. Reducing pernicious polarization: A comparative historical analysis of depolarization. In *Carnegie Endowment for International Peace (CEP). Publications*; Coherent Digital, LLC: Alexandria, VA, USA, 5 May 2022; Available online: https://policycommons.net/artifacts/2392500/reducing-pernicious-polarization/3413931/ (accessed on 10 December 2023).
6. McCoy, J.; Somer, M. (Eds.) Special Issue: Polarization and Democracy: A Janus-faced Relationship with Pernicious Consequences. *Am. Behav. Sci.* **2018**, *62*, 3–145. Available online: https://journals.sagepub.com/toc/absb/62/1 (accessed on 10 December 2023).
7. McCoy, J.; Rahman, T.; Somer, M. Polarization and the global crisis of democracy: Common patterns, dynamics, and pernicious consequences for democratic polities. *Am. Behav. Sci.* **2018**, *62*, 16–42. [CrossRef]

8. Somer, M.; McCoy, J. Déjà vu? Polarization and endangered democracies in the 21st century. *Am. Behav. Sci.* **2018**, *62*, 3–15. [CrossRef]
9. Böttcher, L.; Gersbach, H. The great divide: Drivers of polarization in the US public. *EPJ Data Sci.* **2020**, *9*, 32. [CrossRef] [PubMed]
10. Schaeffer, K. Far more Americans see 'very strong' partisan conflicts now than in the last two presidential election years. *Pew Research Center*, 4 March 2020. Available online: https://pewrsr.ch/2uRKXSs(accessed on 10 December 2023).
11. Kaufman, M.; Kaufman, S.; Diep, H.T. Statistical mechanics of political polarization. *Entropy* **2022**, *24*, 1262. [CrossRef] [PubMed]
12. Manning, J.E. *Membership of the 118th Congress: A Profile*; CRS Report R47470; U.S. Congressional Research Service: Washington, DC, USA, 2023. Available online: https://crsreports.congress.gov/product/pdf/R/R47470 (accessed on 10 December 2023).
13. Williamson, E. Hundreds of Biden nominees stuck in Senate limbo amid G.O.P. blockade. *The New York Times*, 8 January 2022. Available online: https://www.nytimes.com/2022/01/08/us/politics/biden-nominees-senate-confirmation.html(accessed on 10 December 2023).
14. Walter, J.; Helmore, E. Joe Manching hails expansive bill he finally agrees to as "great for America". *The Guardian*, 31 July 2022. Available online: https://www.theguardian.com/us-news/2022/jul/31/joe-manchin-hails-deal-inflation-reduction-act(accessed on 10 December 2023).
15. Newport, F. Update: Partisan gaps expand most on government power, climate. *Gallup*, 7 August 2023. Available online: https://news.gallup.com/poll/509129/update-partisan-gaps-expand-government-power-climate.aspx?version=print&thank-you-subscription-form=1(accessed on 10 December 2023).
16. Doherty, C.; Kiley, J.; Johnson, B. The partisan divide on political values grows even wider. *Pew Research Center*, 5 October 2017. Available online: https://www.pewresearch.org/politics/2017/10/05/the-partisan-divide-on-political-values-grows-evenwider/(accessed on 10 December 2023).
17. Fukuyama, F. Paths to depolarization. *Persuasion*, 3 August 2022. Available online: https://www.persuasion.community/p/fukuyama-paths-to-depolarization(accessed on 10 December 2023).
18. Klein, E.; Robison, J. Like, post, and distrust? How social media use affects trust in government. *Political Commun.* **2020**, *37*, 46–64. [CrossRef]
19. Rekker, R. The nature and origins of political polarization over science. *Public Underst. Sci.* **2021**, *30*, 352–368. [CrossRef]
20. Kubin, E.; von Sikorski, C. The role of (social) media in political polarization: A systematic review. *Ann. Int. Commun. Assoc.* **2021**, *45*, 188–206. [CrossRef]
21. Maes, M.; Flache, A. Differentiation without distancing. Explaining bi-polarization of opinions without negative influence. *PLoS ONE* **2013**, *8*, e74516. [CrossRef] [PubMed]
22. Dandekar, P.; Goel, A.; Lee, D.T. Biased assimilation, homophily, and the dynamics of polarization. *Proc. Natl. Acad. Sci. USA* **2013**, *110*, 5791–5796. [CrossRef] [PubMed]
23. Doise, W. Intergroup relations and polarization of individual and collective judgments. *J. Personal. Soc. Psychol.* **1969**, *12*, 136–143. [CrossRef]
24. Mackie, D.; Cooper, J. Attitude polarization: Effects of group membership. *J. Personal. Soc. Psychol.* **1984**, *46*, 575–585. [CrossRef]
25. Nuesser, A.; Johnston, R.; Bodet, M.A. The dynamics of polarization and depolarization: Methodological considerations and European evidence. In Proceedings of the American Political Association 2014 Annual Meeting and Exhibition, Washington, DC, USA, 28–31 August 2014; Available online: https://ssrn.com/abstract=2452108 (accessed on 10 December 2023).
26. van der Linden, S.; Leiserowitz, A.; Maibach, E.W. Communicating the scientific consensus on human-caused climate change is an effective and depolarizing public engagement strategy: Experimental evidence from a large national replication study. *SSRN* **2016**, 2733956. [CrossRef]
27. Liu, R.; Wang, L.; Jia, C.; Vosoughi, S. Political depolarization of news articles using attribute-aware word embeddings. *Proceed. Intl. AAAI Conf. Weblogs Soc. Media* **2021**, *15*, 385–396. [CrossRef]
28. Ojer, J.; Starnini, M.; Pastor-Satorras, R. Modeling explosive opinion depolarization in interdependent topics. *Phys. Rev. Lett.* **2023**, *130*, 207401. [CrossRef] [PubMed]
29. Sude, D.J.; Knobloch-Westerwick, S. When we have to get along: Depolarizing impacts of cross-cutting social media. *Int. J. Commun.* **2023**, *17*, 5268–5290.
30. Vinokur, A.; Burnstein, E. Depolarization of attitudes in groups. *J. Personal. Soc. Psychol.* **1978**, *36*, 872–885. [CrossRef]
31. Brauer, M.; Judd, C.M. Group polarization and repeated attitude expressions: A new take on an old topic. *Eur. Rev. Soc. Psychol.* **1996**, *7*, 173–207. [CrossRef]
32. McCoy, J.; Somer, M. Toward a theory of pernicious polarization and how it harms democracies: Comparative evidence and possible remedies. *Ann. Am. Acad. Political Soc. Sci.* **2019**, *681*, 234–271. [CrossRef]
33. Somer, M.; McCoy, J. Transformations through polarizations and global threats to democracy. *Ann. Am. Acad. Political Soc. Sci.* **2019**, *681*, 8–22. [CrossRef]
34. Sobkowicz, P. Social depolarization and diversity of opinions—Unified ABM framework. *Entropy* **2023**, *25*, 568. [CrossRef] [PubMed]
35. McCoy, J.; Somer, M. Overcoming polarization. *J. Democr.* **2021**, *32*, 6–21. [CrossRef]
36. The Hyper-Polarization Crisis: A Conflict Resolution Challenge. In *Beyond Intractability/Moving Beyond Intractability. Constructive Conflict Initiative*; Available online: https://www.beyondintractability.org/crq-bi-hyper-polarization-discussion (accessed on 10 December 2023).

37. Diamond, M.J.; Lobitz, W.C. When familiarity breeds respect: The effects of an experimental depolarization program on police and student attitudes toward each other. *J. Soc. Iss.* **1973**, *29*, 95–109. [CrossRef]
38. Currin, C.B.; Vera, S.V.; Khaledi-Nasab, A. Depolarization of echo chambers by random dynamical nudge. *Sci. Rep.* **2022**, *12*, 9234. [CrossRef] [PubMed]
39. Galam, S. *Sociophysics. A Physicist's Modeling of Psycho-political Phenomena*; Springer Science + Business Medai, LLC: New York, NY, USA, 2012; pp. 3–19. [CrossRef]
40. Galam, S. Unanimity, coexistence, and rigidity: Three sides of polarization. *Entropy* **2023**, *25*, 622. [CrossRef] [PubMed]
41. Galam, S. Opinion dynamics and unifying principles: A global unifying frame. *Entropy* **2022**, *24*, 1201. [CrossRef]
42. Diep, H.T.; Kaufman, M.; Kaufman, S. An agent-based statistical physics model for political polarization: A Monte Carlo study. *Entropy* **2023**, *25*, 981. [CrossRef]
43. Epstein, J.M. Agent-based computational models and generative social science. *Complexity* **1999**, *4*, 41–60. [CrossRef]
44. Lempert, R. Agent-based modeling as organizational and public policy simulators. *Proc. Natl. Acad. Sci. USA* **2002**, *99* (Suppl. S3), 7195–7196. [CrossRef] [PubMed]
45. Kaufman, M.; Diep, H.T.; Kaufman, S. Sociophysics of intractable conflicts: Three-group dynamics. *Phys. A* **2019**, *517*, 175–187. [CrossRef]
46. Batty, M.; Torrens, P.M. Modelling complexity: The limits to prediction. *Cybergeo Eur. J. Geogr.* **2001**, *Dossiers*, 201. [CrossRef]
47. Jones, J. U.S Party Preferences Evenly Split in 2022 after Shift to GOP. *Gallup*, 12 January 2023. Available online: https://news.gallup.com/poll/467897/party-preferences-evenly-split-2022-shift-gop.aspx(accessed on 10 December 2023).
48. Gallup. Party Affiliation. Trend since 2004. Available online: https://news.gallup.com/poll/15370/party-affiliation.aspx (accessed on 10 December 2022).
49. Blume, M. Theory of the first-order magnetic phase change in UO_2. *Phys. Rev.* **1966**, *141*, 517–525. [CrossRef]
50. Capel, H.W. On the possibility of first-order phase transitions in Ising systems of triplet ions with zero-field splitting. *Physica* **1966**, *32*, 966–988. [CrossRef]
51. Blume, M.; Emery, V.J.; Griffiths, R.B. Ising model for the λ transition and phase separation in He^3–He^4 mixtures. *Phys. Rev. A* **1971**, *4*, 1071–1077. [CrossRef]
52. Lipiecki, A.; Sznajd-Weron, K. Polarization in the three-state q-voter model with anticonformity and bounded confidence. *Chaos Solitons Fractals* **2022**, *165*, 112809. [CrossRef]
53. Burgess, G.; Burgess, H. Massively Parallel Peacebuilding. *Beyond Intractability*, 2020. Available online: https://www.beyondintractability.org/frontiers/mpp-paper(accessed on 10 December 2023).

physics

MDPI

Article

Phase Transition in Ant Colony Optimization

Shintaro Mori [1,*], Shogo Nakamura [1], Kazuaki Nakayama [2] and Masato Hisakado [3]

1 Department of Mathematics and Physics, Faculty of Science and Technology, Hirosaki University, Bunkyo-cho 3, Hirosaki 036-8561, Japan; h22ms109@hirosaki-u.ac.jp
2 Department of Mathematical Sciences, Faculty of Science, Shinshu University, Asahi 3-1-1, Matsumoto 390-8621, Japan; nakayama@math.shinshu-u.ac.jp
3 Nomura Holdings Inc., Otemachi 2-2-2, Chiyoda-ku, Tokyo 100-8130, Japan; hisakadom@yahoo.co.jp
* Correspondence: shintaro.mori@hirosaki-u.ac.jp; Tel.: +81-0132-39-3628

Abstract: Ant colony optimization (ACO) is a stochastic optimization algorithm inspired by the foraging behavior of ants. We investigate a simplified computational model of ACO, wherein ants sequentially engage in binary decision-making tasks, leaving pheromone trails contingent upon their choices. The quantity of pheromone left is the number of correct answers. We scrutinize the impact of a salient parameter in the ACO algorithm, specifically, the exponent α, which governs the pheromone levels in the stochastic choice function. In the absence of pheromone evaporation, the system is accurately modeled as a multivariate nonlinear Pólya urn, undergoing phase transition as α varies. The probability of selecting the correct answer for each question asymptotically approaches the stable fixed point of the nonlinear Pólya urn. The system exhibits dual stable fixed points for $\alpha \geq \alpha_c$ and a singular stable fixed point for $\alpha < \alpha_c$ where α_c is the critical value. When pheromone evaporates over a time scale τ, the phase transition does not occur and leads to a bimodal stationary distribution of probabilities for $\alpha \geq \alpha_c$ and a monomodal distribution for $\alpha < \alpha_c$.

Keywords: ant colony optimization; Pólya urn process; phase transition

1. Introduction

Sociophysics emerged in the 1970s and has evolved into a captivating research field within statistical physics [1,2]. In particular, herding behavior, or the inclination to follow the majority, has captured the attention of many researchers due to its pivotal role in understanding social phenomena [3–7]. Various probabilistic models have been proposed to describe herding behavior, with one notable example being the ant recruitment model [8,9]. This model explains the intermittent oscillation observed in ants when they are presented with two identical food sources [10,11]. When ants choose a food source from among two food sources, the ant recruitment model incorporates a straightforward herding mechanism in which a randomly selected ant chooses one of the two based on the number of ants that have already made the same choice. Scouts play a crucial role by exploring the terrain to locate food sources [12,13]. When a scout discovers food, it returns to the nest, leaving a pheromone trail in its wake. Other ants are drawn to these pheromone marks and consequently become recruited to forage at the food source.

Ant colony optimization (ACO) is a model-based meta-heuristic inspired by the foraging behavior of ants in their search for the shortest path to food sources [14–18]. While ants may not be highly intelligent individually, they collectively find the shortest path by following pheromone trails left by their fellow ants. The optimal path is determined by the route on which the maximum number of ants travel. Consider a classic problem known as the traveling salesman problem (TSP), which involves finding the shortest possible route that visits each city represented as vertices in a given graph exactly once and returning to the origin city [19]. In ACO, ants make decisions about their next city to visit based on a concept called "pheromone". Pheromone represents the preference for a particular

Citation: Mori, S.; Nakamura, S.; Nakayama, K.; Hisakado, M. Phase Transition in Ant Colony Optimization. *Physics* **2024**, *6*, 123–137. https://doi.org/10.3390/physics6010009

Received: 15 October 2023
Revised: 10 December 2023
Accepted: 28 December 2023
Published: 18 January 2024

choice and is collaboratively learned by the ants during their search process [15]. In the context of the TSP defined on a graph, pheromone values are typically associated with edges between cities, reflecting the preference for traveling from one city to another along the corresponding edges. These pheromone values are learned through a reinforcement strategy, where each ant reinforces its chosen paths based on the quality of the solution constructed. This quality is often determined using the inverse of the total length of the route. ACO has found successful applications in various industrial and academic constraint optimization problems and has become one of the most popular methods [20–23].

ACO has seen significant improvements, and modern ACO algorithms deviate substantially from the original ACO [24]. The fundamental modification lies in controlling the diversity of solutions and achieving convergence [25,26]. In this context, "convergence" refers to the tendency of ants to cluster around similar solutions in the neighborhood and ultimately converge toward the same solution. Early convergence to a small region of the search space leaves large enough sections of the search space unexplored and fails to find suitable solutions. On the other hand, quite slow convergence means that the computational cost required to reach acceptable solutions is high, rendering the search inefficient. Diversity control aims to prevent complete convergence by slowing down the search process.

Many algorithms have been proposed for controlling the diversity of the ACO algorithm. One of the diversity control mechanisms involves modifying the probabilistic decision function [24,27]. Bernd Meyer studied the influence of α, the exponent on the pheromone level in the selection function, and suggested that α qualitatively determines diversity and convergence behavior. Additionally, Meyer introduced a dynamic α that changes throughout the search process to enhance search efficiency, a technique known as α-"annealing". Meyer also emphasized the significance of noise in ACO using stochastic differential equations in both static and dynamic environments [24,28–30]. Ants respond to a two-choice question, and the noisy communication among ants prevents them from selecting suboptimal choice. In this paper, we study a simple model of ACO in which ants sequentially answer a series of two-choice quizzes. We investigate the phase transition and the qualitative change of the convergence behavior by varying α. We start from the discrete time formulation of the problem. Our approach can be straightforwardly generalized to ACO for many-body problems, allowing us to derive the model parameters of the continuous time approach concretely. Additionally, there are various possibilities for the continuous time limit, apart from Wiener noise [31]. In Section 2, we introduce a model and derive stochastic differential equations (SDEs) using diffusion approximation. In Section 3, we investigate the time evolution and examine the effect of α on the convergence properties of the solutions. Section 4 provides a summary of the results. Appendix A explains the estimation of the initial conditions for the SDEs.

2. Method and Definitions

There are N two-choice questions, each of which is answered by quite a large number of ants sequentially [32]. These questions are labeled by $n = 1, \ldots, N$. The answer provided by the tth ant is denoted as $X(n,t) \in \{0,1\}$, where $X(n,t) = 1$ indicates a correct answer, and $X(n,t) = 0$ indicates an incorrect answer. In ACO, multiple ants typically search for the optimal solution simultaneously in each iteration. However, in our model, only one ant conducts the search. Each ant receives 1 point for a correct answer. After ant t has answered all N questions, the total points (TPs) earned by the ant can be calculated using the following equation:

$$\mathrm{TP}(t) = \sum_{n=1}^{N} X(n,t).$$

Ant t deposits pheromones on their answer $X(n,t) \in \{0,1\}$. The amount of the pheromones is $\mathrm{TP}(t)$. We assume that the pheromones evaporate and decrease by $e^{-1/\tau}$ per unit time, where τ represents the time scale of the pheromone evaporation.

Ant $t+1$ observes the total values of the pheromones associated with question m for each choice $X(m,t+1)=\{0,1\}$. We denote the total value of pheromones that remains on all the questions after ant t has answered,

$$S(t) \equiv \sum_{s=1}^{t} \mathrm{TP}(s) e^{-(t-s)/\tau} = \sum_{s=1}^{t}\sum_{n=1}^{N} X(n,s) e^{-(t-s)/\tau}. \tag{1}$$

Then, the remaining pheromone on $X(m,s)=1, 1 \le s \le t$ is

$$S(m,t) \equiv \sum_{s=1}^{t} \mathrm{TP}(s) X(m,s) e^{-(t-s)/\tau} = \sum_{s=1}^{t}\sum_{n=1}^{N} X(n,s) X(m,s) e^{-(t-s)/\tau}. \tag{2}$$

The remaining pheromone on $X(m,s)=0, 1 \le s \le t$ is given by $S(t)-S(m,t)$.

Ants are not highly intelligent, and the probability of them making the correct choice in the two-choice questions by themselves is $1/2$. The information provided by TPs gives them an indirect clue about the correct choice. If $\mathrm{TP}(s) > N/2$, the posterior probability for $X(m,s)=1$ is larger than $1/2$. Similarly, if $S(m,t)$ is greater than $S(t)/2$, the posterior probability for $X(m,t+1)=1$ is greater than $1/2$. In ACO, a decision function with a positive parameters α and K is introduced that uses the values of the pheromones as follows:

$$P(X(m,t+1)=1) = \frac{(S(m,t)+K)^{\alpha}}{(S(m,t)+K)^{\alpha} + (S(t)-S(m,t)+K)^{\alpha}}.$$

Here, the exponent α determines the response of the choice to the values of the pheromones. As K is positive, one can avoid $S(m,t)=0$ and $S(m,t)=S(t)$ becoming the absorbing states of the process. However, instead of K, we adopt the following form for the choice function:

$$P(X(m,t+1)=1) = (1-\epsilon)\frac{S^{\alpha}(m,t)}{S^{\alpha}(m,t) + (S(t)-S(m,t))^{\alpha}} + \frac{1}{2}\epsilon. \tag{3}$$

Here, $\epsilon > 0$ is a small positive constant to avoid the absorbing states $S(m,t)=0$ and $S(m,t)=S(t)$ of the stochastic process [8].

We denote the ratio of the remaining pheromones on the correct choices as $Z(m,t)$,

$$Z(m,t) \equiv \frac{S(m,t)}{S(t)}.$$

We divide both the denominator and numerator of Equation (3) by $S^{\alpha}(t)$, and the probability of the correct choice $X(m,t+1)=1$ is expressed as

$$P(X(m,t+1)=1) = (1-\epsilon)\left(\frac{Z^{\alpha}(m,t)}{Z^{\alpha}(m,t) + (1-Z(m,t))^{\alpha}}\right) + \frac{1}{2}\epsilon = f(Z(m,t)).$$

Here, $f(z)$ is defined as

$$f(z) \equiv (1-\epsilon)\left(\frac{z^{\alpha}}{z^{\alpha} + (1-z)^{\alpha}}\right) + \frac{1}{2}\epsilon. \tag{4}$$

We note that $f(1/2)=1/2$ and $f(1-x)=1-f(x)$. The slope of $f(x)$ at $x=1/2$ is $(1-\epsilon)\alpha$. We also introduce the discount factor $D(t)$ and the ratio of correct answers $Z(t)$ as

$$\begin{aligned} D(t) &= \sum_{n=1}^{N}\sum_{s=1}^{t} e^{-(t-s)/\tau} = N\left(\frac{1-e^{-t/\tau}}{1-e^{-1/\tau}}\right), \\ Z(t) &= \frac{S(t)}{D(t)}. \end{aligned}$$

Stochastic Differential Equations

First, let us derive the recursive relation for $S(t)$ and $D(t)$. According to the definition, $D(t+1)$ and $S(t+1)$ obey the next recursive relations

$$\begin{aligned} D(t+1) &= N + D(t)e^{-1/\tau}, \\ S(t+1) &= \sum_{i=1}^{N} X(i,t+1) + S(t)e^{-1/\tau}. \end{aligned}$$

If τ is finite, for $t \gg \tau \gg 1$, we have $D(t) \simeq N\tau$. In the limit $\tau \to \infty$, the pheromone does not evaporate, and one has $D(t) = Nt$.

For $N \gg 1$, we can replace the sum of $X(i,s)$, $i = 1, \dots, N$, by the sum of the expected values using the law of large numbers; then:

$$S(t+1) \simeq \sum_{i=1}^{N} f(Z(i,t)) + S(t)e^{-1/\tau}.$$

Let us denote the average of $\{f(Z(i,t))\}$ as

$$\overline{f(Z(i,t))} \equiv \frac{1}{N}\sum_{i=1}^{N} f(Z(i,t)).$$

Then,

$$S(t+1) \simeq S(t)e^{-1/\tau} + N\overline{f(Z(i,t))}.$$

The recursive relation for $Z(t)$ is

$$Z(t+1) \simeq \frac{S(t+1)}{D(t+1)} = \frac{D(t+1)-N}{D(t+1)}Z(t) + \frac{N}{D(t+1)}\overline{f(Z(i,t)}.$$

$\Delta Z(t) = Z(t+1) - Z(t)$ is given as

$$\Delta Z(t) \simeq \frac{N}{D(t+1)}(\overline{f(Z(i,t))} - Z(t)).$$

In the continuous time limit $dt = 1 \to 0$, one obtains:

$$\frac{d}{dt}Z(t) = \frac{N}{D(t+1)}(\overline{f(Z(i,t))} - Z(t)). \tag{5}$$

One sees that $Z(t)$ converges to $\overline{f(Z(i,t))}$. When the pheromones evaporate and $\tau < \infty$, $D(t) \simeq N\tau$, and the prefactor of the differential equation is $1/\tau$. If one assumes that the dynamics of $Z(i,t)$ are faster than those of $Z(t)$ (adiabatic approximation), the time scale of the convergence is given by τ as $\overline{f(Z(i,t))} - Z(t) \propto e^{-t/\tau}$. When the pheromone does not evaporate and $\tau \to \infty$, $D(t) \simeq Nt$. The prefactor of the differential equation is $1/t$, and $\overline{f(Z(i,t))} - Z(t) \propto t^{-1}$. The convergence becomes quite slow.

Next, let us study the dynamics of $Z(m,t)$. The recursive relation for $S(m,t)$ is

$$S(m,t+1) = S(m,t)e^{-1/\tau} + X(m,t+1)(\sum_{i=1\neq m}^{N} X(i,t+1) + 1).$$

$Z(m,t+1)$ is then estimated as

$$
\begin{aligned}
Z(m,t+1) &= \frac{S(m,t+1)}{S(t+1)} = \frac{S(t)e^{-1/\tau}}{S(t+1)}Z(m,t) + \frac{\sum_{i\neq m} X(i,t+1)+1}{S(t+1)}X(m,t+1) \\
&\simeq \frac{S(t+1) - N\overline{f(Z(i,t))}}{S(t+1)}Z(m,t) + \frac{N\overline{f(Z(i,t))} + (1-f(Z(m,t))}{S(t+1)}X(m,t+1).
\end{aligned}
$$

Using $S(t+1) = D(t+1)Z(t+1)$, one obtains:

$$
\Delta Z(m,t) = \frac{N\overline{f(Z(i,t))}}{D(t+1)Z(t+1)}\left[\left(1+\frac{1-f(Z(m,t))}{N\overline{f(Z(i,t))}}\right)X(m,t+1) - Z(m,t)\right].
$$

We denote the history of $Z(s), \{Z(i,s)\}$, $s = 1,\ldots,t$, as H_t, and the conditional expected value of $\Delta Z(m,t)$ is estimated as

$$
E(\Delta Z(m,t)|H_t) = \frac{N\overline{f(Z(i,t))}}{D(t+1)Z(t+1)}\left[\left(1+\frac{1-f(Z(m,t))}{N\overline{f(Z(i,t))}}\right)f(Z(m,t)) - Z(m,t)\right]. \quad (6)
$$

Likewise, the conditional variance of $\Delta Z(m,t)$ can be approximated as

$$
V(\Delta Z(m,t)|H_t) \simeq \left(\frac{N\overline{f(Z(i,t)}}{D(t+1)Z(t+1)}\right)^2 f(Z(m,t))[1-f(Z(m,t))].
$$

Here, we neglect the subleading terms in $\left(1+\frac{1-f(Z(m,t))}{N\overline{f(Z(i,t))}}\right)^2$. We read the drift and diffusion term from the results and the SDEs are

$$
dZ(m,t) = E(\Delta Z(m,t)|H_t)dt + \sqrt{V(\Delta Z(m,t)|H_t)}dW(t),\ m = 1,\cdots,N. \quad (7)
$$

Here, $W(t)$ is the Wiener process. Equations (5) and (7) describe the dynamics of the system. The system can be described as a multivariate Pólya urn process.

We note that $\frac{1-f(Z(m,t))}{N\overline{f(Z(i,t))}}$ in Equation (6) breaks the Z_2 symmetry of the system. If one neglects the term, $E(\Delta Z(m,t)|H_t)$ is proportional to $f(Z(m,t)) - Z(m,t)$. As $f(1-x) = 1-f(x)$, $f(1-x)-(1-x) = -(f(x)-x)$ holds. $Z(m,t)$ and $1-Z(m,t)$ obey the same dynamics, and we call the symmetry Z_2 symmetry. The term is always positive and drives $Z(m,t)$ in the positive direction. As the term is proportional to $1/N$, the strength of the Z_2-symmetry-breaking field becomes smaller as N becomes larger.

3. Results

We analyze the SDEs given in Equation (7) and investigate the convergence properties of $Z(m,t)$. As the convergence behavior relies on the initial value of $Z(m,t=t_0)$ in the context of the nonlinear Pólya urn model, we commence by examining the initial distribution of $Z(m,t)$.

3.1. Initial Distribution of $\{Z(m,t)\}$

We assume that ants adopt $\alpha = 0$ and do not respond to the values of the pheromones for $t = 1, \cdots, t_0$. The ants answer the questions independently, and $P(X(i,t) = 1) = 1/2$. We estimate $Z(t)$ and $Z(m,t)$ for $\tau < t \le t_0$ as follows:

$$\begin{aligned} Z(t) &\sim \mathrm{N}\left(\frac{1}{2}, \frac{D_h(t)}{4D^2(t)}\right), \\ Z(m,t) &\sim \mathrm{N}\left(\frac{1}{2} + \frac{1}{2N}, \frac{ND_h(t)}{4D^2(t)}\right), \\ D_h(t) &\equiv N\sum_{s=1}^{t} e^{-2(t-s)/\tau} = N \cdot \frac{1-e^{-2t/\tau}}{1-e^{-2/\tau}}, \end{aligned} \tag{8}$$

where $\mathrm{N}(a,b)$ denotes the normal distribution with the mean a and the variance b. The details of the calculations are given in Appendix A. If τ is finite, one has $D_h(t) \simeq N\tau/2$ for $t \gg \tau \gg 1$. In the limit $\tau \to \infty$, the pheromone does not evaporate and then $D_h(t) = Nt$.

The essential differences between $Z(t)$ and $Z(m,t)$ include a shift in the expected value by $1/2N$ and the presence of a factor of N in the numerator of the variance of $Z(m,t)$. The shift of $1/2N$ arises from the finding that $X(m,s)$ in $S(m,t)$ is 1, which is larger than $E[X(i,s)] = 1/2$ for $i \neq m$. The value of the pheromone contains information about the correct choice, leading to $E[S(m,t)] > \frac{1}{2}E[S(t)]$. However, in the "cheating" process, the variables $X(i,s)$, are combined by $X(m,s)$, as in Equation (2), resulting in a larger variance for $S(m,t)$. The factor of N in the numerator of the variance of $Z(m,t)$ is a consequence of this combination process.

From the distribution of $Z(m,t_0)$, one can determine the values of τ or t_0 that guarantee that $Z(m,t_0)$ is greater than 0.5 for the limits $t \gg \tau \gg 1$ and $\tau \to \infty$, respectively. With a confidence level of 1%, τ and t_0 should satisfy the following conditions:

$$\begin{aligned} t \gg \tau \gg 1 : \frac{1}{2N} \ge \frac{2.58}{2\sqrt{2\tau}} \to \tau \ge 3.33N^2, \\ \tau \to \infty : \frac{1}{2N} \ge \frac{2.58}{2\sqrt{t_0}} \to t_0 \ge 6.66N^2. \end{aligned}$$

3.2. Case with $\tau \to \infty$

We take the limit $\tau \to \infty$ in Equations (5) and (7). We replace $D(t+1)$ and $D_h(t+1)$ with $N(t+1)$ and $N(t+1)$, respectively. This results in the following equations:

$$\begin{aligned} dZ(m,t) &= \frac{\overline{f(Z(i,t))}}{(t+1)Z(t+1)}\left[\left(1 + \frac{1-f(Z(m,t))}{N\overline{f(Z(i,t))}}\right)f(Z(m,t)) - Z(m,t)\right]dt \\ &+ \left(\frac{\overline{f(Z(i,t))}}{(t+1)Z(t+1)}\right)\sqrt{f(Z(m,t)[1-f(Z(m,t))]}dW(t), m = 1, \cdots, N, \\ \frac{d}{dt}Z(t) &= \frac{1}{(t+1)}(\overline{f(Z(i,t))} - Z(t)). \end{aligned}$$

The initial conditions of $Z(t)$ and $Z(m,t)$ at $t = t_0$ are as follows:

$$\begin{aligned} Z(t_0) &\sim \mathrm{N}\left(\frac{1}{2}, \frac{1}{4Nt_0}\right), \\ Z(m,t_0) &\sim \mathrm{N}\left(\frac{1}{2} + \frac{1}{2N}, \frac{1}{4t_0}\right). \end{aligned}$$

In the adiabatic approximation, where the time development of $Z(m,t)$ is much faster than that of $Z(t)$, the structure of the SDE for each $Z(m,t)$, where $m = 1, \cdots, N$, is the same as that of a nonlinear P'olya urn [33,34]. The probability for $Z(m,t)$ to converge to a stable solution of the following equation is positive [35],

$$\left(1 + \frac{1 - f(z)}{N\overline{f(Z(i,t))}}\right) f(z) = z. \tag{9}$$

Here, a stable (unstable) solution of Equation (9) means that the curve of the left-hand-side of the equation crosses the diagonal curve $y = z$ in in the downward (upward) direction [35]. We denote the stable and unstable solutions as z_s and z_u, respectively.

In the limit $N \to \infty$, the positive driving force $\frac{1-f(z)}{N\overline{f(Z(i,t))}}$ in Equation (9) disappears. The system becomes Z_2-symmetric, and $z = 1/2$ is a the solution of $f(1/2) = 1/2$. If N is finite, the positive driving force breaks the Z_2 symmetry. We plot $(1 + b(1 - f(z))f(z)$ in Figure 1. $b > 0$ corresponds to $1/N\overline{f(Z(i,t))}$ and $b \in [0.01, 0.02]$ for $N = 100$ and $\overline{f(Z(i,t)} \in [1/2, 1]$; $b = 0$ and $b = 0.2$ are used in Figure 1, left and right, respectively, along with $\epsilon = 0.1$.

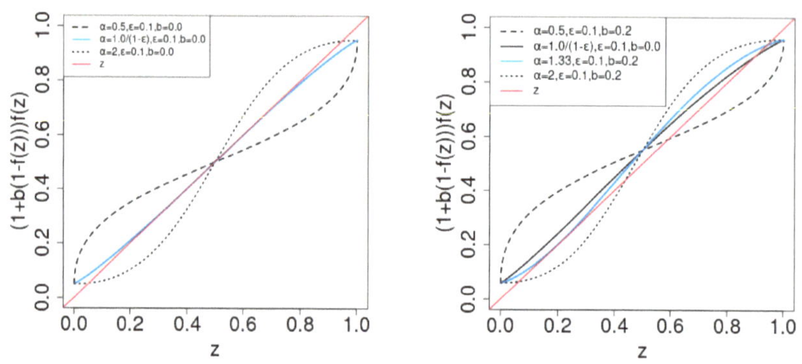

Figure 1. $f(z)$ (4) **(left)** and $(1 + b(1 - f(z))f(z)$ **(right)** versus z for b, α, and ϵ, as indicated. To visualize the fixed point of $f(z)$, the z-function is also shown.

In the Z_2-symmetric case ($b = 0$), the stability of the solution $z = 1/2$ depends on the slope of $f(z)$ at $z = 1/2$. As $f'(1/2) = (1 - \epsilon)\alpha$, where the prime denotes z-derivative, the critical value of α is $\alpha_c = 1/(1 - \epsilon)$. If $\alpha < (>)\alpha_c$, $z = 1/2$ is (un)stable. If $\alpha = \alpha_c$, the curve of $f(z)$ is tangential to the diagonal. $Z(m,t)$ converges to $1/2$ for $\alpha \le \alpha_c$, as $z = 1/2$ is the unique stable solution z_s. For $\alpha > \alpha_c$, there appears two stable solutions z_{s1}, z_{s2}: z_{s1} is within $(0, 0.5)$, and z_{s2} is within $(0.5, 1.0)$. $z = 0.5$ becomes the unstable solution z_u.

The dotted lines in Figure 2 show the solutions versus α for the Z_2-symmetric case. For $\alpha < \alpha_c$, $z = 1/2$ (black dotted line) is the stable solution. For $\alpha > \alpha_c$, $z = 1/2$ (gray dotted line) becomes unstable ($z_u = 1/2$) and two stable solutions z_{s1}, z_{s2} depart from $z = 1/2$ continuously with $\alpha > \alpha_c$. The stable solution to which $Z(m,t)$ converges depends on the initial value of $Z(m,t_0)$. In general, if $Z(m,t_0)$ is greater (smaller) than $1/2$, the probability of the convergence to 1 is greater (smaller) than $1/2$. z_u determines the "attractive domains" for the stable solutions z_{s1}, z_{s2}. The susceptibility of the expected value of $Z(m,t)$ to the initial value $Z(m,t_0)$ is the order parameter of the nonlinear Pólya urn [33,34]. As the order parameter is proportional to the difference in the two stable states z_{s1}, z_{s2}, the order parameter is a continuous function of α, and the phase transition is continuous.

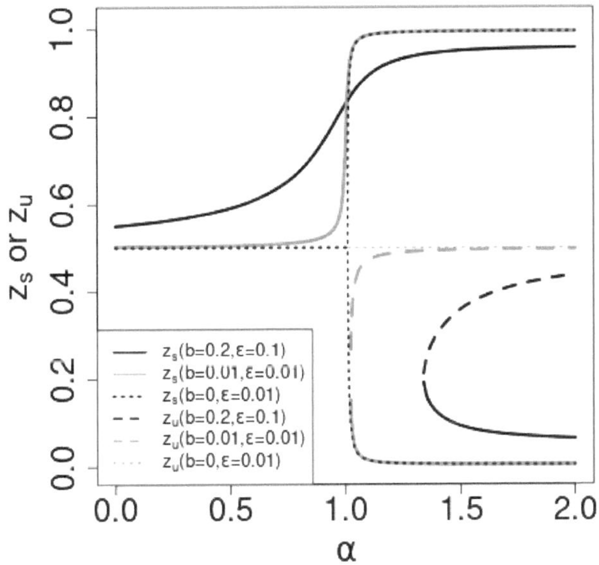

Figure 2. The stable, z_s, and unstable, z_u, solutions for b and ϵ parameters, as indicated. See text for details.

In the Z_2-asymmetric case ($b \neq 0$), for $\alpha < \alpha_c$, there is a stable solution, z_s, in the z-range $(0.5, 1)$. z_s increases with α; at some critical value, α_c, of α, the curve $f(z)(1+b(1-f(z))$ becomes tangential to the diagonal at some z_t. z_t is known as a touchpoint and to be stable [36]. As α continues to increase beyond α_c, two significant changes occur: a new stable solution, z_{s1}, emerges from the touchpoint z_t, while an unstable solution, z_u, also becomes apparent.

When $\alpha < \alpha_c$, only one stable solution, z_s, exists within the range $1/2 < z_s < 1$. Conversely, for $\alpha > \alpha_c$, the specific stable solution to which $Z(m,t)$ converges depends on the initial values of $Z(m,t_0)$. At $\alpha = \alpha_c$, both the stable fixed point z_s and the touchpoint z_t remain stable. The solution to which $Z(m,t)$ converges is determined by the initial values of $Z(m,t_0)$ in this case as well. For $\alpha > \alpha_c$, the situation mirrors that of $\alpha = \alpha_c$. Once $\alpha \geq \alpha_c$, the order parameter becomes positive, and the phase transition becomes discontinuous.

Figure 3 shows the results of the numerical studies in the limit $\tau \to \infty$. We sampled a trajectory of $Z(m,t)$ and $Z(t)$ for $1 \leq t \leq 10^9$ with $N = 10^2$ and $\epsilon = 0.01$. Figure 3 presents the distribution of $Z(m,t_0)$ for two different values of t_0, namely, $t_0 \in 10^3, 10^6$. The mean value of $Z(m,t_0)$ is approximately $1/2 + 1/2N$, which aligns with the theoretical predictions. The variance of $Z(m,t_0)$ is given by $1/4t_0$, so the variance for $t_0 = 10^3$ is about 10^3 times larger than that for $t_0 = 10^6$. Consequently, if $t_0 = 10^3$ is chosen, a significant proportion of $m \in 1, \cdots, N$ has $Z(m,t_0) < z_u \simeq 0.5$, resulting in a high probability that $Z(m,t)$ converges to $z'_s < 1/2$ for $\alpha = 2.0$. In such cases, $Z(t)$ cannot reach 1 due to the convergence of $Z(m,t)$ to z'_s. On the other hand, if $t_0 = 10^5$ is set, the ratio of $m \in 1, \cdots, N$ with $Z(m,t_0) < z_u \simeq 0.5$ is zero, ensuring that $Z(m,t)$ always converges to $z_s > 1/2$. As a result, $Z(t)$ monotonically increases toward 1 for $\alpha = 2.0$. For $\alpha = 1.0 < \alpha_c$, where only one stable state, $z_s \simeq 1$, exists, $Z(m,t)$ consistently converges to z_s.

One can see that $Z(t)$ monotonically approaches z_s with time for both $t_0 = 10^3$ and $t_0 = 10^5$ cases within the range $t \leq 10^9$. In the case of $\alpha = 0.5$, where $z_s \simeq 0.5$, $Z(t)$ experiences relatively little change.

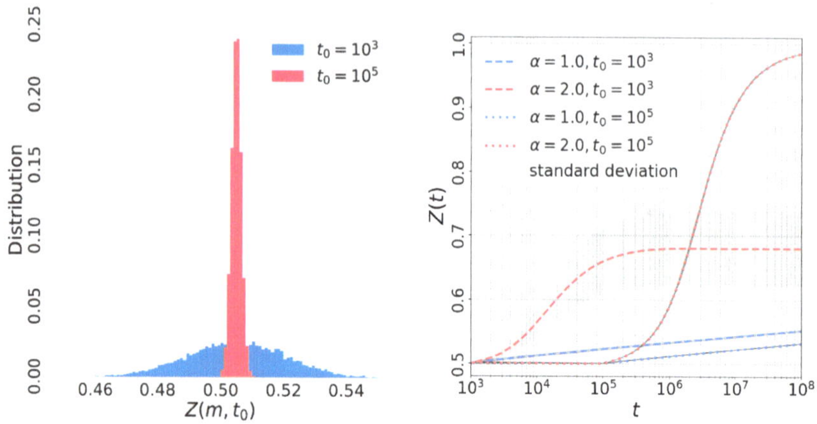

Figure 3. Left: the initial distribution of $Z(m,t_0)$ for the limit $\tau \to \infty$. $\alpha = 0, \epsilon = 0.01$ and $N = 10^2$ are used. **Right**: $Z(t)$ versus t for different α and t_0 as indicated. Gray areas show the standard deviation of $Z(t)$ which is quite large for $\alpha = 2.0, t_0 = 10^3$ and smaller or compatible with the line widths for other cases. See text for details.

3.3. Case with $\tau < \infty$

In the case where τ is finite, let us make the assumption that $t \gg \tau \gg 1$ and replace $D(t+1)$ with $N\tau$ in Equations (5) and (7). This leads to the following equations:

$$\begin{aligned} dZ(m,t) &= \frac{\overline{f(Z(i,t))}}{\tau Z(t+1)}\left[\left(1+\frac{1-f(Z(m,t))}{N\overline{f(Z(i,t))}}\right)f(Z(m,t)) - Z(m,t)\right]dt \\ &+ \frac{\overline{f(Z(i,t))}}{\tau Z(t+1)}\sqrt{f(Z(m,t))(1-f(Z(m,t))}dW(t), \; m = 1,\ldots,N, \\ \frac{d}{dt}Z(t) &= \frac{1}{\tau}(\overline{f(Z(i,t))} - Z(t)). \end{aligned} \tag{10}$$

The initial conditions for $Z(t_0)$ and $Z(m,t_0)$ at $t_0 \gg \tau$ are as follows:

$$\begin{aligned} Z(t_0) &\sim \mathrm{N}\left(\frac{1}{2}, \frac{1}{8N\tau}\right), \\ Z(m,t_0) &\sim \mathrm{N}\left(\frac{1}{2}+\frac{1}{2N}, \frac{1}{8\tau}\right). \end{aligned} \tag{11}$$

The dynamics of $Z(m,t)$, $m = 1,\ldots,N$, are coupled through $Z(t)$ and $\overline{f}(Z(i,t))$. To simplify and analyze this coupled system, we focus on the stationary state of $Z(m,t)$ in the limit $t \to \infty$.

$Z(m,t)$ are expected to fluctuate around the stable fixed points of $f(z)$ in the stationary state. As was observed in Section 3.2 for the case of $\tau \to \infty$, for $\alpha < \alpha_c$, there is only one stable fixed point, and for $\alpha \geq \alpha_c$, two stable fixed points exist, one of which is near one. The stationary distribution is unimodal for $\alpha < \alpha_c$ and bimodal for $\alpha \geq \alpha_c$. Let us denote the stationary distribution and the mean value of $Z(m,t)$ as $P_{\mathrm{st}}(z)$ and μ_{st}, respectively. As $f(Z(m,t))$ is the probability for $X(m,t+1) = 1$, one can assume $\overline{f(Z(i,t)} = Z(t) = \mu_{\mathrm{st}}$ in the stationary state. The SDEs in Equation (10) can be simplified as follows when replacing $\overline{f}(Z(i,t))$ and $Z(t)$ with μ_{st}:

$$
\begin{aligned}
dZ(m,t) &= \frac{1}{\tau}\left[\left(1+\frac{1-f(Z(m,t))}{N\mu_{\rm st}}\right)f(Z(m,1)) - Z(m,t)\right]dt \\
&+ \frac{1}{\tau}\sqrt{f(Z(m,t))(1-f(Z(m,t))}dW(t) \\
&= A(Z(m,t))dt + B(Z(m,t))dW(t), \\
A(z) &= \frac{1}{\tau}\left[\left(1+\frac{1-f(z)}{N\mu_{\rm st}}\right)f(z) - z\right], \\
B(z) &= \frac{1}{\tau}\sqrt{f(z)(1-f(z)}.
\end{aligned} \tag{12}
$$

The stationary state with reflecting boundary conditions is determined by a potential solution [37], which can be expressed as:

$$
\begin{aligned}
P_{\rm st}(z) &\propto \frac{1}{B^2(z)}\exp\left(\int_{1/2}^{z}\frac{2A(y)}{B^2(y)}dy\right), \\
\frac{2A(y)}{B^2(y)} &= 2\tau\frac{f(y)-y}{f(y)(1-f(y))} + \frac{2\tau}{N\mu_{\rm st}}.
\end{aligned} \tag{13}
$$

The second term, $2\tau/N\mu_{\rm st}$, arises from the Z_2 symmetry-breaking field and causes a shift in the stationary distribution in the positive direction.

Figure 4 shows $P_{\rm st}(z)$ (13) for $\epsilon = 0.01$ and $N = 10^2$; $\mu_{\rm st}$ is chosen so that the mean value of $P_{\rm st}(z)$ coincides with

$$\mu_{\rm st} = \int_0^1 P_{\rm st}(z)zdz.$$

The pair of the parameters $(\alpha, \mu_{\rm st})$ are $(0.0, 0.50)$, $(0.5, 0.51)$, $(0.9, 0.54)$, $(0.99, 0.67)$, $(1/0.99, 0.74)$, and $(2.0, 0.54)$. As α increases, the peak position shifts in the positive z-direction, which can be expected by the dependence of the stable solution z_s on α in Figure 2. If $\alpha = 1/(1-\epsilon)$, the peak appears at $z = 1$, since there is only one stable fixed point near $z = 1$ in Figure 1. When $\alpha = 2$, there are two stable fixed point, and the stationary distribution is bimodal.

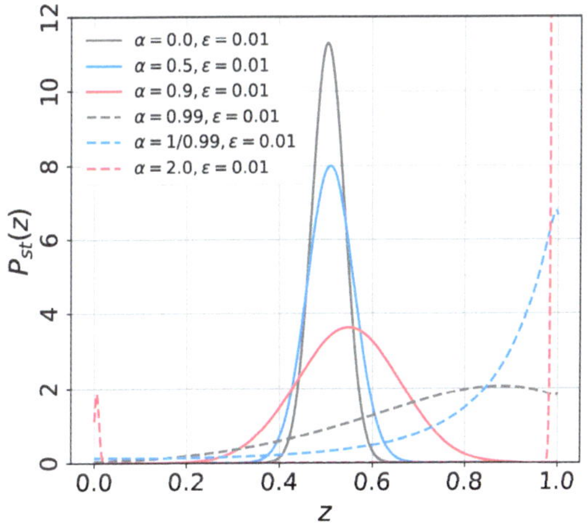

Figure 4. The stationary distribution, $P_{\rm st}$ (13), for $N = 10^2$ and ϵ and α parameters as indicated. See text for details.

In order to derive the dependence of $\mu_{\rm st}$ and the variance of $P_{\rm st}(z)$ on α, let us assume that $Z(m,t)$ fluctuates around $\mu_{\rm st} \simeq \frac{1}{2}$ for $\alpha < \alpha_c$. We linearize $f(z)$ in the vicinity of $z = 1/2$ as

$$f(z) = \frac{1}{2} + (1-\epsilon)\alpha\left(z - \frac{1}{2}\right).$$

Let us also approximate $B^2(z)$ as $\mu_{\rm st}(1-\mu_{\rm st}) = 1/4$, then $P_{\rm st}(z)$ becomes

$$P_{\rm st}(z) \propto \exp\left\{ -\frac{\left[z - \left(\frac{1}{2} + \frac{1}{2N(1-(1-\epsilon)\alpha)}\right)\right]^2}{\frac{2}{8\tau(1-(1-\epsilon)\alpha)}} \right\}.$$

In the case with $\alpha = 0$, the ants do not observe the information of the pheromones and decide by themselves. The expected value and the variance are consistent with the results (11) for the initial state. The expected value and the variance increase with α for $0 \le \alpha < 1/(1-\epsilon)$.

The shape of the stationary distribution changes from the monomodal shape for $\alpha < \alpha_c$ to the bimodal shape for $\alpha \ge \alpha_c$. Figure 5 shows the stationary distribution $P_{\rm st}(z)$ of $Z(m,t) = z$ for $\alpha \in \{0, 0.5, 0.9, 0.99, 1/0.99, 2.0\}, \epsilon = 0.01, \tau = 100$ and $N = 10^2$. We also plot $P_{\rm st}(z)$ in Equation (13) with solid-line curves. Except the $\alpha = 2.0$ case, the numerical results agree with the theoretical ones. As α increases from 0 to $1/0.99$, the mean value and the variance of $Z(m,t)$ increases. For $\alpha = 1/0.99$, the distribution of $Z(m,t)$ has a peak at $z = 1$. The distribution becomes bimodal and has two peaks near $z = 0$ and $z = 1$ for $\alpha = 2$. For $\alpha = 2.0$, $P_{\rm st}(z)$ becomes bimodal and the equilibration time to reach the stationary state becomes quite long. We beleive the latter is the reason for the discrepancy between the numerical and theoretical results.

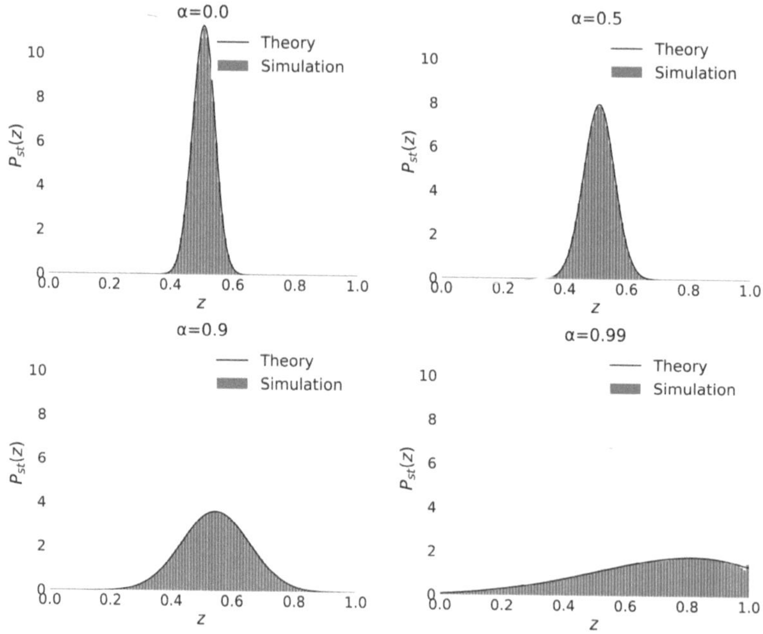

Figure 5. *Cont.*

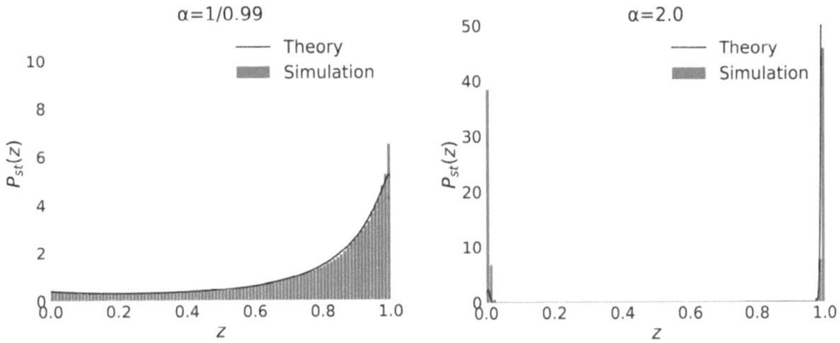

Figure 5. The distributions $P_{\mathrm{st}}(z)$ of $Z(m,t=10^5)=z$ for $\tau=10^2$, $t_0=10^5$ and $\epsilon=0.01$ and initial state α values as indicated.

4. Discussion

In the presented paper, we studied a simple model for ACO and the convergence properties of the solutions. Ants answer many two-choice questions in sequence and deposit pheromone as they choose. As the amount of the pheromones is the number of correct answers, the following ants can receive information or hints about the correct choices. We showed that the model reduces to a multivariate nonlinear Pólya urn process and the pheromones break the Z_2 symmetry of the process. By varying the exponent α of the decision function of the ants, there occurs a phase transition about the convergence of the probability of choosing the correct answer for each question in the limit $\tau\to\infty$. For $\tau<\infty$, the change in the stationary distribution between the monomodal and the bimodal shape occurs as we vary α.

Previous studies adopted values of $\alpha=1$ or smaller in solving actual optimization problems, like the TSP [24]. In α-annealing, α increases gradually, as shown in Ref. [24]. In our study, we showed that the duration of the period $\alpha=0$ should be long enough to ensure that the initial value of $Z(m,t)$ is in the attractive domain of the suitable stable state ($Z(m,t)>z_u$) in the case when $\tau=\infty$. Subsequently, with $\alpha\geq 1$ in effect, $Z(t)$ converges to a value close to 1. In the case of $\tau<\infty$, the timescale for pheromone evaporation, represented by τ, shown to be sufficiently long to maintain the same initial conditions. However, $Z(m,t)$ does not converge to a specific value; instead, it follows a stationary distribution that exhibits both bimodal and monomodal shapes depending on the value of α. To achieve a distribution of $Z(t)$ with a prominent peak near $z=1$, the α-annealing process is an effective strategy. Both the stable solution z_s and the distribution of $Z(m,t)$ suggest that after a lengthy enough period with $\alpha=0$, it is advantageous to gradually increase α from 1. However, the efficiency of the annealing process depends on the specific problem being addressed. Future studies are considered to clarify on an efficient α-annealing schedule.

Author Contributions: Conceptualization, S.M. and M.H.; methodology, S.M. and M.H.; software, S.N.; validation, S.M. and S.N.; formal analysis, S.M. and K.N.; investigation, S.N.; resources, S.M.; data curation, S.N.; writing—original draft preparation, S.M.; writing—review and editing, S.M., S.N., K.N. and M.H.; visualization, S.M. and S.N.; supervision, S.M.; project administration, S.M.; funding acquisition, S.M. All authors have read and agreed to the published version of the manuscript.

Funding: This research was funded by JPSJ KAKENHI (grant No. 22K03445), Japan.

Data Availability Statement: We performed numerical simulations using Julia 1.7.3. The code is available at https://github.com/LABO-M/ACO-PhaseTransition-Sampling.git.

Conflicts of Interest: Author Masato Hisakado was employed by the company Nomura Holdings Inc., Tokyo, Japan.

Appendix A. Initial Conditions of $Z(t)$ and $Z(m,t)$

Let us assume $X(i,s)$ are independent and identically distributed Bernoulli random variable with $P(X(i,s)=1)=1/2$ for $i=1,\cdots,N$ and $s\leq t_0$. As soon as $E(X(i,s))=1/2$ and $V(X(i,s))=1/4$, one finds:

$$\begin{aligned} E[S(t)] &= \sum_{i=1}^{N}\sum_{s=1}^{t} E[X(i,s)]e^{-(t-s)\tau} = \frac{1}{2}D(t), \\ V(S(t)) &= \sum_{i=1}^{N}\sum_{s=1}^{t} V(X(i,s)e^{-(t-s)/\tau}) = \frac{1}{4}N\sum_{s=1}^{t} e^{-2(t-s)/\tau} = \frac{1}{4}D_h(t). \end{aligned}$$

Here, we define $D_h(t)$ as

$$D_h(t) = N\sum_{s=1}^{t} e^{-2(t-s)/\tau}.$$

Applying the central limit theorem, we conclude that $Z(t)=S(t)/D(t)$ behaves like a normal distribution, with its probability density function given by

$$Z(t) \sim \mathrm{N}\left(\frac{1}{2}, \frac{D_h(t)}{4D^2(t)}\right). \tag{A1}$$

$S(m,t)$ in Equation (2) is rewritten as

$$S(m,t) = \sum_{s=1}^{t} X(m,s)(1+\sum_{i\neq m} X(i,s))e^{-(t-s)/\tau}.$$

Then, the conditional and unconditional expected values of $S(m,t)$ read

$$\begin{aligned} E[S(m,t)|\{X(m,s)\}_{s=1,\cdots,t}] &= \sum_{s=1}^{t} X(m,s)(\frac{1}{2}(N+1))e^{-(t-s)/\tau} \\ E[S(m,t)] &= E[E[S(m,t)|\{X(m,s)\}_{s=1,\cdots,t}]] = \left(\frac{1}{4}+\frac{1}{4N}\right)D(t). \end{aligned}$$

The conditional variance of $S(m,t)$ reads

$$V(S(m,t)|\{X(m,s)\}_{s=1,\cdots,t}) = \sum_{s=1}^{t} X(m,s)\frac{1}{4}(N-1)e^{-2(t-s)/\tau}.$$

The unconditional variance reads

$$\begin{aligned} V(S(m,t)) &= E[V(S(m,t)|\{X(m,s)\}_{s=1,\cdots,t})] + V(E[S(m,t)|\{X(m,s)\}_{s=1,\cdots,t}]) \\ &= \left(\frac{1}{16}N+\frac{1}{4}-\frac{1}{16N}\right)D_h(t) \simeq \frac{1}{16}N\cdot D_h(t). \end{aligned}$$

We estimate the variance of $Z(m,t)=S(m,t)/S(t)$ by neglecting the fluctuation of $S(t)$ as

$$V(Z(m,t)) \simeq \frac{V(S(m,t))}{(E[S(t)])^2} = \frac{ND_h(t)}{4D^2(t)}.$$

From the central limit theorem, $Z(m,t)=S(m,t)/S(t)$ behaves as

$$Z(m,t) \sim \mathrm{N}\left(\frac{1}{2}+\frac{1}{2N}, \frac{ND_h(t)}{4D^2(t)}\right). \tag{A2}$$

References

1. Galam, S. Sociophysics: A review of Galam models. *Int. J. Mod. Phys.* **2008**, *19*, 409–440. [CrossRef]
2. Galam, S. *Sociophysics: A Physicist's Modeling of Psycho-Political Phenomena*; Springer Science+Business Media, LLC: New York, NY, USA, 2012. [CrossRef]
3. Galam, S. Majority rule, hierarchical structures, and democratic totalitarianism: A statistical approach. *J. Math. Psychol.* **1986**, *30*, 426–434. [CrossRef]
4. Brian Arthur, W. Competing technologies, increasing returns, and lock-In by historical events. *Econ. J.* **1989**, *99*, 116–131. [CrossRef]
5. Bikhchandani, S.; Hirshleifer, D.; Welch, I. A Theory of fads, fashion, custom, and cultural change as informational cascades. *J. Political Econ.* **1992**, *100*, 992–1026. Available online: https://www.jstor.org/stable/2138632 (accessed on 26 December 2023).
6. Mori, S.; Hisakado, M.; Takahashi, T. Phase transition to a two-peak phase in an information-cascade voting experiment. *Phys. Rev. E* **2012**, *86*, 026109. [CrossRef]
7. Galam, S.; Cheon, T. Asymmetric contrarians in opinion dynamics. *Entropy* **2020**, *22*, 25. [CrossRef]
8. Kirman, A. Ants, rationality, and recruitment. *Quart. J Econ.* **1993**, *108*, 137–156. [CrossRef]
9. Hisakado, M.; Mori, S. Information cascade, Kirman's ant colony model, and kinetic Ising model. *Physica A* **2015**, *417*, 63–75. [CrossRef]
10. Deneubourg, J.L.; Aron, S.; Goss, S.; Pasteels, J.M. Error, communication and learning in ant societies. *Eur. J. Oper. Res.* **1987**, *30*, 168–172. [CrossRef]
11. Pasteels, J.; Deneubourg, J.L.; Goss, S. Transmission and amplification of information in a changing environment: The case of insect societies. In *Law of Nature and Human Conduct*; Prigogine, I., Sanglier, M., Eds.; GORDES: Bruxelles, Belgium, 1987; pp. 129–156.
12. Pasteels, J.; Deneubourg, J.; Detrain, C. (Eds.) *Information Processing in Social Insects*; Birkhäuser Verlag/Springer Basel AG: Basel, Switzerland, 2012. [CrossRef]
13. Camazine, S.; Deneubourg, J.-L.; Franks, N.L.; Sneyd, J.; Theraula, G.; Bonabeau, E. *Self-Organization in Biological Systems*; Princeton University Press: Princeton, NJ, USA, 2001.
14. Dorigo, M. Optimization, Learning and Natural Algorithms. Ph.D. Thesis, Poltecnico di Milan, Milan, Italy, 1992.
15. Dorigo, M.; Gambardella, L.M. Ant colonies for the travelling salesman problem. *Biosystems* **1997**, *43*, 73–81. [CrossRef]
16. Cordón, O.; Herrera, F.; Stützle, T. A review on the ant colony optimization metaheuristic: Basis, models and new trends. *Mathware Soft Comput.* **2002**, *9*, 141–175. Available online: https://eudml.org/doc/39241 (accessed on 26 December 2023).
17. Meuleau, N.; Dorigo, M. Ant colony optimization and stochastic gradient descent. *Artif. Life* **2002**, *8*, 103–121. [CrossRef]
18. Dorigo, M.; Zlochin, M.; Meuleau, N.; Birattari, M. Updating ACO pheromones using stochastic gradient ascent and cross-entropy methods. In *Applications of Evolutionary Computing. EvoWorkshops 2002: EvoCOP, EvoIASP, EvoSTIM/EvoPLAN. Kinsale, Ireland, April 3–4. 2002. Proceedings*; Cagnoni, S., Gottlieb, J., Hart, E., Middendorf, M., Raidl, G.R., Eds.; Springer: Berlin/Heidelberg, Germany, 2002; pp. 21–30. [CrossRef]
19. Lawler, E.L.; Lenstra, J.K.; Rinnooy Kan, A.H.G.; Shmoys, D.B. (Eds.) *The Travelling Salesman Problem. A Guided Tour of Combinatorial Optimization*; John Wiley & Sons, Ltd.: Chichester, UK, 1987. Available online: https://archive.org/details/travelingsalesma00lawl/ (accessed on 26 December 2003).
20. Dorigo, M.; Stützle, T. Ant colony optimization: Overview and recent advances. In *Handbook of Metaheuristics*; Gendreau, M., Potvin, J.Y., Eds.; Springer Science+Business Media, LLC: New York, NY, USA, 2010; pp. 227–263. [CrossRef]
21. Tang, K.; Wei, X.F.; Jiang, Y.H.; Chen, Z.W.; Yang, L. An adaptive Aat colony optimization for solving large-scale traveling salesman problem. *Mathematics* **2023**, *11*, 4439. [CrossRef]
22. Li, W.; Xia, L.; Huang, Y.; Mahmoodi, S. An ant colony optimization algorithm with adaptive greedy strategy to optimize path problems. *J. Ambient. Intell. Humaniz. Comput.* **2022**, *13*, 1557–1571. [CrossRef]
23. Gad, A.G. Particle swarm optimization algorithm and its applications: A systematic review. *Arch. Comput. Methods Eng.* **2022**, *29*, 2531–2561. [CrossRef]
24. Meyer, B. On the convergence behaviour of ant colony search. *Complex. Intl.* **2008**, *12*, 1–15.
25. Gutjahr, W.J. ACO algorithms with guaranteed convergence to the optimal solution. *Inf. Process. Lett.* **2002**, *82*, 145–153. [CrossRef]
26. Nakamichi, Y.; Arita, T. Diversity control in ant colony optimization. *Artif. Life Robot.* **2004**, *7*, 198–204. [CrossRef]
27. Randall, M.; Tonkes, E. Intensification and diversification strategies in ant colony system. *Complex. Intl.* **2002**, *9*, 1–7. Available online: https://www.researchgate.net/publication/246494892 (accessed on 26 December 2023).
28. Meyer, B. A Tale of two wells: Noise-induced adaptiveness in self-organized systems. In Proceedings of the Second IEEE International Conference on Self-Adaptive and Self-Organizing Systems (SASO 2008), Venice, Italy, 20–24 October 2008; Brueckner, S., Robertson, P., Bellur, U., Eds.; IEEE Computer Society: Los Alamos, CA, USA, 2008; pp. 435–444. [CrossRef]
29. Meyer, B. Optimal information transfer and stochastic resonance in collective decision making. *Swarm Intell.* **2017**, *11*, 131–154. [CrossRef]
30. Meyer, B.; Ansorge, C.; Nakagaki, T. The role of noise in self-organized decision making by the true slime mold Physarum polycephalum. *PLoS ONE* **2017**, *12*, e0172933. [CrossRef]

31. Hisakado, M.; Hattori, K.; Mori, S. From the multiterm urn model to the self-exciting negative binomial distribution and Hawkes processes. *Phys. Rev. E* **2022**, *106*, 034106. [CrossRef] [PubMed]
32. Hisakado, M.; Hino, M. Between ant colony optimization and genetic algorithm. *Intl. Proceed. Soc. Jpn. Transact. Math. Model. Appl. (IPSJ TOM)* **2016**, *9*, 8–14.
33. Mori, S.; Hisakado, M. Correlation function for generalized Pólya urns: Finite-size scaling analysis. *Phys. Rev. E* **2015**, *92*, 052112. [CrossRef]
34. Nakayama, K.; Mori, S. Universal function of the nonequilibrium phase transition of a nonlinear Pólya urn. *Phys. Rev. E* **2021**, *104*, 014109. [CrossRef] [PubMed]
35. Hill, B.M.; Lane, D.; Sudderth, W. A Strong law for some generalized urn processes. *Ann. Probab.* **1980**, *8*, 214–226. [CrossRef]
36. Pemantle, R. When are touchpoints limits for generalized polya urns? *Proc. Am. Math. Soc.* **1991**, *113*, 235. [CrossRef]
37. Gardiner, C. *Stochastic Methods: A Handbook for the Natural and Social Science*; Springer: Berlin/Heidelberg, Germany, 2009.

physics

MDPI

Article

Do Successful Researchers Reach the Self-Organized Critical Point?

Asim Ghosh [1,*] and Bikas K. Chakrabarti [2,3]

1 Department of Physics, Raghunathpur College, Raghunathpur 723133, India
2 Saha Institute of Nuclear Physics, Kolkata 700064, India; bikask.chakrabarti@saha.ac.in
3 Economic Research Unit, Indian Statistical Institute, Kolkata 700108, India
* Correspondence: asimghosh066@gmail.com

Abstract: The index of success of the researchers is now mostly measured using the Hirsch index (h). Our recent precise demonstration, that statistically $h \sim \sqrt{N_c} \sim \sqrt{N_p}$, where N_p and N_c denote, respectively, the total number of publications and total citations for the researcher, suggests that average number of citations per paper (N_c/N_p), and hence h, are statistical numbers (Dunbar numbers) depending on the community or network to which the researcher belongs. We show here, extending our earlier observations, that the indications of success are not reflected by the total citations N_c, rather by the inequalities among citations from publications to publications. Specifically, we show that for highly successful authors, the yearly variations in the Gini index (g, giving the average inequality of citations for the publications) and the Kolkata index (k, giving the fraction of total citations received by the top $(1-k)$ fraction of publications; $k = 0.80$ corresponds to Pareto's 80/20 law) approach each other to $g = k \simeq 0.82$, signaling a precursor for the arrival of (or departure from) the self-organized critical (SOC) state of his/her publication statistics. Analyzing the citation statistics (from Google Scholar) of thirty successful scientists throughout their recorded publication history, we find that the g and k for the highly successful among them (mostly Nobel laureates, highest rank Stanford cite-scorers, and a few others) reach and hover just above (and then) below that $g = k \simeq 0.82$ mark, while for others they remain below that mark. We also find that all the lower (than the SOC mark 0.82) values of k and g fit a linear relationship, $k = 1/2 + cg$, with $c = 0.39$, as suggested by an approximate Landau-type expansion of the Lorenz function, and this also indicates $k = g \simeq 0.82$ for the (extrapolated) SOC precursor mark.

Keywords: Lorenz function; Landau-like expansion; citation analysis; Gini index; Kolkata index; Hirsch index; Stanford cite score; Dunbar number; self-organized criticality

Citation: Ghosh, A.; Chakrabarti, B.K. Do Successful Researchers Reach the Self-Organized Critical Point? *Physics* **2024**, *6*, 46–59. https://doi.org/10.3390/physics6010004

Received: 20 September 2023
Revised: 15 November 2023
Accepted: 4 December 2023
Published: 30 December 2023

1. Introduction

Inspiring research in sociophysics (see, e.g., [1–6]) has, in years, led to intense research activities in several statistical and statistical physical models and analysis of sociodynamical problems. For example, the social opinion formation models of Galam (see e.g., [7,8]), of Biswas-Chatterjee-Sen (see e.g., [9,10]), of Minority Games (see e.g., [11]), of Kolkata Paise Restaurant games (see, e.g., [12,13]), and others. In view of the automatically encoded wide range of the citation data of the publications by scientists and their straightforward availability on the internet, we have studied the inequality statistics from Google Scholar data. The presence of ubiquitous inequalities allowed recently the studies of various scaling, etc., properties in their statistics (see e.g., [14,15]) of the Hirsch index [16], or the universal (or limiting) self-organized critical (SOC) behavior (see e.g., [17–19]) and their citation inequality like the century-old Gini (g) [20] and the recently introduced Kolkata (k) [21,22] indices. It may be noted at this stage that while g values measure the overall inequality in the distributions and k gives the fraction of "mass" or of total citations coming from the $(1-k)$ fraction of avalanches or publications, these studies [17–19] indicated that the inequalities in the avalanche size distributions, measured by

g and k, just prior to the arrival of the SOC point in several standard physical models (like the sand-pile models of Bak–Tang–Wiesenfeld (BTW) [23], Manna [24], and others), and in social contexts of citations from publications [18,19] becomes equal: $g = k = 0.84 \pm 0.04$. It may also be noted that $k = 0.80$ corresponds to Pareto's 80/20 law (see e.g., [21,22]). This Pareto principle asserts that 20% of the causes are responsible for 80% of the outcomes. In other words, the principle suggests that a small fraction of the factors contribute in causing a large fraction of major events, from economics to quality management and even in personal development. In business, it is often used to identify the most important areas for improvement. It may be mentioned here that our earlier studies of inequality indices g and k [17–22] corresponded to the cumulative dynamics (as the sand-pile dynamics progress and cluster distributions grow or the publications by the authors or from the institutions progress over time and the citation size distributions grow since the start of the dynamics) as the system approaches the respective SOC states. Our study here is for the same inequality indices, but for small time intervals along growth dynamical paths of individual researchers.

We study here the inequality dynamics measured by the g and k indices of several successful researchers (mostly winners of international prizes, medals, or awards like Nobel, Fields, Boltzmann, Breakthrough, highest level Stanford c-score achievers, etc.), some distinguished sociophysics researchers, along with those of a few high level (but not so high Stanford c-score, though within "top 2%") researchers, for data up to 2022, since their recorded first publication year. We collected the citation data of the publications (from online free Google Scholar, if an individual Google Scholar page exists). We calculate the g and k indices for each year, starting their first publication, by taking the citation statistics today (collected and analyzed in July–August 2023). We extracted the values for g and k for all the recorded publications of the scientist in each overlapping five-year window (since the first publication), where the windows continuously shift by one year till the year 2022 (corresponding to the last central year 2020 of the researcher) in the following figures for each researcher. The choice of five-year window size is found to give optimal stability in statistics (a smaller three-year window size did not give stability to the citation statistics for quite a few of the scientists.)

We find that the majority of the chosen scientists crossed the $g = k \simeq 0.82$ mark (which we interpret here as the precursor level of the SOC point [17]) early in their life and often they hover just above or below but around that level of inequality mark. Some others just touched the precursor mark ($g = k$) once or even multiple times and a few remained below that mark. For other known researchers considered here, the $g = k$ mark occurs marginally but does not cross ever. It is to be noted that this mark of reaching the SOC state (beyond the $g = k \simeq 0.82$) level of inequality is for yearly statistics (within a 5-year window which slides yearly) and not for the overall success measuring indices (in their cumulative citation statistics) studied earlier for the citation statistics of some distinguished researchers (see e.g., [14]), where the SOC mark is observed to be a little higher ($g = k \simeq 0.86$).

As mentioned earlier, the Hirsch index (h) [16], which gives the highest number of publications by a researcher, each of which has received equal or more than that number of citations, does not seem give a most suitable measure [15,25] of the success of individual researchers. It has now been well demonstrated [15] (using the kinetic theoretical exchange model ideas), analyzing the Scopus citation data for the top 120,000 (within the "top 2%") Stanford cites score achievers that statistically $h \sim \sqrt{N_c} \sim \sqrt{N_p}$, where N_c and N_p denote, respectively, the total number of citations and total number of publications by the researcher. This suggests convincingly that the average number of citations per paper (N_c/N_p), and hence h, are statistical numbers (given by the effective Dunbar number [26,27]) depending on the community or network in which the researcher belongs [15,18]. We show here, extending our earlier observations (see, e.g., [14,18]), that the indications of success are not reflected by the total citations N_c, or for that matter by the Hirsch index h, rather than by the inequalities among the citations from publication to publication. Specifically, we show that for highly successful authors, the yearly variations (given by the statistics with overlapping 5-year windows) in the g index (given by the average inequality of the citations for the publications, $0 \leq g \leq 1$) and the k index (giving the fraction of total citations received by the top $(1-k)$ fraction of publications, $0.5 \leq k \leq 1$). In particular, achieving $g = k \simeq 0.82$ signals a precursor to the SOC state in the publication statistics. Analyzing the citation statistics (from the open-access Google Scholar) of thirty successful scientists throughout their recorded publication history, starting from their first recorded publication that the very successful among them (mostly Nobel laureates, higher-ranking Stanford c-scorers, and a few others) reach and hover just above and below that $g = k \simeq 0.82$ mark, characteristic of the SOC state ($k = 0.82$ means 82% of citations come from 18% of publications). Others remain below the (SOC) level of extreme inequality in publication statistics.

2. Socio-Statistical Inequality and Its Measures

In 1905, American economist Lorenz [2,3] developed the Lorenz curve, a graphical representation of the distribution of wealth in a society. To construct this curve (illustrated by the red curve in Figure 1), one organizes the society's population in ascending order of their wealth and then plots the cumulative fraction of wealth, denoted as $L(p)$, held by the poorest p fraction of individuals. One can similarly plot the cumulative fraction of citations against the fraction of papers that attracted those many citations. As indicated in Figure 1, the Gini index is calculated from the area between the equality line and the Lorenz curve, divided by the area (1/2) below the equality line for normalization. As such, $g = 0$ signifies perfect equality and $g = 1$ corresponds to extreme inequality. The index k is given by the fixed point of the complementary Lorentz function $L_c(p) \equiv 1 - L(p)$. As such, k gives the fraction of citations attracted by the top cited k fraction of papers and $k = 0.5$ means perfect equality, while extreme inequality corresponds to $k = 1$.

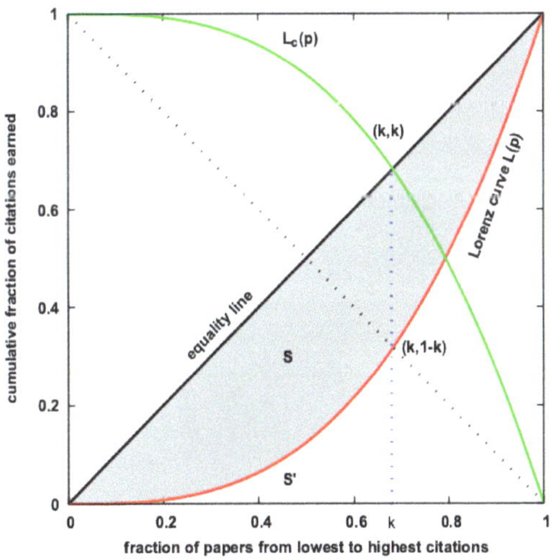

Figure 1. The Lorenz curve, represented by $L(p)$ (in red), denotes the cumulative proportion of total citations possessed by a fraction p of papers, when organized in ascending order of citation counts. Conversely, the black dotted line indicates perfect equality, where each paper receives an equal number of citations. The Gini index (g) is computed from the area (S) between the Lorenz curve and the equality line (the shaded region), normalized by the total area under the equality line ($S + S' = 1/2$). The Kolkata index (k) is obtained by locating the fixed point of the complementary Lorenz function (L_c; shown in green), defined as $L_c(p) \equiv 1 - L(p)$: $L_c(k) = k$. By geometry, the value of k gives the proportion of total citations owned or possessed by $(1-k)$ fraction of the top cited papers.

Landau-Like Expansion of $L(p)$ and g, k Approximate Relation

A minimal expansion [28] of the Lorenz function $L(p)$, employing a Landau-like expansion of free energy, suggests $L(p) = Ap + Bp^2, A > 0, B > 0, A + B = 1$. This gives $L(0) = 0$ and $L(1) = 1$ (with $B = 0$, the Lorenz function can represent only the equality line in Figure 1).

One can then calculate $g = 1 - 2\int_0^1 L(p)dp$, giving $A = 1 - 3g$ and $B = 3g$. Since $L(k) = 1 - k = Ak + Bk^2$, one can obtain a quadratic equation involving g and k. An approximate solution of it, in the $g \to 0$ limit gives

$$k = 1/2 + C \times g, \tag{1}$$

where $C = 3/8$ [28] suggesting that $g = k$ will occur at the Pareto value $k = 0.80$. We see here a little deviation in the value of the constant C in the relation (1), for all the reported observations.

3. Inequality Data Analysis from Google Scholar

We collect the citation data for all the recorded publications in each year since the first entry in the record for thirty successful researchers having individual Google Scholar page and having minimum and maximum number of total publications N_p = 127 and 2954, minimum and maximum number of total citations N_c = 5769 and 463,382, minimum and maximum values of h = 22 and 328, respectively, for all those selected researchers. We considered three Nobel prize winners in each of the science subjects: physics (H. Amano, B. Josephson, A. B. McDonald), chemistry (R. Henderson, J. Frank, J.-P. Sauvage), physiology and medicine (M. Houghton, G. L. Semanza, S. Yamanka), and economics (A. Banerjee, W. Nordhaus, J. Stiglitz). Two Fields medalists (mathematics; S. Smalle, E. Witten), two Boltzmann award winners (statistical physics; D. Dhar, H. E. Stanley), two Breakthrough prize winners (physics; C. Kane, A. Sen), three of the top-most cite-scorers in the Stanford Scopus c-score list (M. Graetzel, R. C. Kessler, and Z. L. Wang; considered for h-index statistics in Ref. [15]), and six known contributors in econophysics and sociophysics: W. Brian Arthur (known for "El Farol Bar Problem" of minority choice, see, e.g., [11]), B. K. Chakrabarti (one of the "fathers of econophysics" [29,30]), R. I. M. Dunbar (known for Dunbar's number of social connectivity, see, e.g., [31]), S. Galam (considered a pioneer of sociophysics), R. Mantegna (one of the "fathers of econophysics" [29,30]), V. M. Yakovenko (pioneer of kinetic exchange models of income/wealth distributions; see, e.g., [32]). We considered three of the highest-ranked Stanford cite-scorers for 2022 (M. Graetzel, R. C. Kessler, and Z. L. Wang [33]), and, for comparison, we also considered three lower-rank holders of the same "top 2% Stanford cite-scores" (I. Fofana, U. Sennur, and N. Tomoyuki [33]).

For studying the growth of inequality in the citation statistics of each of these researchers, we select a 5-year window, starting earliest publication, and note the present-day citations of each of these publications. We then construct the Lorenz function (see Figure 1) and extract the g and k indices as described in Section 4. We associate the g and k values with the middle year of the respective 5-year window and shift the window by one year and get the values of the inequality indices for each of the successive years up to 2020 (considering data up to 2022). These are shown in the following Figures 2–6.

One can see from Figures 2–6, for all the above-mentioned thirty scientists that for many of them (mostly Nobel prize winners and highest rank c-scorers), the g-index value goes over the k-index value in one (or multiple years) by crossing the $k = g \simeq 0.82$ line (see the corresponding insets). These crossings of the indices (at values above 0.80 value) preciously indicates large inequalities and entry into the SOC state [17] of the citation statistics of these scientists (see Table 1 for consolidated results of g and k).

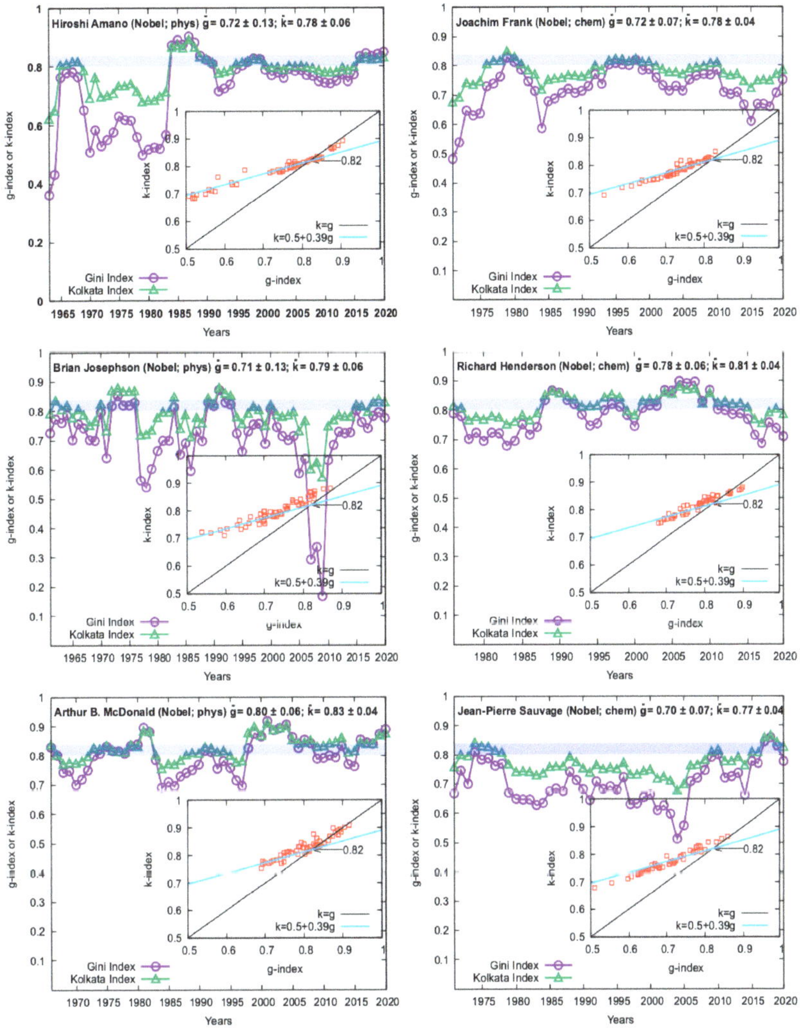

Figure 2. Yearly variations of the citation inequality indices, Gini (g) and Kolkata (k), for three Nobel prize winners in physics and three in chemistry. The indices are calculated using the present citation data for the publications within a 5-year window, starting from first recorded one in Google Scholar, and the window sliding by one year. The corresponding year shown is mid mid-year of the window until 2022 (shown for year 2020 for the last 5-year window). The g value crossing above (and coming down) the k value marks the precursor of onset (leaving) the SOC state with time. The inset shows k versus g (in red) over the entire career of the scientist. It fits well with the linear (Landau-like) relationship, $k = 1/2 + 0.39g$, suggesting a crossing SOC precursor point at $k = g = 0.82 \pm 0.02$.

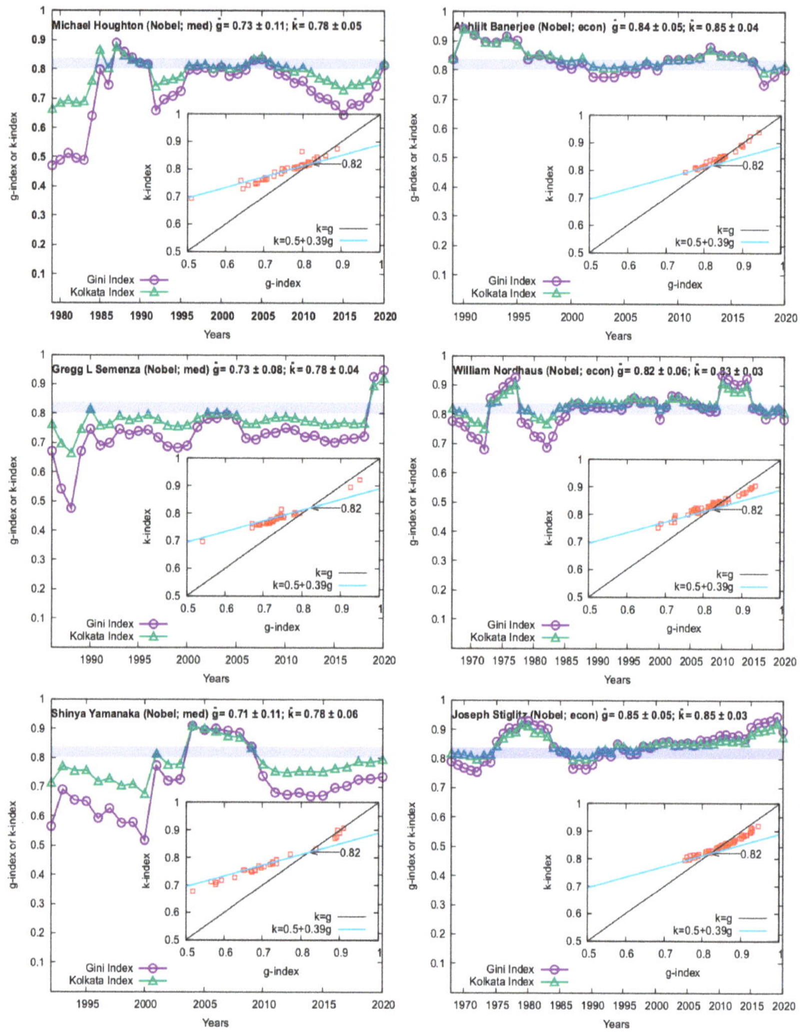

Figure 3. Yearly variations of the citation inequality indices, Gini (g) and Kolkata (k), for three Nobel prize winners in physiology-medicine and three in economics. The indices are calculated using the present citation data for the publications within a 5-year window, starting from the first recorded one in Google Scholar, and the window sliding by one year. The corresponding year shown is mid-year of the window until 2022 (shown for the year 2020 for the last 5-year window). The g value crossing above (and coming down) the k value marks the precursor of onset (leaving) the SOC state with time. The inset shows k versus g (in red) over the entire career of the scientist. It fits well with the linear (Landau-like) relationship, $k = 1/2 + 0.39g$, suggesting a crossing SOC precursor point at $k = g = 0.82 \pm 0.02$.

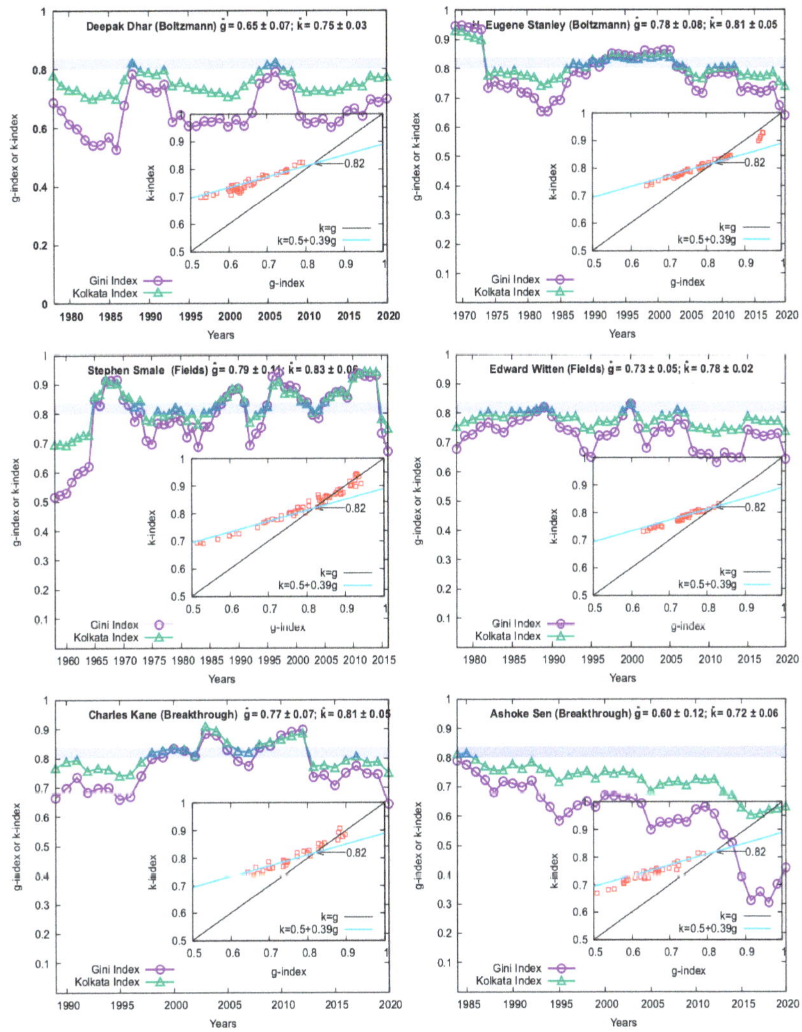

Figure 4. Yearly variations of the citation inequality indices, Gini (g) and Kolkata (k), for two winners each of Fields medal (mathematics), Boltzmann prize (statistical physics) and Breakthrough prize (physics). The indices are calculated using the present citation data for the publications within a 5-year window, starting from first recorded one in Google Scholar, and the window sliding by one year. The corresponding year shown is mid year of the window until 2022 (shown for year 2020 for the last 5-year window). The g value crossing above (and coming down) the k value marks the precursor of onset (leaving) the SOC state with time. The inset shows k versus g (in red) over the entire career of the scientist. It fits well with the linear (Landau-like) relationship, $k = 1/2 + 0.39g$, suggesting a crossing SOC precursor point at $k = g = 0.82 \pm 0.02$.

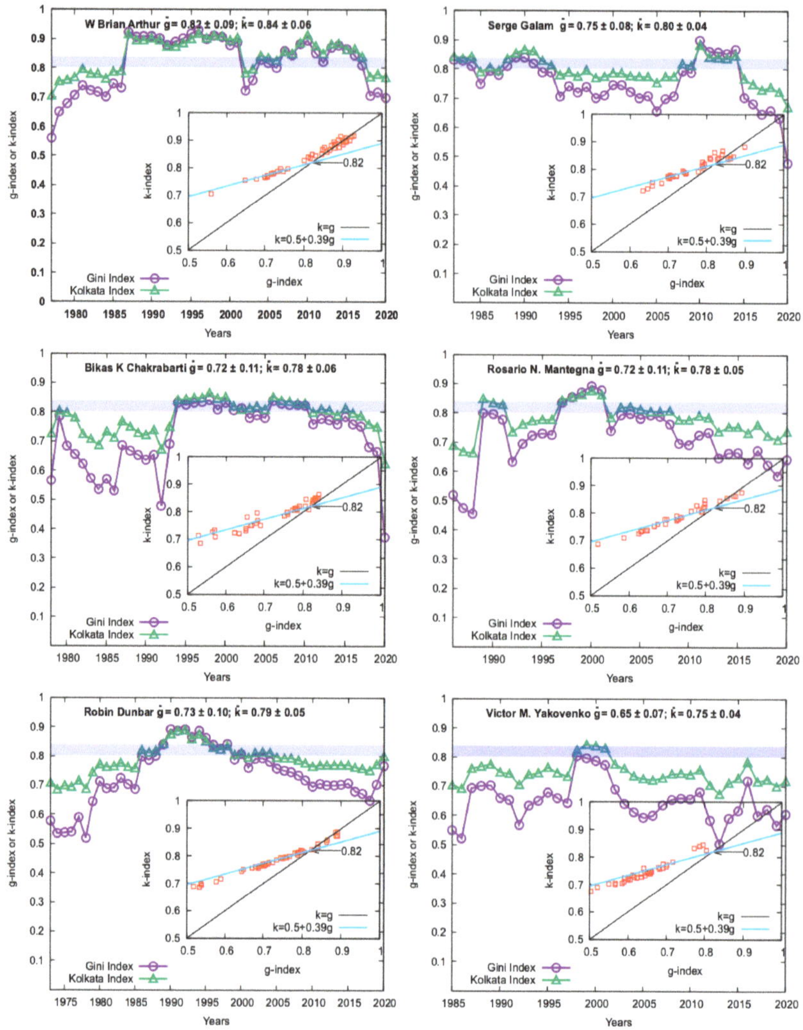

Figure 5. Yearly variations of the citation inequality indices, Gini (g) and Kolkata (k), for six distinguished researchers in econophysics and sociophysics. The indices are calculated using the present citation data for the publications within a 5-year window, starting from the first recorded one in Google Scholar, and the window sliding by one year. The corresponding year shown is mid-year of the window until 2022 (shown for year 2020 for the last 5-year window). The g value crossing above (and coming down) the k value marks the precursor of onset (leaving) the SOC state with time. The inset shows k versus g (in red) over the entire career of the scientist. It fits well with the linear (Landau-like) relationship, $k = 1/2 + 0.39g$, suggesting a crossing SOC precursor point at $k = g = 0.82 \pm 0.02$.

Although the study of the time variations of the g and k indices (as shown in Figures 2–6) and checking if g value ever goes over the k value by crossing the k versus g line (as shown in the insets) is indispensable for detecting if the SOC state has arrived or not, one can also have a straightforward (but only approximate) indication of the SOC state by looking at the ratio R of the citation number $n_C^{\max}$ of the highest cited paper and the effective Dunbar number D given by the average citation N_c/N_p of the researcher. In Table 2, we compare these $R = n_C^{\max}/D$ values (where $D = N_c/N_p$) and see how its higher values compare with the observation of SOC (when k versus g line is crossed affirmatively). We find, for $R \geq 40$, more than 94% of cases correspond to the SOC level.

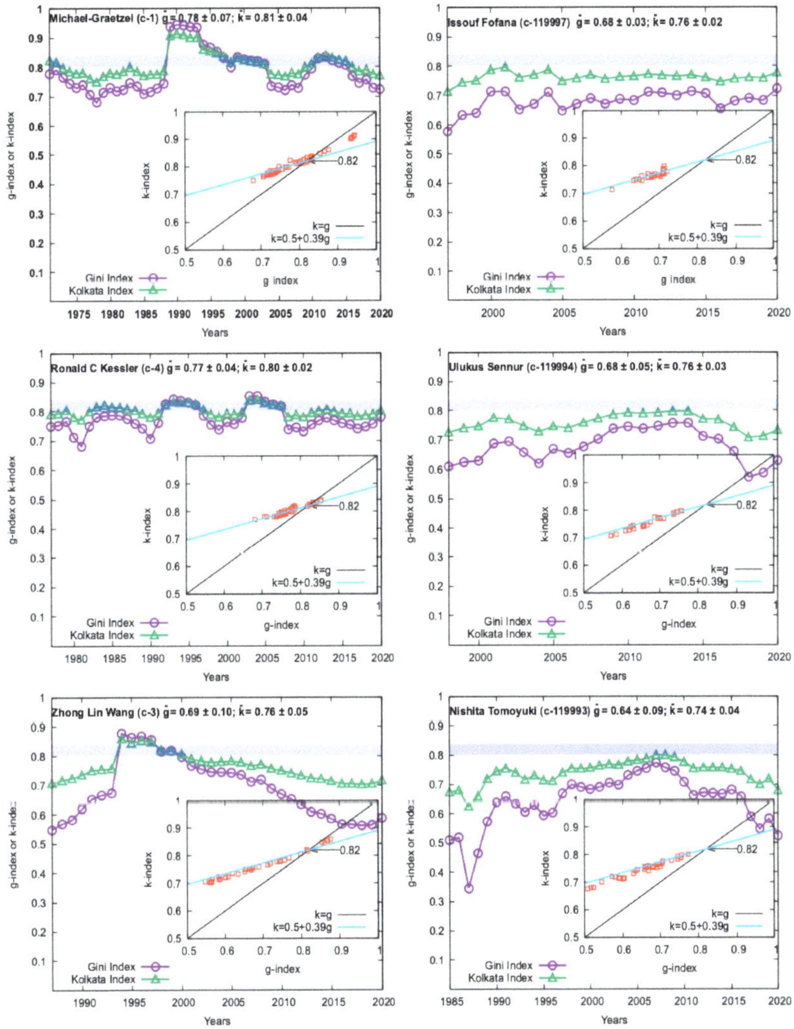

Figure 6. Yearly variations of the citation inequality indices, Gini (g) and Kolkata (k), for three top-most Stanford cite-scores and three lower rank entries from the "top 2% Stanford cite-scores" [15,33]. The indices are calculated using the present citation data for the publications within a 5-year window, starting from first recorded one in Google Scholar, and the window sliding by one year. The corresponding year shown is mid mid-year of the window until 2022 (shown for year 2020 for the last 5-year window). The g value crossing above (and coming down) the k value marks the precursor of onset (leaving) the SOC state with time. The inset shows k versus g (in red) over the entire career of the scientist. It fits well with the linear (Landau-like) relationship, $k = 1/2 + 0.39g$, suggesting a crossing SOC precursor point at $k = g = 0.82 \pm 0.02$.

Table 1. Consolidated inequality index (g, k) results for the citation statistics (from Figures 2–6) of the thirty chosen scientists (including twelve Nobel prize winners, two Fields medalists, two Boltzmann award winners, two Breakthrough prize winners, six distinguished sociophysics and econophysics researchers, three from the top and three from the bottom of the "top 2% Stanford cite-score scientists" (2022 list). NP(P) stands for Nobel Prize in physics, NP(C) stands for Nobel Prize in chemistry, NP(M) stands for Nobel Prize in physiology or medicine and NP(E) stands for Nobel Prize in economics, FM stands for Fields Medal in mathematics, BA stands for Boltzmann Award in statistical physics, BP(P) stands for Breakthrough Prize in physics, FEP stands for "fathers of econoophysics" [29,30], EFBP stands for "El Farol Bar problem" (see, e.g., [11]), DN stands for "Dunbar Number" (see, e.g., [31]), FSP stands for "father of sociophysics", PKEM stands for pioneer in kinetic exchange modeling of Wealth distribution [32], SCS-x stands for Stanford Cite Score rank ('x' denoting the rank) among the "top 2%" scientists in 2022 [33], and "yearly-av." stands for yearly-averaged publication in 5-year window; see text for more details.

					Inequality Indices: Hirsch (h), Gini (g), Kolkata (k)				$g = k$ Line
Researcher Name	Award/Prize/ Known for	N_p	N_c	h	g (overall)	k (overall)	g (yearly-av.)	k (yearly-av.)	Crossed Near 0.82?
H Amano	NP(P)	2161	57,281	106	0.84	0.83	0.72 ± 0.13	0.78 ± 0.06	Yes
B Josephson	NP(P)	127	11,685	22	0.94	0.92	0.71 ± 0.13	0.79 ± 0.06	No
AB McDonald	NP(P)	437	25,111	53	0.91	0.88	0.80 ± 0.06	0.83 ± 0.04	Yes
J Frank	NP(C)	686	50,518	116	0.77	0.80	0.72 ± 0.07	0.78 ± 0.04	No
R Henderson	NP(C)	267	31,822	65	0.85	0.84	0.78 ± 0.06	0.81 ± 0.04	Yes
JP Sauvage	NP(C)	655	61,572	114	0.70	0.76	0.70 ± 0.07	0.77 ± 0.04	Marginally
M Houghton	NP(M)	529	59,029	102	0.85	0.84	0.73 ± 0.11	0.78 ± 0.05	Yes
GL Semenza	NP(M)	682	192,246	196	0.80	0.81	0.73 ± 0.08	0.78 ± 0.04	Yes
S Yamanaka	NP(M)	345	124,106	125	0.85	0.84	0.71 ± 0.11	0.78 ± 0.06	Yes
A Banerjee	NP(E)	524	79,076	106	0.86	0.86	0.84 ± 0.05	0.85 ± 0.04	Yes
W Nordhaus	NP(E)	647	101,219	124	0.87	0.86	0.82 ± 0.06	0.83 ± 0.03	Yes
J Stiglitz	NP(E)	2408	364,237	235	0.89	0.87	0.85 ± 0.05	0.85 ± 0.03	Yes
D Dhar	BA	209	8299	44	0.76	0.80	0.65 ± 0.07	0.75 ± 0.03	No
HE Stanley	BA, FEP	2070	225,169	204	0.83	0.83	0.78 ± 0.08	0.81 ± 0.05	Yes
S Smale	FM	346	48,084	85	0.87	0.85	0.79 ± 0.11	0.83 ± 0.06	Yes
E Witten	FM, BP(P)	620	242,911	206	0.79	0.81	0.73 ± 0.05	0.78 ± 0.02	Marginally
C Kane	BP(P)	189	80,714	75	0.88	0.87	0.77 ± 0.07	0.81 ± 0.05	Yes
A Sen	BP(P)	401	37,065	103	0.69	0.76	0.60 ± 0.12	0.72 ± 0.06	No
WB Arthur	EFBP	196	52,545	56	0.91	0.89	0.82 ± 0.09	0.84 ± 0.06	Yes
BK Chakrabarti	FEP	390	12,596	47	0.81	0.82	0.72 ± 0.11	0.78 ± 0.06	Marginally
RIM Dunbar	DN	857	85,486	141	0.79	0.81	0.73 ± 0.10	0.79 ± 0.05	Yes
S Galam	FSP	252	8828	42	0.81	0.83	0.75 ± 0.08	0.80 ± 0.04	Yes
RN Mantegna	FEP	259	28,561	68	0.84	0.84	0.72 ± 0.11	0.78 ± 0.05	Yes
VM Yakovenko	PKEM	171	9076	44	0.73	0.78	0.65 ± 0.07	0.75 ± 0.04	No
M Graetzel	SCS-1	2282	463,382	295	0.82	0.82	0.78 ± 0.07	0.81 ± 0.04	Yes
RC Kessler	SCS-4	1829	523,835	328	0.83	0.83	0.77 ± 0.04	0.80 ± 0.02	Yes
ZL Wang	SCS-3	2954	394,080	299	0.71	0.77	0.69 ± 0.10	0.76 ± 0.05	Yes
I Fofana	SCS-119997	353	5759	39	0.73	0.78	0.68 ± 0.03	0.76 ± 0.02	No
N Tomoyuki	SCS-119993	304	9073	51	0.75	0.79	0.64 ± 0.09	0.74 ± 0.04	No
U Sennur	SCS-119994	455	18,987	63	0.73	0.78	0.68 ± 0.05	0.76 ± 0.03	No

Table 2. An approximate indicator, $R = n_C^{\max}/D$, where the effective Dunbar number $D = N_c/N_p$ (N_c denotes the total number of citations for N_p papers by the researcher) and $n_C^{\max}$ denotes the citation of the most-cited paper by the researcher, to check if the researcher has achieved the SOC level or not. We find, for $R \geq 40$, the corresponding researchers certainly belong to the SOC level (94% success rate).

Researcher Name	Award/ Known for	N_p	N_c	$n_C^{\max}$	h	$D = N_C/N_P$	$R = n_C^{\max}/D$	SOC Level Achieved (Table 1)	Comments
H Amano	NP(P)	2161	57,281	3154	106	27	119	Yes	Out of the eighteen researchers with $R \geq 40$, one failed in achieving the SOC level (crossing the $k = g \simeq 0.82$ line in Figures 2–6; see Table 1). $R \geq 40$ therefore indicates SOC level for the researcher with more than 94% success rate.
B Josephson	NP(P)	127	11,685	6554	22	92	71	No	
AB McDonald	NP(P)	437	25,111	5375	53	57	94	Yes	
J Frank	NP(C)	686	50,518	2299	116	74	31	No	
R Henderson	NP(C)	267	31,822	3681	65	119	31	Yes	
JP Sauvage	NP(C)	655	61,572	1805	114	94	19	Marginally	
M Houghton	NP(M)	529	59,029	9952	102	112	89	Yes	
GL Semenza	NP(M)	682	192,246	12,229	196	282	43	Yes	
S Yamanaka	NP(M)	345	124,106	30,735	125	360	85	Yes	
A Banerjee	NP(E)	524	79,076	9254	106	151	61	Yes	
W Nordhaus	NP(E)	647	101,219	19,605	124	156	125	Yes	
J Stiglitz	NP(E)	2408	364,237	23,844	235	151	158	Yes	
D Dhar	BA	209	8299	1182	44	40	30	No	
HE Stanley	BA, FEP	2070	225,169	14,348	204	109	132	Yes	
S Smale	FM	346	48,084	7912	85	139	57	Yes	
E Witten	FM, BP(P)	620	242,911	14,380	206	392	37	Marginally	
C Kane	BP(P)	189	80,714	19,504	75	427	46	Yes	
A Sen	BP(P)	401	37,065	1443	103	92	16	No	
WB Arthur	EFBP	196	52,545	15,227	56	268	57	Yes	
BK Chakrabarti	FEP	390	12,596	730	47	32	23	Marginally	
RIM Dunbar	DN	857	85,486	5312	141	100	53	Yes	
S Galam	FSP	252	8828	653	42	35	19	Yes	
RN Mantegna	FEP	259	28,561	5796	68	110	53	Yes	
VM Yakovenko	PKEM	171	9076	920	44	53	17	No	
M Graetzel	SCS-1	2282	463,382	35,789	295	203	176	Yes	
RC Kessler	SCS-4	1829	523,835	35,079	328	286	122	Yes	
ZL Wang	SCS-3	2954	394,080	8120	299	133	61	Yes	
I Fofana	SCS-119997	353	5759	333	39	16	20	No	
N Tomoyuki	SCS-119993	304	9073	467	51	30	16	No	
U Sennur	SCS-119994	455	18,987	1198	63	42	29	No	

4. Summary and Discussion

Our earlier analysis [15] of the Scopus citation data for the 120,000 top Stanford cite-score scientists showed that the Hirsch index $h \sim \sqrt{N_c} \sim \sqrt{N_p}$, where N_c and N_p denote, respectively, the total number of citations and the total number of publications by the researcher. This, in turn, says that the average number of citations per paper (N_c/N_p), and hence h, are statistical numbers (determined by the effective Dunbar num ber [26,31]) of the community or network (coauthors and followers) in which the researcher belongs [15,18]. Indeed, the anticipated increase in research impacts through collaboration (by increasing the number of coauthors) has been studied in Ref. [34], by looking at the average value of the community Dunbar number or N_c/N_p. Also, a detailed study from Google Scholar data on the relation between the Hirsch index of individual scientists with their average number of co-authors per paper has been reported in Ref. [35]. Our study here shows that the Hirsch index can not be a suitable measure of success for the researchers (even in Table 1; the highest $h = 328$ does not correspond to a Nobel Prize winner, while the least one with $h = 22$ do).

In an earlier study [18], we proposed that the citation inequality indices Gini ($0 \leq g \leq 1$) and Kolkata ($0.5 \leq k \leq 1$) to give better measures of success of the scientist (not N_c or h) and g and k both expected to approach equality at $g = k \simeq 0.86$ for successful researchers. It may be mentioned here that we used the entire citation data (over all the years) to get the Lorenz curve and the overall values of g and k of the researcher, and this gave a little higher value of $g = k \simeq 0.86$ point. Indeed, our numerical study [17] of the overall or cumulative inequality statistics of the avalanches or cluster sizes in some mostly studied and mstly established SOC models also suggested the arrival of the equality point of the avalanche size inequality indices ($g = k \simeq 0.86$) just appears as a precursor of the SOC point of the respective sand-pile or SOC models. In other words, as mentioned already, the SOC points in sandpile models (like BTW [23], Manna [24], and other models) of physics signifies a critical state where sand grain avalanches of all sizes occur following a power law distribution. As shown in Ref. [17], even in these physics SOC models, the inequality statistics (Gini and Kolkata indices) corresponding to the avalanche size statistics reach similar values for the inequality indices of the unequal citations (considered here equivalent to the sand mass avalanches in sand piles).

We analyzed the citation data for all the recorded publications in each year since the first entry in the record for the chosen thirtky successful scientists, each having an individual Google Scholar page. They have the minimum and maximum number of total publications N_p = 127 and 2954, and minimum and maximum number of total citations, N_c = 5769 and 463,382, respectively. For studying the growth of inequality in the citation statistics of each of these scientists, we select 5-year windows, where the central year of each window moves every year. We construct the Lorenz functions for each of these windows (see Figure 1) and extract the yearly values (corresponding to the central year of the window) of g and k indices. We have plotted these yearly g and k values for all the working years, starting the recorded first year and for the third year from there and continued for successive years up to 2022 (by considering data up to 2022) for each of these chosen thirty scientists. These are then shown in Figures 2–6. The insets in each Figure show the corresponding plot of k versus g (disregarding the yearly sequence). These plots in all thirty cases of the researchers show quite a good linear fit to $k = 0.5 + 0.39g$ (cf. Equation (1)), as obtained approximately using a (Landau-like) minimal polynomial expansion of the Lorenz function (see Section 2). The insets also show the actual or extrapolated (precursor of sand-pile SOC) point at $k = g = 0.82 \pm 0.02$. As we can see from Figures 2–6, for ten of the twelve Nobel prize winners, several of the other international prize winners are considered here, known sociophysicists, econophysicists, and all three of the highest rank Stanford cite-scorers, the crossing(s) of k versus g (often at multiple years), do take place convincingly. The same is also true (often marginally), for several others. The three lower rank (yet from the "top 2%") Stanford cite-scorers did not come up to $g = k$ point. There are certainly a few notable anomalies in this analysis of the data set; e.g., B. Josephson, J. Frank (both Nobel laureates), D. Dhar (Boltzmann award winner), and A. Sen (Breakthrough prize winner) do not fit this picture of indeed reaching the SOC point. These anomalies may indicate some shortcomings of this kind of analysis. On the other hand, noting that out of twenty seven of the researchers have chosen here (neglecting the three lower rank, though from the "top 2%" Stanford cite-scorers), the visible evidence of SOC are seen for nineteen (neglecting the "no" and "marginal" entries in the last column of Table 1 for these twenty seven researchers), indicating a success rate more than 70% for identifying the outstanding researchers. In Table 2, we give quite a simple (though approximate) indicator $R = n_C^{max}/D$ (where n_C^{max} denotes the maximum citation of any paper and D the effective Dunbar number of the researcher) to check if the researcher has achieved the SOC level or not. We see that the SOC level is achieved for $R \geq 40$, with a more than 94% coincidence rate.

In summary, as the Hirsch index h of a prolific researcher grows with the total citations N_c as $h = 0.5\sqrt{N_c}$ [15] and N_c grows linearly with the total number N_p of publications by the researcher, $N_c = DN_p$ (see [15,18]), where the effective Dunbar number D (~75 [15]) of the network community in which the scientist belongs, h and N_c can only give some average measures of success. In fact, most appreciated members of the community can in principle have uniformly high citations of order D for each of their publications and hence $h \geq D \simeq 75$. Though such uniformly appreciated or cited scientists will have significantly low values of Gini and Kolkata indices, $g \simeq 0$ and $k \simeq 0.5$. Our study here shows, notwithstanding some anomalies, most successful researchers have large fluctuations in the citations of one or more of their publications (presumably due to uneven but accurate appreciations from the usual Dunbar network or community and also possibly from outside the usual Dunbar community), which do not increase directly the D or h values, but lead to larger values of their inequality indices g and k, which may then hover around the SOC level value $g = k \simeq 0.82$, a little above the Pareto value (k = 0.80).

Author Contributions: Conceptualization, A.G. and B.K.C.; methodology, A.G. and B.K.C.; software, A.G.; formal analysis, A.G.; investigation, A.G.; writing—original draft, B.K.C.; writing—review and editing, A.G.; supervision, B.K.C. Chakrabarti. All authors have read and agreed to the published version of the manuscript.

Funding: B.K.C. is grateful to the Indian National Science Academy for their Senior Scientist Research Grant.

Data Availability Statement: The data that support the findings of this study are available on request from the corresponding author.

Acknowledgments: We are thankful to Soumyajyoti Biswas, Arnab Chatterjee, and Arnab Das for careful readings of the manuscript and useful criticisms and suggestions.

Conflicts of Interest: The authors declare no conflicts of interest.

References

1. de Oliveira, S.M.; de Oliveira, P.M.C.; Stauffer, D. *Evolution, Money, War and Computers. Non-Traditional Applications of Computational Statistical Physics*; B. G. Teubner Stuttgart: Leipzig, Germany; Springer Fachmedien Wiesbaden GmbH: Wiesbaden, Germany, 1999. [CrossRef]
2. Chakrabarti, B.K.; Chakraborti, A.; Chatterjee, A. (Eds.) *Econophysics and Sociophysics: Trends and Perspectives*; Wiley-VCH Verlag GmbH and Co. KGaA: Weinheim, Germany, 2006. [CrossRef]
3. Castelano, C.; Fortunato, S.; Loreto, V. Statistical physics of social dynamics. *Rev. Mod. Phys.* **2009**, *81*, 591–646. [CrossRef]
4. Helbing, D. *Quantitative Sociodynamics: Stochastic Methods & Models of Social Interaction Processes*; Springer: Berlin/Heidelberg, Germany, 2010. [CrossRef]
5. Galam, S. *Sociophysics: A Physicist's Modeling of Psycho-Political Phenomena*; Springer Science+Business Media, LLC: New York, NY, USA, 2012. [CrossRef]
6. Sen, P.; Chakrabarti, B.K. *Sociophysics: An Introduction*; Oxford University Press: Oxford, UK, 2014. Available online: https://archive.org/details/sociophysicsintr0000senp (accessed on 2 December 2023).
7. Galam, S. Sociophysics: A review of Galam models. *Int. J. Mod. Phys. C* **2008**, *19*, 409–440. [CrossRef]
8. Galam, S.; Cheon, T. Tipping points in opinion dynamics: A universal formula in five dimensions. *Front. Phys.* **2020**, *8*, 566580. [CrossRef]
9. Biswas, S.; Chatterjee, A.; Sen, P.; Mukherjee, S.; Chakrabarti, B.K. Social dynamics through kinetic exchange: The BChS model. *Front. Phys.* **2023**, *11*, 1196745. [CrossRef]
10. Filho, E.A.; Lima, F.W.; Alves, T.A.; Alves, G.A.; Plascak, J.A. Opinion dynamics systems via Biswas–Chatterjee–Sen model on Solomon networks. *Physics* **2023**, *5*, 873–882. [CrossRef]
11. Challet, D.; Marsili, M.; Zhang, Y.-C. *Minority Games. Interacting Agents in Financial Markets*; Oxford University Press: Oxford, UK, 2005.

12. Martin, L. Extending Kolkata Paise Restaurant problem to dynamic matching in mobility markets. *Jr. Manag. Sci.* **2019**, *4*, 1–34. [CrossRef]
13. Harlalka, A.; Belmonte, A.; Griffin, C. Stability of dining clubs in the Kolkata Paise Restaurant Problem with and without cheating. *Phys. A Stat. Mech. Appl.* **2023**, *620*, 128767. [CrossRef]
14. Ghosh, A.; Chakrabarti, B.K.; Ram, D.R.S.; Mitra, M.; Maiti, R.; Biswas, S.; Banerjee, S. Scaling behavior of the Hirsch index for failure avalanches, percolation clusters, and paper citations. *Front. Phys.* **2022**, *10*, 1019744. [CrossRef]
15. Ghosh, A.; Chakrabarti, B.K. Scaling and kinetic exchange like behavior of Hirsch index and total citation distributions: Scopus-CiteScore data analysis. *Phys. A Stat. Mech. Appl.* **2023**, *626*, 129061. [CrossRef]
16. Hirsch, J.E. An index to quantify an individual's scientific research output. *Proc. Natl. Acad. Sci. USA* **2005**, *102*, 16569–16572. [CrossRef] [PubMed]
17. Manna, S.S.; Biswas, S.; Chakrabarti, B.K. Near universal values of social inequality indices in self-organized critical models. *Phys. A Stat. Mech. Appl.* **2022**, *596*, 127121. [CrossRef]
18. Ghosh, A.; Chakrabarti, B.K. Limiting value of the Kolkata index for social inequality and a possible social constant. *Phys. A Stat. Appl.* **2021**, *573*, 125944. [CrossRef]
19. Banerjee, S.; Biswas, S.; Chakrabarti, B.K.; Challagundla, S.K.; Ghosh, A.; Guntaka, S.R.; Koganti, H.; Kondapalli, A.R.; Maiti, R.; Mitra, M.; et al. Evolutionary dynamics of social inequality and coincidence of Gini and Kolkata indices under unrestricted competition. *Int. J. Mod. Phys. C* **2023**, *34*, 2350048. [CrossRef]
20. Gini, C. Measurement of inequality of incomes. *Econ. J.* **1921**, *31*, 124–126. [CrossRef]
21. Ghosh, A.; Chattopadhyay, N.; Chakrabarti, B.K. Inequality in societies, academic institutions and science journals: Gini and k-indices. *Phys. A Stat. Mech. Appl.* **2014**, *410*, 30–34. [CrossRef]
22. Banerjee, S.; Chakrabarti, B.K.; Mitra, M.; Mutuswami, S. Social inequality measures: The Kolkata index in comparison with other measures. *Front. Phys.* **2020**, *8*, 562182. [CrossRef]
23. Bak, P.; Tang, C.; Wiesenfeld, K. Self-organized criticality: An explanation of the $1/f$ noise. *Phys. Rev. Lett.* **1987**, *59*, 381–384. [CrossRef]
24. Manna, S.S. Two-state model of self-organized criticality. *J. Phys. A Math. Gen.* **1991**, *24*, L363–L370. [CrossRef]
25. Yong, A. A critique of hirsch's citation index: A combinatorial fermi problem. *Not. Am. Math. Soc.* **2014**, *61*, 1040–1050. [CrossRef]
26. Dunbar, R.I.M. Neocortex size as a constraint on group size in primates. *J. Hum. Evol.* **1992**, *22*, 469–493. [CrossRef]
27. Dunbar, R. *How Many Friends Does One Person Need? Dunbar's Number and Other Evolutionary Quirks*; Faber and Faber Limited: London, UK, 2010. Available online: https://archive.org/details/howmanyfriendsdo0000dunb (accessed on 2 December 2023).
28. Joseph, B.; Chakrabarti, B.K. Variation of Gini and Kolkata indices with saving propensity in the Kinetic Exchange model of wealth distribution: An analytical study. *Phys. A Stat. Mech. Appl.* **2022**, *594*, 127051. [CrossRef]
29. Jovanovic, F.; Schinckus, C. *Econophysics and Financial Economics: An Emerging Dialogue*; Oxford University Press: Oxford, UK, 2017; p. 178. [CrossRef]
30. Schinckus, C. When Physics Became Undisciplined: An Essay on Econophysics. Ph.D. Thesis, University of Cambridge, Cambridge, UK, 2018; pp. 15–16. [CrossRef]
31. Dunbar, R.I.M. Coevolution of neocortical size, group size and language in humans. *Behav. Brain Sci.* **1993**, *16*, 681–735. [CrossRef]
32. Yakovenko, V.M.; Rosser, J.B. Statistical mechanics of money, wealth, and income. *Rev. Mod. Phys.* **2009**, *81*, 1703–1725. [CrossRef]
33. Ioannidis, J.P.A. *September 2022 Data-Update for "Updated Science-Wide Author Databases of Standardized Citation Indicators". Version 5*; Elsevier Data Repository, Elsevier Inc.: New York, NY, USA, 2022. [CrossRef]
34. Katz, J.S.; Hicks, D. How much is a collaboration worth? A calibrated bibliometric model. *Sociometrics* **1997**, *40*, 541–554. [CrossRef]
35. Arnaboldi, V.; Dunbar, R.I.M.; Passarella, A.; Conti, M. Analysis of co-authorship ego networks. In *Advances in Network Science. Proceedings of the 12th International Conference and School, NetSci-X 2016, Wroclaw, Poland, 11–13 January 2016*; Wierzbicki, A., Brandes, U., Schweitzer, F., Pedreschi, D., Eds.; Springer International Publishing: Cham, Switzerland, 2016; pp. 82–96.

physics

Article

Graph-Based Generalization of Galam Model: Convergence Time and Influential Nodes

Sining Li * and Ahad N. Zehmakan

School of Computing, Australian National University, Canberra, ACT 2601, Australia; ahadn.zehmakan@anu.edu.au
* Correspondence: sining.li@anu.edu.au

Abstract: We study a graph-based generalization of the Galam opinion formation model. Consider a simple connected graph which represents a social network. Each node in the graph is colored either blue or white, which indicates a positive or negative opinion on a new product or a topic. In each discrete-time round, all nodes are assigned randomly to groups of different sizes, where the node(s) in each group form a clique in the underlying graph. All the nodes simultaneously update their color to the majority color in their group. If there is a tie, each node in the group chooses one of the two colors uniformly at random. Investigating the convergence time of the model, our experiments show that the convergence time is a logarithm function of the number of nodes for a complete graph and a quadratic function for a cycle graph. We also study the various strategies for selecting a set of seed nodes to maximize the final cascade of one of the two colors, motivated by viral marketing. We consider the algorithms where the seed nodes are selected based on the graph structure (nodes' centrality measures such as degree, betweenness, and closeness) and the individual's characteristics (activeness and stubbornness). We provide a comparison of such strategies by conducting experiments on different real-world and synthetic networks.

Keywords: sociophysics; Galam model; graph theory; Markov chain; social networks; convergence time; opinion formation; influential nodes; viral marketing

Citation: Li, S.; Zehmakan, A.N. Graph-Based Generalization of Galam Model: Convergence Time and Influential Nodes. *Physics* **2023**, *5*, 1094–1108. https://doi.org/10.3390/physics5040071

Received: 15 September 2023
Revised: 4 November 2023
Accepted: 9 November 2023
Published: 28 November 2023

1. Introduction

Humans' decisions and opinions can be influenced by others. When forming an opinion or making a decision on a subject such as a new product or a political election, people usually seek advice from their families, friends, and other people with whom they often keep in contact with. In addition, people also consider the opinions of the figures that they value or respect; for instance, figures in specific fields or celebrities that people look up to. Therefore, people often keep updating their decisions and opinions by interacting with their connections.

Furthermore, the prevalence of online social media and social networking platforms, such as Facebook, Instagram, and WeChat, contribute to the increasing pace of opinion formation. These online platforms provide people with a convenient and fast way to communicate with friends, obtain information, and express opinions. Consequently, they significantly accelerate the information spreading and opinion formation and exchange process.

In recent decades, there has been a growing demand for a deeper understanding of how opinions form and how information spreads in social networks. A variety of opinion diffusion models, which aim to simulate the process of information spreading in a society, have thus been developed and studied within sociophysics using tools and concepts from statistical physics [1]. More recently, the topic has been enriched by contributions from theoretical computer science [2,3].

The studied models typically involve a graph G and an initial coloring of its nodes, with each node being either blue or white. The graph represents a social network, in which each node corresponds to an individual. The edges of the graph represent relationships between individuals, such as personal contacts, friendships, or followers. The color assigned to each node, say white or blue, represents the individual's positive or negative opinion on a subject. Following the initialization, a group of nodes in the graph update their colors based on a predefined rule in each round.

The majority model is an instance of the aforementioned basic model [4,5] and it has become popular within the community of social networks. In this model, all nodes in the graph G simultaneously update their color to the most frequent color among their neighbors. If there is a tie, a node does not change its current color. It should be noted that the number of neighbors of each node can be quite large, which makes the model somewhat unrealistic. For example, it is almost impossible for a person who has 50 friends to discuss a topic with all the friends at the same time.

Another established model is the Galam model [6], which describes the dynamics of the spreading of the minority opinion in a social debate. In this model, all individuals (nodes) are randomly assigned to groups of different sizes, based on the geometry of social life of people, with the discussing groups composed of small sizes from 1 person up to 5 or 6 people. Only the people in the same group can discuss the topic and update their opinion based on the local majority opinion in the group. Therefore, the Galam model is more realistic. In the general majority-based model, the nodes with a higher degree can influence more nodes in each round. But, in the Galam model, the node can only influence other nodes in the same group. Moreover, at a tie, the Galam model introduces the possibility of a tie breaking determined using the unconscious prejudices of the agents [7].

In the classic Galam model, each individual has the same probability of grouping with any other individual. However, it is unusual that two people get together and discuss on some topics if they do not have any relationships. To add another layer of realism to the model, we leverage the underlying graph structure of a social network to introduce a generalization of the Galam model in the present paper. In the Graph-based Galam model (GGM), only the nodes that form a clique in the graph can be assigned to a group.

In the present paper, we focus on two problems: (1) How long does it take for the GGM to converge? This is arguably the most well-studied question for different dynamic systems. In our model, the answer highly depends on the underlying graph structure. (2) What is the subset of nodes which maximizes the final expected number of blue nodes, if initially blue, for a given fixed budget? This is also another popular question investigated for various models, motivated from viral marketing. For example, when a company plans to launch a new product on to the market, it is often intended to persuade a specific group of people into using the product to maximize the final cascade of the adoptions of the product. Another example is when a political party aims to convince a subset of individuals to adopt a positive opinion about a topic with the aim of a large further adoption of the positive opinion. The main problem here is to find a subset of nodes (individuals) of a given size whose adoption of the desired product/opinion results in its maximum adoption at the end.

We study the convergence time of GGM on different synthetic and real-world graph data. Experiments conducted on cycle and complete graphs show that the convergence times of GGM on cycle and complete graphs are in $\mathcal{O}(n^2)$ and $\mathcal{O}(\log n)$, respectively, where n denotes the number of nodes. We also investigate how adding new edges can influence the convergence time of the process and observe that it essentially depends on the manner that the edges are added. Particularly, if the added edges increase the expansion (connectivity) of the graph, the convergence time usually decreases.

Furthermore, the experiments on complete graphs and real-world social networks indicate that the initial blue nodes must be more than half to ensure that the blue nodes "dominate" at the end, if one selects the initial blue nodes randomly. If the initial blue nodes are selected based on some smarter strategies such as centrality-based measures (such as degree, betweenness, and closeness), then a much smaller fraction of initial blue nodes is required for the blue color to be the winning color at the end.

Inspired by some prior work [8,9], we also study the setup where the nodes have different levels of activeness and stubbornness; that is, some nodes are more probable to participate in the gatherings and some nodes are less inclined to follow the majority opinion. We then introduce some strategies for choosing initial seed nodes, which take these parameters into account, in addition to the graph structural properties.

The paper is organized as follows. We first give some basic definitions and an overview of some prior work in the rest of this Section. Our findings, which address questions (1) and (2), are presented in Sections 2 and 3, respectively.

1.1. Preliminaries

1.1.1. Graph Definitions

- Graph
  Let $G = (V, E)$ be a simple connected undirected graph. Let $n := |V|$ and $m := |E|$ represent the number of nodes and the number of edges, respectively. For a given node, $v \in V$, $N(v) := \{u \in V : \{u, v\} \in E\}$ represents the neighborhood of v. Furthermore, let $d(v) := |N(v)|$ represent the degree of node v.
- Expander Graph
  An expander graph is expected a graph with good connectivity. There are three parameters often being used to measure the expansion of a graph: vertex expansion, edge expansion, and spectral expansion. A graph will be called here an expander graph if it has strong expansion properties.
  (1) Vertex Expansion. Let $S \subseteq V$ be a subset of nodes, and $\overline{S} := V \backslash S$. Let $\partial S_{\text{out}} := \{v \in \overline{S} | \exists u \in N(v), \text{such that } u \in S\}$, which is the outer boundary of set S. The vertex expansion is defined as $h_{\text{out}}(G) := \min_{0<|S|<\frac{n}{2}} |\partial S_{\text{out}}|/|S|$. For nodes in any "small" subset S (with a size less than $n/2$) of V, the greater the $h_{\text{out}}(G)$ is, the larger the number of their neighbors outside S is, and the better connected the graph G is. Therefore, a graph G with a greater $h_{\text{out}}(G)$ has stronger expansion properties
  (2) Edge Expansion. Let $S \subseteq V$ be a subset of nodes, and $\overline{S} := V \backslash S$. Let $\partial S := \{\{u, v\} \in E | u \in S, v \in \overline{S}\}$, which is the boundary of set S. The edge expansion is defined as $h(G) := \min_{0<|S|<\frac{n}{2}} |\partial S|/|S|$. For the number of edges from the nodes inside any "small" subset (with a size less than $n/2$) of V to the nodes outside this subset, the greater the $h(G)$ is, the larger the number of edges is, and the better connected the graph G is. Therefore, a graph G with a greater $h(G)$ has stronger expansion properties
  (3) Spectral Expansion. Let A be the adjacency matrix of the graph G. Since A is symmetric and real, it has n real-valued eigenvalues. Let $\lambda(G)$ represent the second-largest absolute eigenvalue of A. A graph G with a smaller value of $\lambda(G)$ has stronger expansion properties.

1.1.2. Model

- Majority Model and Random Majority Model
  Consider a graph G that represents a social network. Each node in the graph is either blue or white at the initial state, which represents a person holding a positive or negative opinion about a product or a topic. In each round, all the nodes simultaneously update their color to the most frequent color among their neighbors. If there is a tie, the node keeps its color. This is known as the Majority Model. The Random Majority Model is the same as the Majority Model, except for the tie-breaking rule. In the Random Majority Model, a node chooses blue or white with an equal probability of 0.5 in case of a tie.
- Galam Model
  Consider a population of n individuals (nodes) who can hold a positive or negative opinion on a subject (i.e., are blue or white). Furthermore, consider a set of rooms of various sizes, such that the summation of the size of the rooms is equal to n. In each

round, all individuals are randomly assigned to these rooms. Then, all individuals simultaneously update their opinion to the most frequent opinion in the room. If there is a tie, then a tie-breaking rule is applied to handle the situation. In the original description of the model, individuals choose negative in case of a tie [6]. However, other variants are considered, for example, random tie-breaking rules [7].

- Graph-based Galam Model
  In the present paper, we introduce a graph-based generalization of the Galam model, GGM. A graph $G = (V, E)$ is used to represent the society to be considered. The nodes and edges in the graph represents the individuals and the relationship between them, respectively. We define a coloring function $\mathcal{C}\colon V \rightarrow \{w, b\}$, where w and b represent white and blue, respectively, which correspond to negative and positive.
  There are two steps in each round of this model: (1) randomly assign all nodes into groups with different sizes; and (2) update the color of each node following the local majority-based rule in the group.

  (1) Group Allocation. All the cliques of size 1 to R in the graph are collected and stored in a list. In our setup, we use $R = 3$, but the model is well defined for larger values of R. Each time, one clique in the list is randomly picked with an equal likelihood. For each node in the clique that is picked, the remaining cliques that contain the node is removed from the clique list. This continues until the nodes are partitioned into cliques of size 1, 2, and 3 (the clique list is empty).
  (2) Color Updating. All the nodes simultaneously update their color to the most frequent color in their group. If there is a tie, a binary random number (0 or 1) is generated with equal probability. The nodes becomes blue if the random number is 1 and white otherwise.

  An example is given in Figure 1. The graph on the left-hand side is the initial state. The list of all cliques of this graph is {{1}, {2}, {3}, {4}, {5}, {6}, {1, 2}, {1, 3}, {1, 5}, {2, 5}, {3, 4}, {4, 5}, {4, 6}, {5, 6}, {1, 2, 5}, and {4, 5, 6}}. One possible outcome of group allocation is {{1, 2}, {3}, and {4, 5, 6}}. In this case, node 1 and node 2 have the same color and keep their color. Node 3 is in a group of itself and it also keeps the color. Nodes 4, 5, and 6 are in the same group, and the major color in this group is blue. Hence, node 4 updates its color from white to blue. The state of the graph after this round is illustrated in the graph in Figure 1, right.

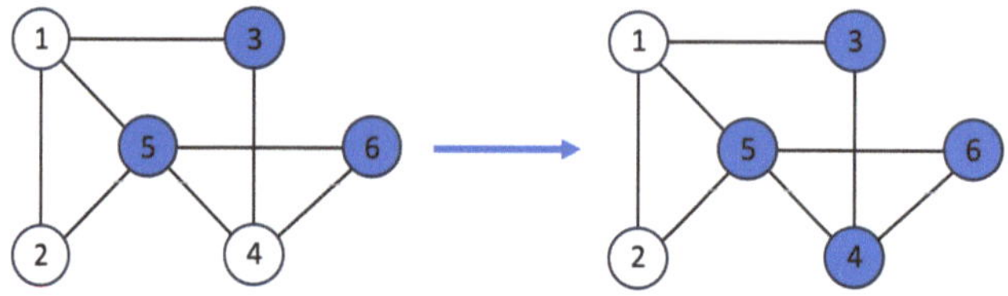

Figure 1. An example of one round of the Graph-based Galam model: the initial state of the graph (**left**) and the updated graph after one round (**right**).

1.2. Prior Studies

- Models
  Numerous models that simulate the spreading of information and the formation of opinions have been introduced and studied, such as the Independent Cascade (IC) model [10], the Linear Threshold (LT) model [10,11], and majority-based models [12–15]. The Galam model is one of the most studied models in the area of sociophysics. The model was originally designed to explain how an initial minority can finally win the debate [6]. This model has become an established model in sociophysics and various variants of this model have been investigated. Some of these variants involve considering three different opinions [16], changing the biased tie-breaking rule to a random one [7], adding inflexible individuals [8], and introducing the level

of activeness [9]. The model studied in this paper falls under the umbrella of this line of models.

- Convergence time
  Svatopluk Poljak and Daniel Turzík proved that the convergence time of the majority model is upper bounded by a linear function of the number of edges; that is, $\mathcal{O}(m)$ [17]. Furthermore, some stronger bounds have been proven for some special types of graphs; see [18]. Studies on convergence property have also been conducted for other majority-based models [19,20]. For the random majority model on a cycle graph, the convergence time is proven to be in $\mathcal{O}(n^2)$ [18]. For the classic Galam model, Bernd Gärtner and one of the authors of this paper proved that the convergence time is in $\mathcal{O}(\log\log n)$ when all groups in the model have a size greater or equal to 3 [21].
- Influential nodes
  The research about viral marketing in social networks has become popular in recent decades. A small set of seed individuals who hold a positive (blue) opinion on some subject needs to be selected to maximize the final number of individuals holding a positive opinion in the social network. Such a problem about how to select seed individuals is typically known as influence maximization (IM). David Kempe, Jon Kleinberg and Éva Tardos provided a foundation for this problem in their seminal paper in 2003 [10]. After that, there has been extensive study on the IM problem [11,22–24].
  The above IM problem can be modeled as a discrete optimization problem and it is usually proven to be computationally hard to solve for various models; more precisely, it is known to be non-deterministic polynomial-time hard (NP-hard); see [10,24]. Therefore, several approximation, randomized, and heuristic algorithms, such as centrality-based algorithms [25,26] (which choose nodes with the highest degree, betweenness, closeness, or pagerank) and greedy approaches [27–29], have been proposed. In recent years, machine learning techniques have become popular, which leads to the development of some machine learning-based algorithms for the IM problem [30–32].

1.3. Experimental Setup

All graph data for the experiments are from Stanford Network Analysis Project (SNAP) [33]. Experiments were conducted on Facebook (FB) and Twitch Spain (T-ES)'s social networks. Some basic properties of these graphs are listed in Table 1. Furthermore, several experiments focus on cycle graphs and complete graphs to examine the relationship between the number of nodes and the convergence time.

Table 1. Basic properties of the two examined social networks. Here, n, m and "AvgDegree" denote the number of nodes, the number of edges and the average degree, respectively.

Social Network Name	n	m	Avg. Degree
Facebook	4039	88,234	43.69
Twitch Spain	4638	59,382	25.55

Since our model is a random process, all experiments were executed 20 times, unless otherwise specified, and the average output was considered. All experiments were conducted on an Apple M1 Pro chip, with 32 GB RAM and MacOS Ventura (Apple Inc., Cupertino, CA, USA).

2. Convergence Time

Some experiments were conducted to study the convergence time of the GGM in cycle graphs, complete graphs, and some other types of special graphs in this Section. Before discussing the outcomes of these experiments, we show that the GGM always converges on a connected graph.

Convergence to an Absorbing State. For a connected graph G with n nodes, the model on such a graph can be regarded as a Markov chain. Since there are 2^n possible colorings in the model, this Markov chain has 2^n states. If the probability of transition from state s to s' is non-zero, then there is a directed edge from s to s'. A state is called an absorbing state if there is no edge going out, and a Markov chain is called an absorbing Markov chain if every state in the Markov chain can reach an absorbing state.

In GGM, the two states where all nodes have the same color (blue or white) are two absorbing states. Since each node in the graph keeps its color in such two states, there is no outgoing edge from the two states. Except these two states, any other state has at least two edges going out from it and there is a path from the state to the absorbing states. Consider a state that there are i blue nodes ($i \neq 0$ and $i \neq n$, i.e., there are not n blue nodes or n white nodes). Since the graph is connected, there must be at least two adjacent nodes with different colors. In the coming round, there is a non-zero probability that these two nodes are assigned to the same group and all other nodes are assigned to other groups of size one. Then, these two nodes become both blue or white and all other nodes keep their color at the end of this round. Therefore, the state that has i blue nodes must have at least two outgoing edges: one edge goes to a state that has $(i+1)$ blue nodes and the other one goes to a state that has $(i-1)$ blue nodes. As the process keeps going, the state of i blue nodes finally reach the state that all the nodes are blue (or white). Hence, the GGM on a connected graph is an absorbing Markov chain with two absorbing states where all nodes are blue or white. Furthermore, the process eventually converges to one of the absorbing states.

2.1. Cycle Graph

As mentioned in Section 1.2, prior studies have shown that the convergence time of the random majority model on a cycle graph with n nodes is in $\mathcal{O}(n^2)$. Since one can think of the GGM somewhat as a room-based random majority-based model, let us speculate that the convergence time of the model in a cycle graph with n nodes is also of order $\mathcal{O}(n^2)$ for GGM.

The experiments were conducted on 10 cycle graphs with the number of nodes selected from 100 to 1000 every 100. The number of initial blue nodes and how to set these blue nodes should be determined to maximize the convergence time. One can expect that it would take more time for the GGM to converge when the number of blue and white nodes are equal. If blue nodes are more (or less) than white nodes, the process tends to converge to a state of all nodes being blue (or white) more quickly. Therefore, the initial number of blue nodes were set to be half of the total nodes. In addition, the blue nodes were selected in two ways. The first way is that all the blue nodes are next to each other to form a path. The other way is that all blue nodes and white nodes are alternating in the cycle. An example of these two types of initial state can be seen in Figure 2. For the graph in Figure 2a, only four nodes have a non-zero probability to change their color in the coming round. However, all the nodes in the graph shown in Figure 2b have a non-zero probability to change their color. All the other configurations of initial blue nodes are expected to place between these two special cases, with regard to convergence time.

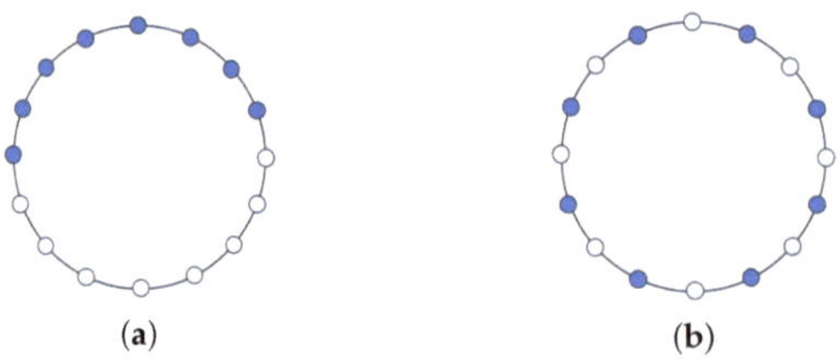

Figure 2. An example of the initial state of a cycle: (**a**) all blue nodes are next to each other; (**b**) white and blue nodes are alternating in the cycle.

Figure 3 shows the outcome of the experiments. Cycle graph (1) and cycle graph (2) correspond to the two ways of initializing. The experiment result indicates that the convergence time of the GGM on cycle graphs is of order $\mathcal{O}(n^2)$. When half of the nodes are blue, the two different ways of setting these blue nodes have little impact on the convergence time.

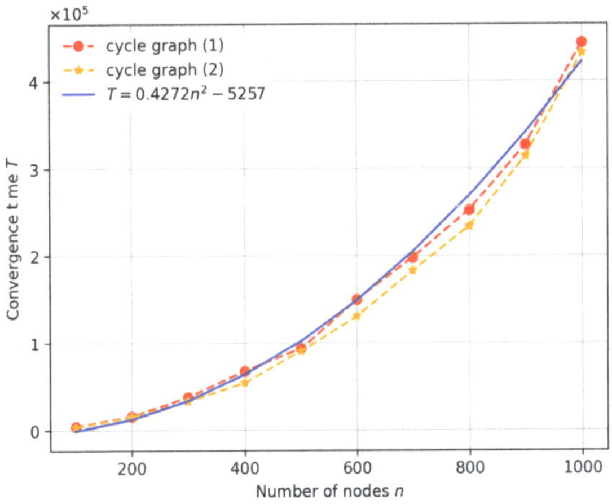

Figure 3. Convergence time in cycle graph.

2.2. Complete Graph

Prior study [21] has shown that the convergence time of the original Galam model with all groups of a size greater or equal to 3 is of order $\mathcal{O}(\log\log n)$. In the original Galam model, any two individuals in the social network can be assigned in the same group, which means that there exists a relationship between them. If a graph is used to represent the social network, there must be an edge between any two nodes in this graph. Then, the original Galam model can be regarded as a complete graph. Therefore, one might speculate that the convergence time of the GGM on complete graphs with n nodes is of order $\mathcal{O}(\log\log n)$.

The experiment was conducted on 10 complete graphs with the number of nodes selected in the set {50, 100, 200, 500, 1000, 2000, 3000, 4000, 5000, and 10,000}. At the initial state, half of the nodes were blue, since as expected, this would result in the largest convergence time. Since each node in a complete graph is connected to all other nodes and there is no structural difference between the nodes, the blue nodes were just randomly selected.

The outcome of the experiment is depicted in Figure 4a. It should be noted that the convergence time of the GGM on complete graphs is of order $\mathcal{O}(\log n)$, but not $\mathcal{O}(\log\log n)$. The main reason for the difference is more expected to be the tie-breaking rule and the number of groups of even size. In the original Galam model, all the people in the group hold a negative (white) opinion if there is a tie. It is a biased tie-breaking rule, which can lead to fast convergence if there are more groups with a size of an even number. In addition, the convergence time $\mathcal{O}(\log\log n)$ has been proved when all groups have a size of equal to or greater than 3. However, the group size in the GGM in the present paper is less than or equal to 3, which might slow down the diffusion of opinions.

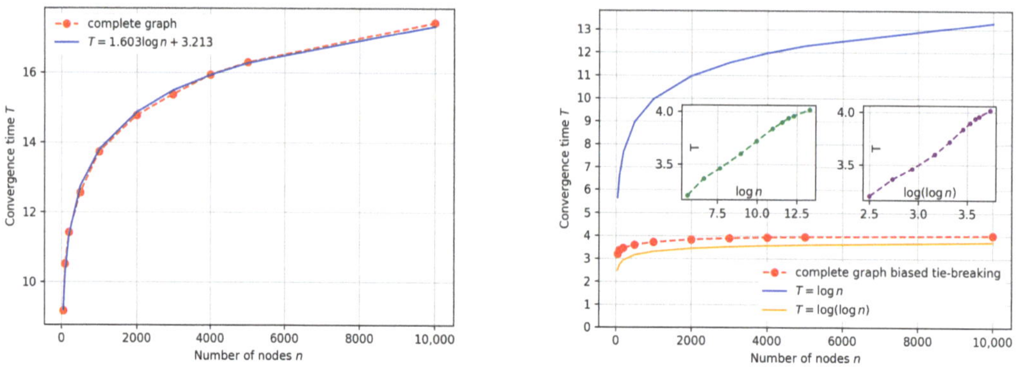

(a) (b)

Figure 4. Convergence time in complete graph with (**a**) random tie-breaking rule and (**b**) biased tie-breaking rule.

We conducted another experiment on those 10 complete graphs with the biased tie-breaking rule. The maximum size of a group was increased from 3 to 4. Figure 4b shows the result of this experiment. There are two insets in Figure 4b, which illustrate the plots of the convergence time, T, against $\log n$ and $\log(\log n)$. The insets show that the convergence time is a linear function of both the logarithm and double logarithm of n. The reason is that $\log(\log n)$ is a linear function of $\log n$ when n is not big enough. Thus, experiments should be executed for n that is large enough to determine the order of the convergence time. However, we cannot run the experiments for so large n due to the limitation of computing power. In such a case, we plot the curves of the convergence time T against the number of nodes n, $T = \log(\log n)$ and $T = \log n$ in Figure 4b to make a direct comparison. It can be observed that the curve of T against n is much closer to the curve of $T = \log(\log n)$. It indicates that the convergence time of the GGM on a complete graph looks to be rather of order $\mathcal{O}(\log\log n)$ if the biased tie-breaking rule is followed and the maximum group size is increased and is an even number.

2.3. Some Other Special Graphs

The outcomes of the experiments on the cycle and complete graphs from the previous sections show that the GGM converges much faster on a complete graph than on a cycle graph if they have the same n. For an n-node graph, the complete graph has $\binom{n}{2} = n(n-1)/2$ edges, but a cycle graph only has n edges. One may expact that adding more edges to a graph can increase the convergence rate.

An experiment was executed on two-cycle graphs (Figure 5a) and random cycle graphs (Figure 5b). A two-cycle graph can be generated by taking a cycle graph and adding an edge between every two nodes with a distance of 2. To build a random cycle graph, two randomly selected extra edges should be added to each node in a cycle graph.

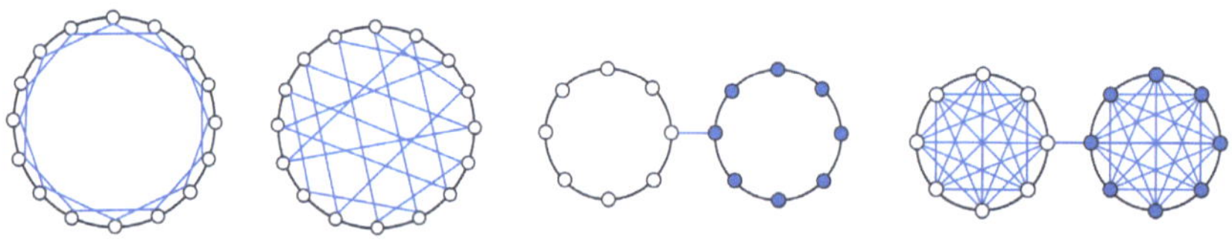

(a) (b) (c) (d)

Figure 5. (**a**) Two-cycle, (**b**) random cycle, (**c**) two-cycle-connected and (**d**) two-complete-connected graphs.

Figure 6a,b depicts the convergence time of the GGM on two-cycle graphs and random cycle graphs, respectively. It can be observed that the process converges faster on two-cycle graphs than on a cycle graph, but the convergence time is still of order $\mathcal{O}(n^2)$. However, the convergence time is of order $\mathcal{O}(n)$ on the random cycle graph. With the same n and m (number of edges), a two-cycle graph converges much slower than a random cycle graph. Therefore, the increase in the convergence speed is not merely the result of adding extra edges, but rather how to add these edges.

It should also be noted that adding extra edges improperly can even slow down the convergence of the process. Figure 5c,d illustrates the two-cycle-connected (TCC) graph and the two-complete-connected (TKC) graph, respectively. These types of graphs can be built by connecting two cycle or complete graphs of the same size with one edge. It is obvious that the number of edges in a TKC graph is greater than that in a TCC graph if they have the same n. We run experiments on these two types of graphs. The outcome shows that the convergence time of the GGM on a TCC graph is $\mathcal{O}(n^2)$ (Figure 6c). However, the TKC graph has a convergence time of $\mathcal{O}(2^{n/2})$ (Figure 6d) even though it has more edges than a TCC graph with the same n. It takes much longer for a TKC graph to converge (i.e., more than 200,000 rounds for only 30 nodes).

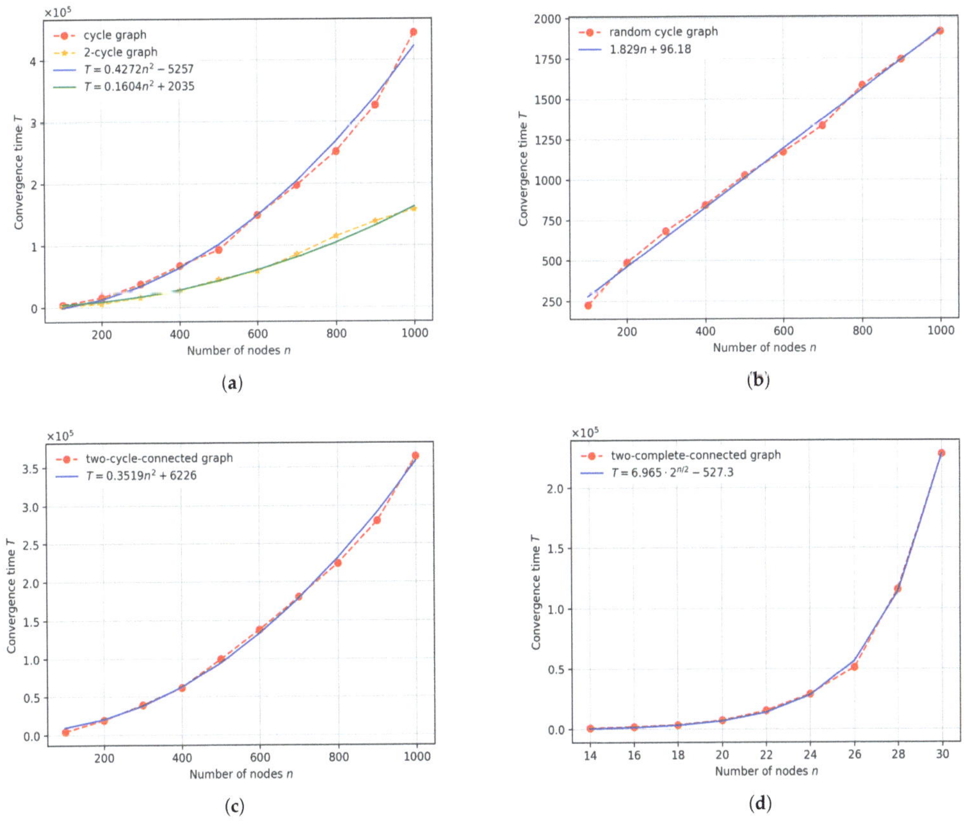

Figure 6. Convergence time in (**a**) two-cycle, (**b**) random cycle, (**c**) two-cycle-connected, and (**d**) two-complete-connected graphs.

Considering the expansion property of TCC and TKC graphs, both of them have an edge expansion of $2/n$ and a node expansion of $2/n$. It means these two types of graphs have a poor connectivity (removing only one edge or one node can make them become disconnected). The spectral expansion of a TCC graph with 50 nodes is only 2.236 and it is much smaller than the spectral expansion of a 50-node TKC graph, which is 23.962. Therefore, the TCC graph has better expansion properties than the TKC graph.

Furthermore, in the case of a two-cycle and random cycle graph, while they have the same number of nodes and edges, the random cycle graph has much stronger expansion properties since the randomly added edges significantly improve the connectivity of the graph. Strong expansion and connectivity properties can increase the flow of information in the network. Consequently, the process is much faster on a random cycle graph than on a two-cycle graph, as observed in our experiments.

From the results of these experiments, it can be observed that the expansion property has an impact on the convergence time. For instance, a complete graph or random cycle graph is a good expander and the process in such a graph converges fast. One can expect that the convergence time of the GGM on a graph with strong expansion properties to be less.

3. Influential Nodes

As discussed in Section 2, the original Galam model can be modeled by a complete graph. Since there is no difference between each node in a complete graph, only the number of initial blue nodes influences the total number of blue nodes at the end. Thus, the initial blue nodes are randomly selected. Experiments were conducted for the GGM on a 1000-node complete graph with different initial ratios of blue nodes selected in the set $\{0.1, 0.2, 0.3, 0.4, 0.45, 0.5, 0.55, 0.6, 0.7, 0.8, \text{ and } 0.9\}$. Experiments were executed 1000 times for each initial ratio of blue nodes. When the process converges to an absorbing state, the average final ratio of blue nodes is checked. Figure 7 shows that the GGM converges to the absorbing state that all nodes are blue (white) eventually if the initial ratio of blue nodes is just over (under) half. This looks to be true in any graph, which possesses some level symmetry (such as vertex-transitive graphs).

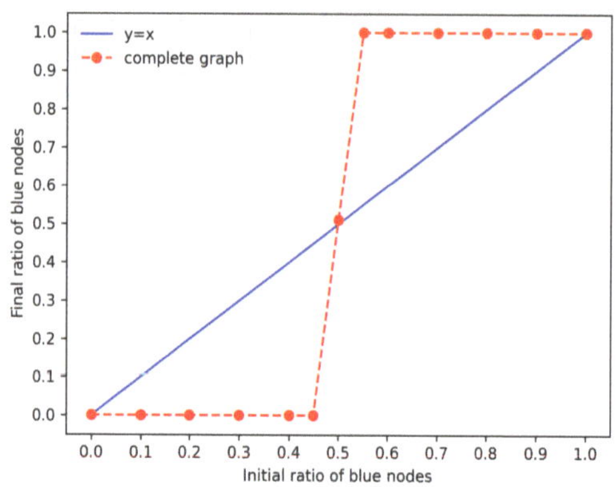

Figure 7. Randomly select initial blue nodes on a 1000-node complete graph.

However, most real-world social networks cannot be modeled as a complete graph. For instance, some individuals may have more connections to a social network than others. Such individuals are modeled as nodes with a higher degree. Selecting these nodes as initial blue nodes may lead to different results compared to selecting nodes with a lower degree. Therefore, a strategy should be developed to smartly choose the initial blue nodes rather than just randomly choosing them. One of the most often used approach is to use different centrality based measures.

3.1. Centrality-Based Influential Nodes

The centrality of a node can be used to identify the "importance" of the node [34]. There are many definitions of centrality, and the degree, betweenness, and closeness centrality are considered in the present paper.

Degree. As defined in Section 1.1.1, it is the number of neighbors of a node:

$$d(v) := |N(v)|.$$

Betweenness. If a node lies on as many shortest paths between other nodes in a graph, it is more probable that this node is placed in the center of the graph. Based on this idea, the betweenness centrality is defined as

$$b(v) := \sum_{s \neq v \neq t} \frac{\sigma_{st}(v)}{\sigma_{st}}.$$

where σ_{st} is the total number of shortest paths from node s to node t, and $\sigma_{st}(v)$ is the number of those paths passing through the node v.

Closeness. If the total distance between a node and all other nodes is short, then the node is close to all other nodes. The closer a node to all other nodes, the more central the node is. The closeness centrality comes from this idea and it is defined as

$$c(v) := \frac{n-1}{\sum_u d(u,v)}.$$

where $d(u,v)$ is the distance (shortest path) between node u and node v.

To compare the difference between randomly selecting initial blue nodes and choosing initial blue nodes based on centrality, we created three lists for the graph on which our model ran. Each list stores all the nodes of the graph. Then, these lists were sorted in descending order of the degree, betweenness, and closeness centrality of the nodes, respectively. When setting the initial blue nodes, we choose the first $\lceil \alpha n \rceil$ (α is the ratio initial of blue nodes) nodes in the list as blue nodes.

We ran our model on the T-ES and FB social networks. Four experiments were designed on each social network. One experiment is for randomly selecting the initial blue nodes and the other three are for selecting the initial blue nodes according to the three types of centrality. It should be mentioned that there is often a time limit for a product to take over a market in the real world. We assumed that each person takes part in three discussions every day on average, and the time limit is 100 days. Therefore, we ran the GGM 300 rounds in each experiment instead of waiting for it to converge. Note that running the process until the final convergence can also be quite expensive computationally for these networks.

Figure 8 shows the outcomes of these experiments. When the initial blue nodes are randomly selected, the outcomes of the experiments on these two real-world social networks are quite similar to that on the complete graph, where 0.5 is a transition point. The final ratio is more than 0.9 (less than 0.1) and tends to be 1 (0) when the initial ratio is greater (less) than 0.5. On the other hand, if the initial blue nodes are selected according to the centrality-based measures, then even a smaller fraction of initial blue nodes can result in a final blue ratio that is greater than 0.5.

For the T-ES social network, there is not much difference between choosing the initial blue nodes based on degree, betweenness, and closeness centrality. When the initial ratio is equal to or greater than 0.4, the final ratio is close to 1. For the FB social network, the difference between degree-based and betweenness-based strategies is not big. However, the initial and final ratios of blue nodes are almost the same when the initial blue nodes are selected according to their closeness. In fact, the number of blue nodes just oscillated around the initial number during the experiment. This may be attributed to the structure of FB's social network.

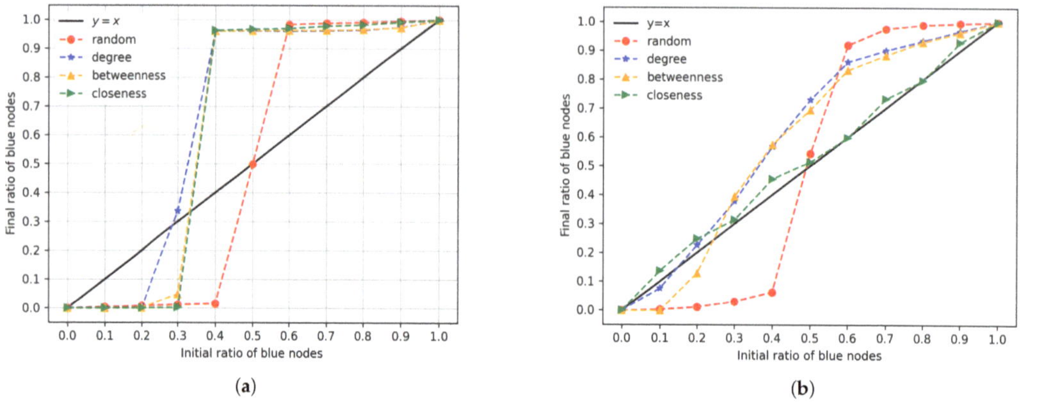

Figure 8. Centrality-based influential nodes selection on (**a**) Twitch Spain's and (**b**) Facebook's social networks.

3.2. Personality-Based Influential Nodes

In addition to considering how "central" individuals in a social network are, one also needs to take into account their personality when picking up the initial blue nodes. For instance, some people are more active and they spend more time on social communication. These people are more expectedly to influence their connections than those who communicate little with others. In addition, some individuals are inflexible and they firmly stick to their vision. To maximize the final ratio of blue nodes, it might be a reasonable way to choose the initial blue nodes that are more active and inflexible.

In Refs. [8,9], the "inflexible agent" and "activeness" have been introduced. In the present paper, we re-introduce these two parameters, called "activeness" and "stubbornness", for each node in a graph and define them in a slightly different way (which we believe is more self-explaining). Each node is assigned a value between 0 and 1 for its activeness and a value between 0 and 1 for its stubbornness. In our experiments, two random numbers were generated from the uniform distribution (0, 1) independently and they were set as activeness and stubbornness, respectively. For a node, the greater the activeness (stubbornness) is, the more active (inflexible) is the node. More precisely, a node v with activeness $act(v)$ decides to participate in the opinion exchange process with probability $act(v)$. Furthermore, for stubbornness $stub(v)$, it decides not to follow the majority updating rule and keep its opinion unchanged.

The GGM needs to be modified after these two parameters are introduced. Before each round of the model, a random number following uniform distribution (0, 1) to be generated for each node in the graph. If the random number is greater than the activeness of a node, then the node to be removed from the graph in this round. When each node is updating their colors, a random number following uniform distribution (0, 1) is generated for each group. If the stubbornness of a node in the group is greater than the random number, this node keeps its color, even though its color is not the majority in the group.

Before experiments on the modified model were executed, three lists of all nodes were created. These three lists are sorted in descending order of the degree, activeness, and stubbornness, respectively. The first $\lceil \alpha n \rceil$ was selected as the initial blue nodes.

The results are illustrated in Figure 9. It can be observed that selecting the initial blue nodes according to the stubbornness of nodes always maximizes the final ratio of blue nodes in T-ES social networks. In FB social network, the stubbornness-based strategy maximizes the final ratio of blue nodes when the initial ratio is greater than or equal to 0.4.

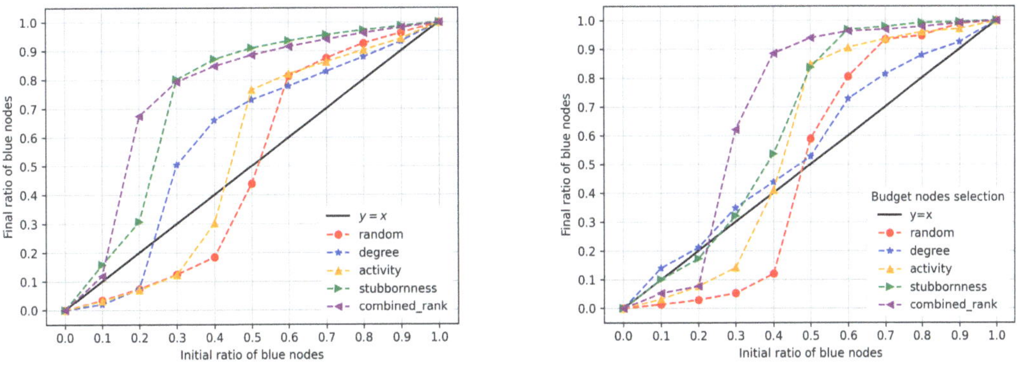

(**a**) (**b**)

Figure 9. Personality-based influential node selection on (**a**) Twitch Spain's and (**b**) Facebook's social networks.

The results indicate that the stubbornness is the dominant parameter in this model. But, what about selecting the initial nodes based on a rank that combines the centrality and personality of a node? A new parameter called combined rank is introduced, considering $d(v)$, $act(v)$, and $stub(v)$. As stubbornness has more of an impact on the final ratio of blue nodes than the other two factors, more weight are added to the $stub(b)$ in the combined rank. Thus, the combined rank $r(v)$ is defined as

$$r(v) = \sqrt{d(v)} \cdot \sqrt{act(v)} \cdot (stub(v))^3.$$

We ran another experiment in which the initial blue nodes were picked up based on the combined rank. The outcomes are the purple line in Figure 9a,b. It can be observed that only 20% of the initial blue nodes will result in about 70% of blue nodes at the end in the T-ES social network. For FB social network, 30% of the initial blue nodes result in about 60% of blue nodes.

4. Conclusions

In the current paper, we generalized the classic Galam opinion formation model by using a graph structure to represent a social network. Conducting several experiments on some special types of graphs, we showed that the convergence time of the GGM on the cycle and complete graphs are in $\mathcal{O}(n^2)$ and $\mathcal{O}(\log n)$, respectively, and it is probable that the expansion of a graph can influence the convergence time of this model. The model on a graph with strong expansion properties (e.g., any "small" subset of vertices of the graph has a great amount of connections with the complement of the subset, or the second-largest eigenvalue of the adjacency matrix of the graph is small) is expected to converge faster. Finding an explicit relationship between the expansion of a graph and the convergence time of GGM could be an interesting problem to tackle in the future.

Furthermore, experiments on real-world social networks indicate that selecting the initial blue nodes based on their centrality properties and personalities can lead to more final blue nodes compared with choosing the initial blue nodes randomly. It should be noted that the number of blue nodes just oscillated around the initial number when we select the initial blue nodes according to the closeness centrality on FB's social network. What types of graph structure can lead to such an outcome? Why does the structure not affect the outcomes of degree-based and betweenness-based strategies? This is left for future studies. In addition, selecting the initial blue nodes after considering all factors (using combined rank) can obtain even better results. A further study of such hybrid strategies can be a potential avenue for future research.

Author Contributions: Conceptualization, A.N.Z.; methodology, A.N.Z.; software, S.L.; formal analysis, S.L.; investigation, S.L. and A.N.Z.; resources, A.N.Z.; data curation, S.L.; writing—original draft preparation, S.L.; writing—review and editing, A.N.Z.; visualization: S.L.; supervision, A.N.Z.; project administration, A.N.Z. All authors have read and agreed to the published version of the manuscript.

Funding: This research received no external funding.

Data Availability Statement: The data supporting this study are available on request.

Acknowledgments: We express our deepest appreciation to Serge Galam for his invaluable advice and feedback on this paper. This paper would not have been possible without his guidance and support.

Conflicts of Interest: The authors declare no conflict of interests.

References

1. Yin, X.; Wang, H.; Yin, P.; Zhu, H. Agent-based opinion formation modeling in social network: A perspective of social psychology. *Phys. A Stat. Mech. Its Appl.* **2019**, *532*, 121786. [CrossRef]
2. Bredereck, R.; Elkind, E. Manipulating opinion diffusion in social networks. In Proceedings of the Twenty-Sixth International Joint Conference on Artificial Intelligence, Melbourne, Australia, 19–25 August 2017; Sierrs, C., Ed.; International Joint Conference on Artificial Intelligence (IJCAI)/Information Sciences Institute: Marina del Rey, CA, USA, 2017; pp. 894–900. [CrossRef]
3. Huang, P.-Y.; Liu, H.-Y.; Chen, C.-H.; Cheng, P. The impact of social diversity and dynamic influence propagation for identifying influencers in social networks. In Proceedings of the WI-IAT'13: 2013 IEEE/WIC/ACM International Joint Conferences on Web Intelligence (WI) and Intelligent Agent Technologies (IAT), Atlanta, GA, USA, 17–20 November 2013; IEEE Computer Society: NW Washington, DC, USA, 2013; Volume 1, pp. 410–416. [CrossRef]
4. Auletta, V.; Caragiannis, I.; Ferraioli, D.; Galdi, C.; Persiano, G. Minority becomes majority in social networks. In *Web and Internet Economics*: Proceedings of the 11th International Conference WINE 2015, Amsterdam, The Netherlands, 9–12 December 2015; Markakis, E., Schäfer, G., Eds.; Springer: Berlin/Heidelberg, Germany, 2015; pp. 74–88. [CrossRef]
5. Auletta, V.; Ferraioli, D.; Greco, G. Reasoning about consensus when opinions diffuse through majority dynamics. In Proceedings of the Twenty-Seventh International Joint Conference on Artificial Intelligence, Stockholm and Twenty-Third European Conference on Artificial Intelligence (IJCAI-ECAI 2018), Sweden, 13–19 July 2018; Lang, J., Ed.; International Joint Conference on Artificial Intelligence (IJCAI)/Information Sciences Institute: Marina del Rey CA, USA, 2018; pp. 49–55. [CrossRef]
6. Galam, S. Minority opinion spreading in random geometry. *Eur. Phys. J. B* **2002**, *25*, 403–406. [CrossRef]
7. Galam, S. Heterogeneous beliefs, segregation, and extremism in the making of public opinions. *Phys. Rev. E* **2005**, *71*, 046123. [CrossRef] [PubMed]
8. Galam, S.; Jacobos, F. The role of inflexible minorities in the breaking of democratic opinion dynamics. *Phys. A* **2007**, *381*, 366–376. [CrossRef]
9. Qian, S.; Liu, Y.; Galam, S. Activeness as a key to counter democratic balance. *Phys. A* **2015**, *432*, 187–196. [CrossRef]
10. Kempe, D.; Kleinberg, J.; Tardos, É. Maximizing the spread of influence through a social network. In KDD-2003: Proceedings of the Ninth ACM SIGKDD International Conference on Knowledge Discovery and Data Mining, Washington, DC, USA, 24–27 August 2003; Dominigos, P., Faloutsos, C., Senator, T., Kargupta, H., Getoor, L., Eds.; Association for Computing Machinery (ACM): New York, NY, USA, 2003; pp. 137–146. [CrossRef]
11. Talukder, A.; Alam, M.G.R.; Tran, N.H.; Niyato, D.; Park, G.H.; Hong, C.S. Threshold estimation models for linear threshold-based influential user mining in social networks. *IEEE Access* **2019**, *7*, 105441–105461. [CrossRef]
12. Zhuang, Z.; Wang, K.; Wang, J.; Zhang, H.; Wang, Z.; Gong, Z. Lifting majority to unanimity in opinion diffusion. In Proceedings of the ECAI 2020: 24th European Conference on Artificial Intelligence, Santiago de Compostela, Spain, 29 August–8 September 2020; De Giacomo, G., Catala, A., Dilkina, B., Milano, M., Barro, S., Bugarin, A., Lang, J., Eds.; IOS Press: Amsterdam, The Netherlands, 2020; pp. 259–266. [CrossRef]
13. Avin, C.; Lotker, Z.; Mizrachi, A.; Peleg, D. Majority vote and monopolies in social networks. In ICDCN'19: Proceedings of the 20th International Conference on Distributed Computing and Networking, Bangalore, India, 4–7 January 2019; Hansdah, R.C., Krishnaswamy, D., Vaidya, N., Eds.; Association for Computing Machinery (ACM): New York NY, USA, 2019; pp. 342–351. [CrossRef]
14. Amir, G.; Baldasso, R.; Beilin, N. Majority dynamics and the median process: Connections, convergence and some new conjectures. *Stoch. Process Their Appl.* **2023**, *155*, 437–458. [CrossRef]
15. Anagnostopoulos, A.; Becchetti, L.; Cruciani, E.; Pasquale, F.; Rizzo, S. Biased opinion dynamics: When the devil is in the detail. In Proceedings of the Twenty-Ninth International Joint Conference on Artificial Intelligence (IJCAI-20), Yokohama, Japan, 7–15 January 2021; Bessiere, C., Ed.; International Joint Conferences on Artifical Intelligence (IJCAI)/Information Sciences Institute: Marina del Rey CA, USA, 2021; pp. 53–59. [CrossRef]
16. Gekle, S.; Peliti, L.; Galam, S. Opinion dynamics in a three-choice system *Eur. Phys. J. B* **2005**, *45*, 569–575. [CrossRef]
17. Poljak, S.; Turzík, D. On pre-periods of discrete influence systems. *Discret. Appl. Math.* **1986**, *13*, 33–39. [CrossRef]

18. Zehmakan, A.N. Random majority opinion diffusion: Stabilization Time, absorbing states, and influential nodes. In Proceedings of the 22nd International Conference on Autonomous Agents and Multiagent Systems (AAMAS 2023), London, UK, 29 May 2023–2 June 2023; Ricci, A., Yeoh, W., Agmon, N., An, B., Eds.; International Foundation for Autonomous Agents and Multiagent Systems (IFAAMAS): Liverpool, UK, 2023; pp. 2179–2187. Available online: https://www.ifaamas.org/Proceedings/aamas2023/forms/contents.htm#6E (accessed on 7 November 2023).
19. Abdullah, M.A.; Draief, M. Global majority consensus by local majority polling on graphs of a given degree sequence. *Discret. Appl. Math.* **2020**. *180*, 1–10. [CrossRef]
20. Cruise, J.; Ganesh, A. Probabilistic consensus via polling and majority rules. *Queueing Syst.* **2014**, *78*, 99–120. [CrossRef]
21. Gärtner, B.; Zehmakan, A.N. Threshold behavior of democratic opinion dynamics. *J. Stat. Phys.* **2015**, *178*, 1442–1466. [CrossRef]
22. Auletta, V.; Ferraioli, D.; Grece, G. On the effectiveness of social proof recommendations in markets with multiple products. In Proceedings of the ECAI 2020: 24th European Conference on Artificial Intelligence, Santiago de Compostela, Spain, 29 August–8 September 2020; De Giacomo, G., Catala, A., Dilkina, B., Milano, M., Barro, S., Bugarin, A, Lang, J., Eds.; IOS Press: Amsterdam, The Netherlands, 2020; pp. 19–26. [CrossRef]
23. Karia, N.; Mallick, F.; Dey, P. How hard is safe bribery? In Proceedings of the 21st International Conference on Autonomous Agents and Multiagent Systems (AAMAS 2022), Online, 9–13 May 2022; Faliszewski, P., Mascardi, V., Pelachaud, C., Taylor, M.E., Eds.; International Foundation for Autonomous Agents and Multiagent Systems (IFAAMAS): Liverpool, UK, 2022; pp. 714–722. Available online: https://www.ifaamas.org/Proceedings/aamas2022/forms/contents.htm#1 (accessed on 7 November 2023).
24. Schoenebeck, G.; Tao, B.; Yu, F.-Y. Limitations of greed: Influence maximization in undirected networks re-visited. In Proceedings of the 19th International Conference on Autonomous Agents and Multiagent Systems (AAMAS 2020), Auckland, New Zealand, 9–13 May 2020; An, B., Yorke-Smith, N., El Fallah Seghrouchni, A., Sukthankar, G., Eds.; International Foundation for Autonomous Agents and Multiagent Systems (IFAAMAS): Liverpool, UK, 2020; pp. 1224–1232. Available online: https://www.ifaamas.org/Proceedings/aamas2020/forms/contents.htm#RTP (accessed on 7 November 2023).
25. Kundu, S.; Murthy, C.A.; Pal, S.K. A new centrality measure for influence maximization in social networks. In *Pattern Recognition and Machine Intelligence:* Proceedings of the 4th International Conference PReMI 2011, Moscow, Russia, 27 June–1 July 2011; Kuznetsov, S.O., Mandal, D.P., Kundu, M.K., Pal, S.K., Eds.; Springer: Berlin/Heidelberg, Germany, 2011; pp. 242–247. [CrossRef]
26. Chen, W.; Wang, Y.; Yang, S. Efficient influence maximization in social networks. In Proceedings of the 15th ACM SIGKDD International Conference on Knowledge Discovery and Data Mining (KDD'09), Paris, France, 28 June–1 July 2009; Elder, J, Soulié Fogelman, F., Flach, P., Zaki, M., Eds.; Association for Computing Machinery (ACM): New York, NY, USA, 2009; pp. 199–208. [CrossRef]
27. Leskovec, J.; Krause, A.; Guestrin, C.; Faloutsos, C.; VanBriesen, J.; Glance, N. Cost-effective outbreak detection in networks. In Proceedings of the 13th ACM SIGKDD International Conference on Knowledge Discovery and Data Mining KDD-2007), San Jose, CA, USA, 12–15 August 2007; Berkhin, P., Caruana, R., Wu, X., Gaffney, S., Eds.; Association for Computing Machinery (ACM): New York, NY, USA, 2007; pp. 420–429. [CrossRef]
28. Goyal, A.; Lu, W.; Lakshmanan, L.V.S. CELF++: Optimizing the greedy algorithm for influence maximization in social networks. In WWW'11: Proceedings of the 20th International Conference Companion on World Wide Web, Hyderabad, India, 28 March 2011–1 April 2011; Sadagopan, S., Ramamritham, K., Kumar, A., Ravindra, M.P., Bertino, E., Kumar, R., Eds.; Association for Computing Machinery (ACM): New York, NY, USA, 2011; pp. 47–48. [CrossRef]
29. Wu, H.; Liu, W.; Yue, K.; Huang, W.; Yang, K. Maximizing the spread of competitive influence in a social network oriented to viral marketing. In *Web-Age Information Management: Proceedings of 16th International Conference WAIM 2015, Qingdao, China, 8–10 June 2015;* Dong, X., Yu, X., Li, J., Sun, Y., Eds.; Springer: Cham, Switzerland, 2015; pp. 516–519. [CrossRef]
30. Kamarthi, H.; Vijayan, P.; Wilder, B.; Ravindran, B.; Tambe, M. Influence maximization in unknown social networks: Learning policies for effective graph sampling. In Proceedings of the 19th International Conference on Autonomous Agents and MultiAgent Systems (AAMAS 2020), Auckland, New Zealand, 9–13 May 2020; An, B., Yorke-Smith, N., El Fallah Seghrouchni, A., Sukthankar, G., Eds.; International Foundation for Autonomous Agents and Multiagent Systems (IFAAMAS): Liverpool, UK, 2020; pp. 575–583. Available online: https://www.ifaamas.org/Proceedings/aamas2020/forms/contents.htm#RTP (accessed on 7 November 2023).
31. Zhao, G.; Jia, P.; Huang, C.; Zhou, A.; Fang, Y. A machine learning based framework for identifying influential nodes in complex networks. *IEEE Access* **2020**, *8*, 65462–65471. [CrossRef]
32. Li, Y.; Gao, H.; Gao, Y.; Guo, J.; Wu, W. A survey on influence maximization: From an ML-based combinatorial optimization. *ACM Trans. Knowl. Discov. Data* **2023**, *17*, 1–50. [CrossRef]
33. Leskovec, J.; SNAP: Standford Large Network Dataset Collection. Available online: http://snap.stanford.edu/data (accessed on 7 November 2023).
34. Newman, M.E.J. *Networks: An Introduction*; Oxford University Press: Oxford, UK, 2018; Ch. 7. [CrossRef]

physics

MDPI

Article

Phase Diagram for Social Impact Theory in Initially Fully Differentiated Society

Krzysztof Malarz [1,*] and Tomasz Masłyk [2]

[1] Faculty of Physics and Applied Computer Science, AGH University of Krakow, al. Mickiewicza 30, 30-059 Kraków, Poland
[2] Faculty of Humanities, AGH University of Krakow, al. Mickiewicza 30, 30-059 Kraków, Poland
* Correspondence: malarz@agh.edu.pl

Abstract: The study of opinion formation and dynamics is one of the core topics in sociophysics. In this paper, the results of computer simulation of opinion dynamics based on social impact theory are presented. The simulations are based on Latané theory in its computerised version proposed by Nowak, Szamrej and Latané. The active parameters of the model describe the volatility of the actors (social temperature T) and the effective range of interaction (governed by an exponent α in a scaling function of distance between actors). Initially, every actor i has his/her own opinion. Our results indicate that ultimately at least 90% of the initial opinions available are removed from the society. For a low social temperature and a long range of interaction, only one opinion survives. Also, a rough sketch of the system phase diagram is presented. It indicates a set of (α, T) leading either to (1) the dominance of the unanimity of the opinions or (2) mixtures of unanimity and polarisation, or (3) taking random opinions by actors, or (4) a mixture of the final fates of the systems. The drastic reduction of finally observed opinions vs. their initial variety may be generic for many sociophysical models of opinions formation but masked by assuming an initially small pool of available opinions (in the worst case, in models with only binary opinions).

Keywords: sociophysics; social impact; opinion dynamics; social temperature; clustering and polarisation

Citation: Malarz, K.; Masłyk, T. Phase Diagram for Social Impact Theory in Initially Fully Differentiated Society. *Physics* **2023**, *5*, 1031–1047. https://doi.org/10.3390/physics5040067

Received: 30 June 2023
Revised: 2 October 2023
Accepted: 18 October 2023
Published: 27 October 2023

1. Introduction

The studies of opinion formation and dynamics are one of the core topics in sociophysics [1–6]. For example, Galam models of opinion dynamics [7–20] are based on the reaction–diffusion model: the dynamics operates via local update rules and reshuffling. In these models, three kinds of actors correspond to floaters, contrarians, and inflexibles. The models assume two or three opinions available in the society [21]. Among other discrete models of opinion formation, one should mention the majority rule [8,10,22,23], voter [24–27], and Sznajd [28–32] models. In these models, usually only binary opinions are considered, which naturally causes society polarisation. However, modifications that allow for multiple opinions were also studied [33–45].

The Nowak–Szamrej–Latané model [46] is based on the Latané social impact theory [47–49]. Latané himself defined his theory as a "bulb theory" of social impact. According to this physical analogy, every actor plays simultaneously the role of an isotropic single wavelength light emitter and a multi-wavelength light detector. We assume that every actor can emit and detect easily distinguished K various light colours. Every discrete time step t actor i switches the emitted wave length (colour) $\lambda_i(t)$ to that perceptible illuminance is detected in his/her position as the strongest. The decision of which colour $\lambda_i(t+1)$ will be emitted by actor i depends on (i) the number of each colour sources, (ii) the distance from this point to every other source of light, (iii) and intensities (illuminance flux, "bulb" power) of each light source.

The earlier attempts to employ this model for sociophysical studies included: observing the influence of the strong leader on opinion formation [50]; studying the noise-induced order/disorder phase transition [35,51]; searching for self-organised criticality in opinion systems [36]; observing the disappearance of some opinions [37] (see Ref. [52] for review).

In this paper, with a computer simulation based on the Nowak–Szamrej–Latané model [46], we check how the initial diversity of opinions influences the possibility of reaching a unanimity of opinions. Namely, we build a phase diagram in the social temperature and the effective range of the interaction space based on the number of surviving opinions; numbers of clusters of opinions; and the probability distribution of the size of the largest cluster. Unlike previous works—where the number of available opinions was usually small (two [50,51], up to three [36], or up to five [35,37])—we proposed as a starting point a situation in which each actor has their own opinion.

2. Model Formalisation

Every actor i at time t has an opinion $\lambda_i(t)$. The social impact $\mathcal{I}_{i,k}(t)$ exerted in time t on an actor i by all actors who share opinions Λ_k is calculated as

$$\mathcal{I}_{i,k}(t) = \sum_{j=1}^{L^2} \frac{4s_j}{g(d_{i,j})} \cdot \delta(\Lambda_k, \lambda_j(t)) \cdot \delta(\lambda_j(t), \lambda_i(t)) \tag{1}$$

or

$$\mathcal{I}_{i,k}(t) = \sum_{j=1}^{L^2} \frac{4p_j}{g(d_{i,j})} \cdot \delta(\Lambda_k, \lambda_j(t)) \cdot [1 - \delta(\lambda_j(t), \lambda_i(t))], \tag{2}$$

where s_j is j-th actor supportiveness, p_j is j-th actor persuasiveness, $d_{i,j}$ stands for Euclidean distance between actors i and j, $g(\cdot)$ is an arbitrary distance scaling function, and Kronecker delta $\delta(x,y) = 0$ when $x \neq y$ and $\delta(x,y) = 1$ when $x = y$. The sum in Equations (1) and (2) reflects the increase in impact $\mathcal{I}_{i,k}$ by increasing the number of "bulbs" (point (i) of the model description in Section 1), the decrease in impact with the distance between "bulbs" (fraction denominator, point (ii) of the model description in Section 1), and the fraction nominator corresponds to the intensities of "bulbs" (point (iii) of the model description in Section 1). The terms with Kronecker's delta are equal to either zero or to one:

- Equation (1) applies to the calculation of the impact of actors currently sharing the opinion of actor i;
- while Equation (2) allows the calculation of the impact of all other opinions.

The parameters supportiveness s_i and persuasiveness p_i describe i-th actor intensity of interaction with actors sharing their opinions or with believers in opposite opinions, respectively. We decided to use $\forall i : p_i = s_i = 1/2$ (as in Ref. [35]) because whether it is easier to stick to our opinion or change it depends on numerous factors, such as the social context, emotions, beliefs, authorities, and persuasion strategies. People's decisions on this matter can vary widely and depend on individual circumstances and preferences. On the one hand, there are theories that explain the tendency to change opinions, such as: social influence theory and conformity [53]; motivation and belief theory [54]; authority influence theory [55] or persuasion theory [56]. On the other hand, there are theories that point to an advantage in trying to keep our opinions, such as cognitive consistency theory [57] or cognitive dissonance theory [58]. Furthermore, keeping the supportiveness and persuasiveness equal for each actor makes the initial variety of opinions (next to social temperature and range of interactions) the dominant factor in the results of our studies. With such assumption, the social impact (1) and (2) may be reduced to

$$\mathcal{I}_{i,k}(t) = \sum_{j=1}^{L^2} \frac{2 \cdot \delta(\Lambda_k, \lambda_j(t))}{g(d_{i,j})}. \tag{3}$$

To ensure a lower impact on the opinions of actors from a more distant neighbour, the distance scaling function $g(\cdot)$ must be an increasing function of its argument. Here, we assume that

$$g(x) = 1 + x^{\alpha}, \tag{4}$$

where the exponent α is a model control parameter while the first addition component ensures finite self-supportiveness $\mathcal{I}_{i,i}$. The parameter α qualitatively describes the effective range of interaction between the actors. Its quantitative meaning was delivered recently in Ref. [37] where it was shown that for $\alpha = 2$, about 25% of the impact comes from only nine nearest neighbours. This ratio increases to approximately 59%, 80% and 96% for $\alpha = 3, 4$ and 6, respectively. Calculating the relative impact exerted by 25 nearest neighbours gives about 39%, 76%, 92%, and 99% of the total social impact for $\alpha = 2, 3, 4$, and 6, respectively (see Ref. [37], Figure 2, Table 1). In Ref. [37], it is concluded that "the parameter α says how influential the nearest neighbours are with respect to the entire population: the larger α, the more influential the nearest neighbours are".

The example of calculating the social impact of nine actors and three colours is available as supporting information in Ref. [36].

In the deterministic version of the algorithm [37], the actor's opinion $\lambda_i(t+1) = \Lambda_k$, when $\mathcal{I}_{i,k}(t)$ has the maximum value among all impacts $\mathcal{I}_{i,j}(t)$ for $j = 1, \cdots, K$ (3). Following our previous studies [35–37], we employ the actors with "free will" by allowing them to avoid taking the opinions that believers exert the greatest impact on them. This scenario is realised in a probabilistic way by introducing a parameter T often termed "information noise" or "social temperature". Quoting Ref. [59]: "Using the statistical mechanical foundation, [$\cdots$] the most probable collective behaviour depends on a group's social temperature, a measure of the group's decision-making volatility. The extreme of zero temperature leads to stable, unchanging collective behaviour with pockets of minority and majority opinions. As group temperatures increase, the model's collective behaviour tends toward a uniform decision without clustering of minority opinions. When the social temperature exceeds a certain limit, the group will have a well defined average opinion, but individuals are no longer stable and vacillate in a nearly random manner between different possible opinions". Quoting Ref. [60] by the same authors: "Individuals are influenced by the group's temperature. When a group's social temperature is high, very little provocation is necessary to induce an individual to change opinion. At low group temperatures, individuals appear more phlegmatic or stubborn, and much greater provocation is required to induce a change in opinion. High social temperatures amplify the slightest excuse for change, whereas low temperatures diminish the arguments for change. Note that each individual's opinion strength is unaltered, but as the group's temperature changes, so does an individual's decision making abilities."

In our recent studies [36,37], we observed the exact same effects on the system evolution as described above, when social temperature T is introduced as a parameter in probabilities in time t of taking in the next time step opinion Λ_k by actor i based on Boltzmann-like factors,

$$p_{i,k}(t) = \begin{cases} 0 & \iff \mathcal{I}_{i,k} = 0, \\ \exp\left(\frac{\mathcal{I}_{i,k}(t)}{T}\right) & \iff \mathcal{I}_{i,k} > 0, \end{cases} \tag{5}$$

which yet requires proper normalisation,

$$P_{i,k}(t) = \frac{p_{i,k}(t)}{\sum_{j=1}^{K} p_{i,j}(t)}. \tag{6}$$

In contrast to our earlier attempt, due to a huge number of available opinions in the system, we reset to zero probability $p_{i,k}(t)$ when the impact from opinion Λ_k holders is zero, i.e., when believers of this opinion vanished.

Initially, every actor i has his/her opinion $\lambda_i(t = 0) = \Lambda_i = i$ among the $K = L^2$ opinions available in the system (see the Listing A1 in Appendix B). The time evolution for $L^2 = 21^2$ actors who decorate the nodes of the square network takes $\tau = 10^5$ time steps. The single step is completed when all L^2 actors attempt to change their opinions. The results are averaged over $R = 10^2$ independent simulations. The averaging procedure is marked by a $\langle \cdots \rangle$. The open boundary conditions are assumed.

To learn more about the spatial distribution of opinions, we detect, count, and measure the sizes of clusters of agents sharing the same opinion. To this end, we apply the Hoshen–Kopelman algorithm [61] (see also [62] (pp. 59–60), [63,64]) allowing the labelling of every actor in such a way that actors who share the same opinions in various clusters are labelled with various labels and actors belonging to a given cluster are labelled with the same label. The number of clusters and the size of the largest cluster at time $t = \tau$ are indicated as n_c and $S_{\max}$, respectively.

3. Results

In this Section, we present the results of computer simulation—based on a computer programme written in Fortran—for $\tau = 10^5$ and $R = 100$. Typically, simulation for this set of parameters τ and R and a single pair of (α, T) takes around 3 days of Central Processing Unit (CPU) time on the Dell Precision Rack 7920 Workstation with a 3.20 GHz CPU clock.

In Figure 1, the results of simulations concerning the average final (at $t = \tau$): number of opinions, $\langle n_o \rangle$ (Figure 1a), number of clusters $\langle n_c \rangle$ (Figure 1b), and their largest size, $\langle S_{\max} \rangle$ (Figure 1c), are presented. The values of $\langle n_o \rangle$ and $\langle n_c \rangle$ are normalised to the number K of opinions available in the system while $\langle S_{\max} \rangle$ is normalised to the system size L^2. Thus, these numbers are presented as percentages.

In Figure 2, examples of probability distribution function, F, for the size of the largest clusters $\langle S_{\max} \rangle$ are presented. Figure 2a shows the initial distribution (i.e., at $t = 0$, when $n_o = K$) of the largest cluster size, while Figure 2b–j show examples of the typical distribution obtained at the end of simulations, i.e., at $t = \tau$. In Figure 2b,c,h,i, the probability distribution of the largest cluster size for the "corners" of the parameter system plane (α, T)—see Figure 1—are presented. Furthermore, in Figure 2j, we present the function F for the limiting case $T \to \infty$, $\alpha \to \infty$.

In the system of opinions studied, independently of the control parameters of the model, at least about 90% initially available opinions are removed from the system. For low social temperature (small T) and effectively long range of interactions (small α), only a single opinion (when consensus on common opinion in society occurs) or two opinions survive (when system polarisation takes place). For high social temperature (large T) and effectively short range of interactions (large α), no clustering of opinions is observed (with their number reduced as mentioned above). Unlike many binary models, these effects are not embedded in the model rules themselves. With computer simulations of the opinion dynamics model, we have shown that successive disappearing of opinions are naturally associated with social impact theory, and the initial diversity of opinion vanishes in several time steps of system evolution. As mentioned in Section 2, the disappearing of any opinion in a given simulation is irreversible—this is not different from the sociological equivalent of Muller's ratchet [65] observed also in Eigen's quasi-species [66].

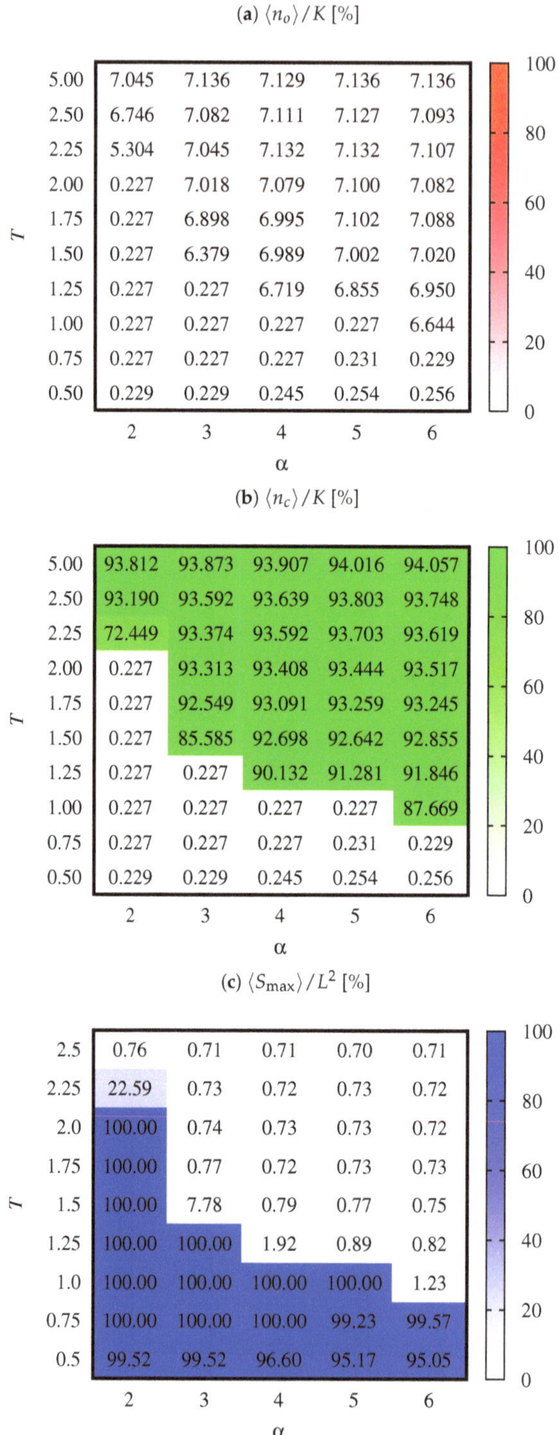

Figure 1. Ultimate (at time $t = 10^5$ steps) averaged (over $R = 10^2$ simulations) numbers (**a**) of observed opinions, $\langle n_o \rangle$, (**b**) clusters of opinions, $\langle n_c \rangle$, and (**c**) the largest cluster size, $\langle S_{\max} \rangle$. The data are normalised to the initial available number of opinions, K (**a**,**b**), and the system size, L^2 (**c**).

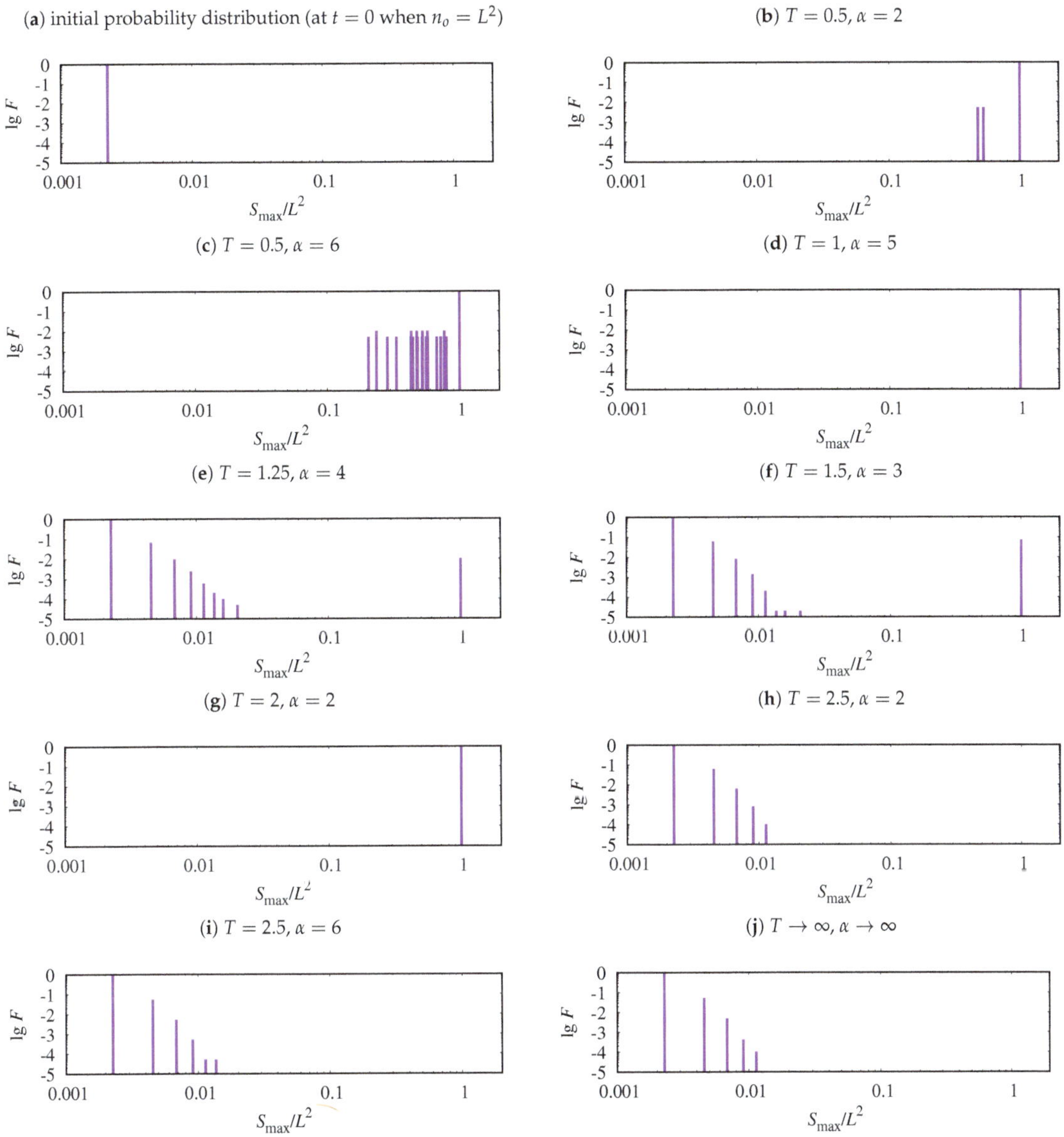

Figure 2. Examples of probability distribution function, F, of the size of the largest clusters $\langle S_{\max}\rangle$ for (**a**) initial configuration (at $t = 0$, $n_o = L^2$), (**b**) $T = 0.5$, $\alpha = 2$ (the lowest left corner of Figure 3), (**c**) $T = 0.5$, $\alpha = 6$ (the lowest right corner of Figure 3), (**d**) $T = 1$, $\alpha = 5$, (**e**) $T = 1.25$, $\alpha = 4$, (**f**) $T = 1.5$, $\alpha = 3$, (**g**) $T = 2$, $\alpha = 2$, (**h**) $T = 2.5$, $\alpha = 2$ (the highest left corner of Figure 3), (**i**) $T = 2.5$, $\alpha = 6$ (the highest right corner of Figure 3), (**j**) $T \to \infty$, $\alpha \to \infty$.

In Appendix A, we show six examples of the time evolution of the number n_o of opinions observed in the system. The presented results are for $T = 0.5$, $\alpha = 2$ (Figure A1a), $T = 2.5$, $\alpha = 6$ (Figure A1b), $T = 0.5$, $\alpha = 2$ (Figure A1c), and $T = 2.5$, $\alpha = 6$ (Figure A1d).

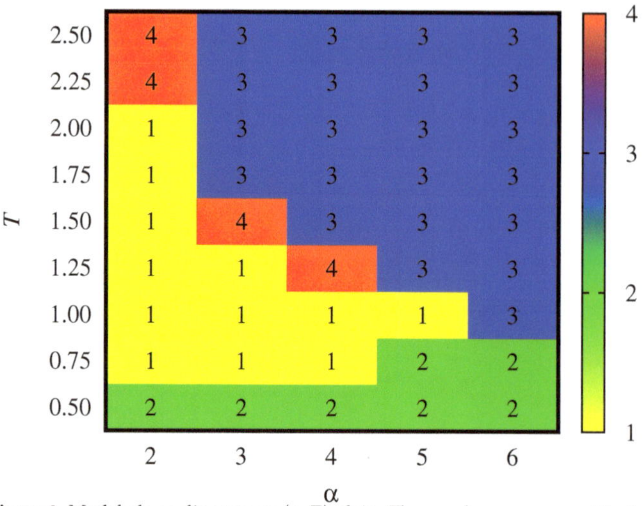

Figure 3. Model phase diagram on (α, T) plain. The numbers correspond to the list enumerator given in Section 4.

4. Discussion

In a single simulation, the number $n_o(t)$ is always a monotonically nonincreasing function of time (see Figure A1). The assumed simulation time $\tau = 10^5$ seems to be long enough to ensure reaching a plateau in the time evolution of n_o. Please note that vanishing during evolution any of the initially available opinion Λ_v, i.e., when at time t_v none of the actors share this opinion Λ_v, then for any $t \geq t_v$, this opinion Λ_v will not be restored. Here, we deal with the sociological equivalent of the famous Muller's ratchet [65] known in evolutionary genetics.

Changes in the number of observed opinions (surviving temporal evolution) $\langle n_o \rangle$ (Figure 1a) are accompanied by changes in the number of clusters $\langle n_c \rangle$ (Figure 1b) and their largest size $\langle S_{\max} \rangle$ (Figure 1c). These numbers bring complementary quantitative information on the system: for instance, for $\alpha = 2$ and $T = 0.5$, one simultaneously has $\langle n_o \rangle = 1$, $\langle n_c \rangle = 1$ and $\langle S_{\max} \rangle = L^2$—which are straightforward signatures of unanimity of opinion.

The analyses of the averages $\langle n_o \rangle$, $\langle n_c \rangle$, and $\langle S_{\max} \rangle$ presented in Figure 1 together with the analyses of the probability distribution function $P(S_{\max})$ shown in Figure 2 allow the identification of four possible phases observed in the system. These phases correspond to:

1. reaching unanimity of opinions ($\langle n_o \rangle = 1$, $\langle n_c \rangle = 1$, $\langle S_{\max} \rangle / L^2 = 1$, probability distribution function of the largest cluster size as in Figure 2d,g);
2. reaching unanimity of opinions or society polarisation (probability distribution function of the largest cluster size as in Figure 2b,c);
3. taking random opinions by actors (probability distribution function of the largest cluster size as in Figure 2h,i);
4. mixture of the phases mentioned above (probability distribution function of the largest cluster size as in Figure 2e,f).

These four scenarios, observed after system time evolution up to $\tau = 10^5$ time steps, can be mapped into (α, T) space to create the phase diagram of the computerised model of the Nowak–Szamrej–Latané social impact theory [46]. This diagram is presented in Figure 3 and the numbers there correspond to the list enumerator given above.

Looking for sociological theories that would explain the disappearance of some of the available opinions, one can refer to Nan Lin's hypothesis on the theory of social capital [67,68]. According to Lin's concept, opinions may be treated as a resource in a social network. The process by which resources in social networks become meaningful and valuable for members of these networks can be considered in relation to several principles ([67], pp. 30–33):

- The first principle has to do with consensus or influence developed or exercised within a group. Consent as to whether a resource is valuable or not can be achieved as a result of persuasion (communication and interaction without sanctions or penalties lead to the formation of an internal conviction in individuals as to the value of a given resource), request (appeals or lobbying result in recognising a given resource as valuable even when the individual does not understand its meaning but wants to remain a member of the group or identify with it), or coercion (an alternative to not recognising the value of a given resource is the threat of sanctions or penalties).
- The second mechanism that allows one to assign value to resources boils down to taking actions by all actors aimed at promoting their own interests by protecting or acquiring valuable resources. For example, it is in the community's well-understood interest to give a higher status to those who, in the opinion of its members (between whom consensus is reached), have valuable resources (knowledge, physical strength, knowledge of members of other communities, etc.). In this sense, the self-interest of individual members of the community becomes convergent with the collective interest (development, security, and cooperation). The devaluation of a given asset requires more than individual effort—it requires the consent of others who make similar demands.
- The third principle regarding valuable resources assumes that their maintenance and acquisition are the two basic motives of individuals' actions, although the former is more important than the latter. Only when the group's resources are secured can its members make an effort to acquire additional resources.

In the case of social resources, two types of mechanisms can be distinguished: network resources to which an individual has access by virtue of membership in that network and contact resources that an individual actually uses in the course of action. The first of them represents constantly available resources due to the durability of social relations in the network, the second represents resources that can be mobilised in order to achieve specific benefits. The nature of the resources contained in the social network to which an individual has access is determined by several factors. First, the range of resources in the network is important, that is, the "distance" between the most valuable and least valuable resource. Second, the most valuable resource available to an individual within the entire hierarchy of resources contained in the network is of importance. Third, the diversity or heterogeneity of resources in a social network plays an important role, and fourth, the composition of resources shaped by those of them that are average or the most typical composition is also significant [67] (p. 37), [68]. In the field of social sciences, Lin's theory is one of the most coherent and well-established theories of social capital. It deals with the exchange of resources in social networks. Like most sociological theories, it does not attempt to indicate how exchange occurs (in quantitative terms), but rather why it occurs. Therefore, it creates a context for understanding the complexity of interpersonal relationships in their social dimension.

The existence of these two mechanisms was confirmed by Luca Valori and coleeagues research [69], which used a large and detailed data set [69]. They have characterised the empirical properties of the large-scale distribution of individuals in multidimensional cultural space. By using simple models, they showed that ultrametricity has profound and nontrivial consequences on short- and long-term cultural dynamics. In the short term, they found the existence of a symmetry-breaking phase transition where collective behaviour arises out of purely local interactions. However, in the long term, the same ultrametric property suppresses cultural convergence by restricting it within disjoint domains, implying a strong sensitivity to the initial conditions. Thus, the apparent paradox of the coexistence of short-term collective social behaviour and long-term cultural diversity might have, as a simple and parsimonious explanation, the empirically observed hierarchical distribution of individuals in cultural space.

The character of actors' connections in social networks determines the availability of social resources and their size. If one treats information or opinion as a social network resource, then some members of the social network (opinion leaders) play a greater role in its propagation [70] than the sender of the message itself. In the first model (two-step flow), the key role is played by opinion leaders who mediate between the sender (mass media) and the rest of society. In this model, unlike the one-step or "hypodermic" model [71], individuals are not treated as atomised recipients of media influence.

Depending on whether the actors in social networks are similar or different from each other, the links between them can be bonding or bridging [72]. Ronald Burt [73,74] characterised two mechanisms of social contagion in the diffusion of innovation (opinions) in social networks depending on their structure: cohesion-induced contagion and equivalence-induced contagion. Cohesion-induced contagion occurs in cohesive networks between actors that maintain frequent and emphatic relationships. It is based on socialising communication. However, equivalence-induced contagion occurs in bridging networks as a result of competition between two actors who have similar relationships with other people. This applies to the competition of people who just use each other to evaluate their relative adequacy. Quoting Ref. [73] (p. 1291): "The more similar ego's and alter's relations with other persons are—that is, the more that alter could substitute for ego in ego's role relations, and so the more intense that ego's feelings of competition with alter are—the more likely it is that ego will quickly adopt any innovation perceived to make alter more attractive as the object or source of relations." Ultimately, a large number of bridges connecting diverse groups is essential for reducing opinion fractionalisation within societies [75]. A large number of bridges also has the effect of reducing distances between unconnected citizens [76].

5. Conclusions

In this paper, the Latané social impact theory is employed to build a model of opinion formation. With computer simulation, we investigate how the initial variety of opinions assigned to actors in such a way that initially every actor has his/her own opinion influences the final opinions number and their spatial distribution. The latter may be, to some extent, automatically checked (without direct analysis of snapshots from simulations) by means of techniques known from studies of site percolation phenomena.

As was pointed out in Refs. [36,37], a small noise dose (not too high a social temperature T) helps to reach consensus (not necessarily observed for the deterministic version of this model, i.e., for $T = 0$, cf. Figure 3 in Ref. [37] and Figure 7 in Ref. [36], where, however, the number of available opinions was restricted to several, namely $K = 3$ [36] and $K = 5$ [37]). This is well seen in Figure 1 and also in Figure 3 for $T = 1$ and $\alpha \le 5$.

Independently of the model control parameters, at least 90% of the initially available opinions are removed from the system. In some cases, only two opinions (when society polarisation occurs) or even a single opinion (when consensuses on a common opinion takes place) survive. As explained by Lin [68], there are various mechanisms that connect the individual to the group around shared resources [67,68]. The group provides the individual with a more effective way of pursuing their interests than if the individual were to act individually. In order to remain a member of the group, one must agree to a consensus on the value of the resources held by the group. This consensus also applies to opinions. In Burt's theory [73,74], opinion reduction is caused by the action of opinion leaders. Opinion leaders are the people whose conversations trigger contagion across the social boundaries between status groups. As a consequence of such actions, groups can become more similar in terms of opinions.

Finally, it would be interesting to investigate if the observed vanishing of opinions is generic, i.e., if it may also be observed in other discrete models of opinion formation. Further studies of the model may also include investigating the influence of the network topology on obtained results: the studies may either deal with regular lattices—triangular (six neighbours) or honeycomb (three neighbours)—or complex networks (including small world networks). The latter requires, however, a redefinition of the distance, $d_{i,j}$, from its Euclidean definition to the shortest paths between actors.

Author Contributions: Conceptualization, K.M. and T.M.; methodology, K.M.; software, K.M.; validation, K.M. and T.M.; formal analysis, K.M. and T.M.; investigation, K.M.; writing—original draft preparation, K.M.; writing—review and editing, K.M. and T.M.; visualization, K.M.; supervision, K.M. All authors have read and agreed to the published version of the manuscript.

Funding: This research received no external funding.

Data Availability Statement: The data generated by simulations are available from the authors upon a reasonable request.

Acknowledgments: K.M. is grateful to Serge Galam for his hospitality at Pierre and Maria Curie University in Paris in July 2004 within the framework of European Cooperation in Science and Technology (COST) Action "Physics of risk" and a fruitful discussion on application of percolation theory for social systems [77,78].

Conflicts of Interest: The authors declare no conflict of interest.

Appendix A. Examples of Time Evolution of the Observed Number of Opinions

In Figure A1 the time evolution of the observed number of opinions $n_o(t)$ for $T = 0.5$, $\alpha = 2$ (Figure A1a), $T = 0.5$, $\alpha = 6$ (Figure A1b), $T = 2.5$, $\alpha = 2$ (Figure A1c) and $T = 2.5$, $\alpha = 6$ (Figure A1d).

(**a**) $T = 0.5$, $\alpha = 2$

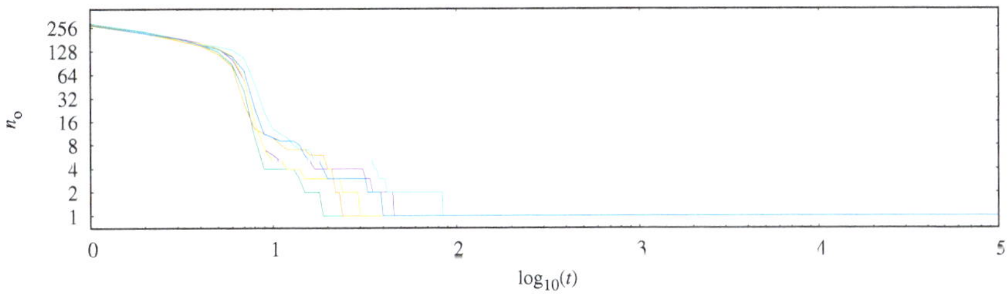

(**b**) $T = 0.5$, $\alpha = 6$

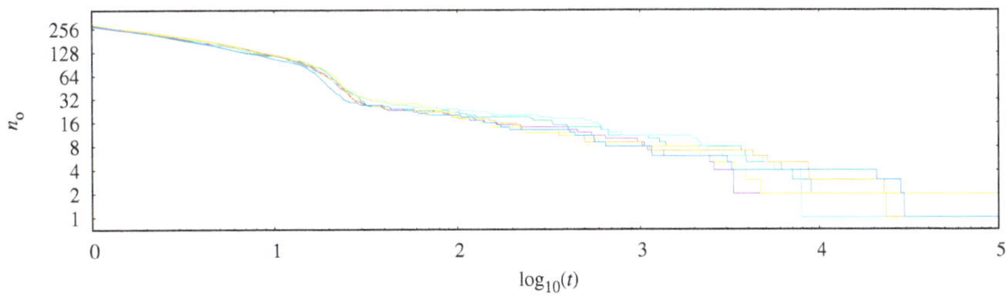

Figure A1. *Cont.*

(**c**) $T = 2.5, \alpha = 2$

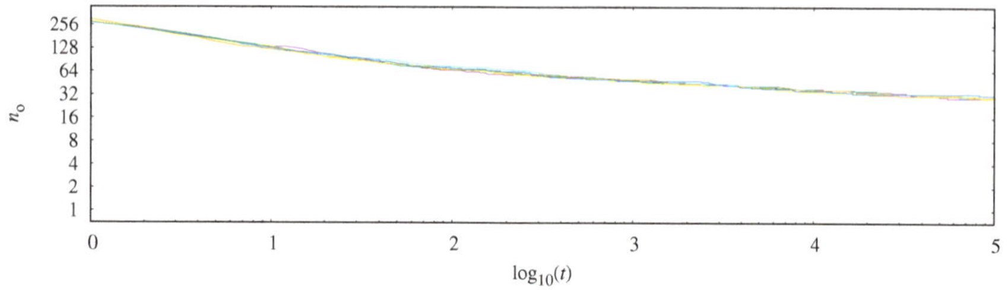

(**d**) $T = 2.5, \alpha = 6$

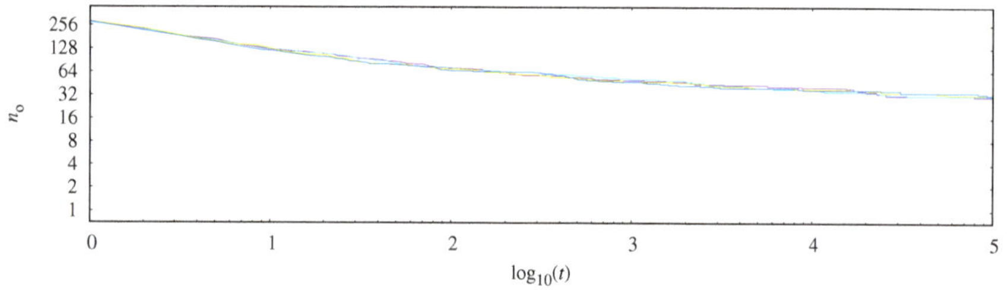

Figure A1. Time evolution of the observed number of opinions $n_o(t)$ for (**a**) $T = 0.5, \alpha = 2$, (**b**) $T = 0.5, \alpha = 6$, (**c**) $T = 2.5, \alpha = 2$ and (**d**) $T = 2.5, \alpha = 6$.

Appendix B. Examples of Spatial Opinion Distribution

An initial state of the system for $L = 21$ and $K = L^2$ is presented in Listing A1. In Listings A2–A7, examples of the final state of the system evolution for $L = 21$ after $\tau = 10^5$ time steps are presented. The numbers represent opinions. The examples are associated with four phases identified and presented in Figure 3. In Listing A2, the case of unanimity of opinions is presented. In Listings A3–A5, three variants of society polarization (with $n_o = 2$ and $n_c = 2$) are presented. In Listings A6 (with $n_o = 30$, $n_c = 399$) and A7 (with $n_o = 32$, $n_c = 415$), snapshots of the (still dynamical) state of the system are presented.

Listing A1. An initial state of the system for $L = 21$ and $K = L^2$. The numbers represent opinions. Every agent starts with his/her own opinion, which is different from the opinions of any other actor.

```
  1   2   3   4   5   6   7   8   9  10  11  12  13  14  15  16  17  18  19  20  21
 22  23  24  25  26  27  28  29  30  31  32  33  34  35  36  37  38  39  40  41  42
 43  44  45  46  47  48  49  50  51  52  53  54  55  56  57  58  59  60  61  62  63
 64  65  66  67  68  69  70  71  72  73  74  75  76  77  78  79  80  81  82  83  84
 85  86  87  88  89  90  91  92  93  94  95  96  97  98  99 100 101 102 103 104 105
106 107 108 109 110 111 112 113 114 115 116 117 118 119 120 121 122 123 124 125 126
127 128 129 130 131 132 133 134 135 136 137 138 139 140 141 142 143 144 145 146 147
148 149 150 151 152 153 154 155 156 157 158 159 160 161 162 163 164 165 166 167 168
169 170 171 172 173 174 175 176 177 178 179 180 181 182 183 184 185 186 187 188 189
190 191 192 193 194 195 196 197 198 199 200 201 202 203 204 205 206 207 208 209 210
211 212 213 214 215 216 217 218 219 220 221 222 223 224 225 226 227 228 229 230 231
232 233 234 235 236 237 238 239 240 241 242 243 244 245 246 247 248 249 250 251 252
253 254 255 256 257 258 259 260 261 262 263 264 265 266 267 268 269 270 271 272 273
274 275 276 277 278 279 280 281 282 283 284 285 286 287 288 289 290 291 292 293 294
295 296 297 298 299 300 301 302 303 304 305 306 307 308 309 310 311 312 313 314 315
316 317 318 319 320 321 322 323 324 325 326 327 328 329 330 331 332 333 334 335 336
337 338 339 340 341 342 343 344 345 346 347 348 349 350 351 352 353 354 355 356 357
358 359 360 361 362 363 364 365 366 367 368 369 370 371 372 373 374 375 376 377 378
379 380 381 382 383 384 385 386 387 388 389 390 391 392 393 394 395 396 397 398 399
400 401 402 403 404 405 406 407 408 409 410 411 412 413 414 415 416 417 418 419 420
421 422 423 424 425 426 427 428 429 430 431 432 433 434 435 436 437 438 439 440 441
```

Listing A2. $\alpha = 2$, $T = 0.5$, $n_o = 1$, $n_c = 1$, $S_{\max}/L^2 = 100\%$.

```
# irun=          10
# t=        100001, lambda:
284  284  284  284  284  284  284  284  284  284  284  284  284  284  284  284  284  284  284  284  284
284  284  284  284  284  284  284  284  284  284  284  284  284  284  284  284  284  284  284  284  284
284  284  284  284  284  284  284  284  284  284  284  284  284  284  284  284  284  284  284  284  284
284  284  284  284  284  284  284  284  284  284  284  284  284  284  284  284  284  284  284  284  284
284  284  284  284  284  284  284  284  284  284  284  284  284  284  284  284  284  284  284  284  284
284  284  284  284  284  284  284  284  284  284  284  284  284  284  284  284  284  284  284  284  284
284  284  284  284  284  284  284  284  284  284  284  284  284  284  284  284  284  284  284  284  284
284  284  284  284  284  284  284  284  284  284  284  284  284  284  284  284  284  284  284  284  284
284  284  284  284  284  284  284  284  284  284  284  284  284  284  284  284  284  284  284  284  284
284  284  284  284  284  284  284  284  284  284  284  284  284  284  284  284  284  284  284  284  284
284  284  284  284  284  284  284  284  284  284  284  284  284  284  284  284  284  284  284  284  284
284  284  284  284  284  284  284  284  284  284  284  284  284  284  284  284  284  284  284  284  284
284  284  284  284  284  284  284  284  284  284  284  284  284  284  284  284  284  284  284  284  284
284  284  284  284  284  284  284  284  284  284  284  284  284  284  284  284  284  284  284  284  284
284  284  284  284  284  284  284  284  284  284  284  284  284  284  284  284  284  284  284  284  284
284  284  284  284  284  284  284  284  284  284  284  284  284  284  284  284  284  284  284  284  284
284  284  284  284  284  284  284  284  284  284  284  284  284  284  284  284  284  284  284  284  284
284  284  284  284  284  284  284  284  284  284  284  284  284  284  284  284  284  284  284  284  284
284  284  284  284  284  284  284  284  284  284  284  284  284  284  284  284  284  284  284  284  284
284  284  284  284  284  284  284  284  284  284  284  284  284  284  284  284  284  284  284  284  284
284  284  284  284  284  284  284  284  284  284  284  284  284  284  284  284  284  284  284  284  284
# histogram of cluster sizes:
#          441            1
### Smax=          441
###   nc=            1
###   no=            1
```

Listing A3. $\alpha = 6$, $T = 0.5$, $n_o = 2$, $n_c = 2$, $S_{\max}/L^2 \approx 52\%$.

```
# irun=           2
# t=        100001, lambda:
 50   50   50   50   50   50   50   50   50   50   50  387  387  387  387  387  387  387  387  387  387
 50   50   50   50   50   50   50   50   50   50   50  387  387  387  387  387  387  387  387  387  387
 50   50   50   50   50   50   50   50   50   50  387  387  387  387  387  387  387  387  387  387  387
 50   50   50   50   50   50   50   50   50   50  387  387  387  387  387  387  387  387  387  387  387
 50   50   50   50   50   50   50   50   50   50  387  387  387  387  387  387  387  387  387  387  387
 50   50   50   50   50   50   50   50   50   50  387  387  387  387  387  387  387  387  387  387  387
 50   50   50   50   50   50   50   50   50   50  387  387  387  387  387  387  387  387  387  387  387
 50   50   50   50   50   50   50   50   50   50  387  387  387  387  387  387  387  387  387  387  387
 50   50   50   50   50   50   50   50   50   50  387  387  387  387  387  387  387  387  387  387  387
 50   50   50   50   50   50   50   50   50   50  387  387  387  387  387  387  387  387  387  387  387
 50   50   50   50   50   50   50   50   50   50  387  387  387  387  387  387  387  387  387  387  387
 50   50   50   50   50   50   50   50   50   50  387  387  387  387  387  387  387  387  387  387  387
 50   50   50   50   50   50   50   50   50   50  387  387  387  387  387  387  387  387  387  387  387
 50   50   50   50   50   50   50   50   50   50  387  387  387  387  387  387  387  387  387  387  387
 50   50   50   50   50   50   50   50   50   50  387  387  387  387  387  387  387  387  387  387  387
 50   50   50   50   50   50   50   50   50   50  387  387  387  387  387  387  387  387  387  387  387
 50   50   50   50   50   50   50   50   50   50  387  387  387  387  387  387  387  387  387  387  387
 50   50   50   50   50   50   50   50   50   50  387  387  387  387  387  387  387  387  387  387  387
 50   50   50   50   50   50   50   50   50   50  387  387  387  387  387  387  387  387  387  387  387
 50   50   50   50   50   50   50   50   50   50  387  387  387  387  387  387  387  387  387  387  387
 50   50   50   50   50   50   50   50   50   50  387  387  387  387  387  387  387  387  387  387  387
# histogram of cluster sizes:
#          212            1
#          229            1
### Smax=          229
###   nc=            2
###   no=            2
```

Listing A4. $\alpha = 6$, $T = 0.5$, $n_o = 2$, $n_c = 2$, $S_{\max}/L^2 \approx 52\%$.

```
# irun=          23
# t=      100001, lambda:
 40   40   40   40   40   40   40   40   40   40   40   40   40   40   40   40   40   40   40   40   40
 40   40   40   40   40   40   40   40   40   40   40   40   40   40   40   40   40   40   40   40   40
 40   40   40   40   40   40   40   40   40   40   40   40   40   40   40   40   40   40   40   40   40
 40   40   40   40   40   40   40   40   40   40   40   40   40   40   40   40   40   40   40   40   40
 40   40   40   40   40   40   40   40   40   40   40   40   40   40   40   40   40   40   40   40   40
 40   40   40   40   40   40   40   40   40   40   40   40   40   40   40   40   40   40   40   40   40
 40   40   40   40   40   40   40   40   40   40   40   40   40   40   40   40   40   40   40   40   40
 40   40   40   40   40   40   40   40   40   40   40   40   40   40   40   40   40   40   40   40   40
 40   40   40   40   40   40   40   40   40   40   40   40   40   40   40   40   40   40   40   40   40
 40   40   40   40   40   40   40   40   40   40   40   40   40   40   40   40   40   40   40   40   40
 40   40   40   40   40   40   40   40   40   40   40   40   40   40   40   40   40   40   40   40   40
120  120  120  120  120  120  120  120  120  120  120  120  120  120  120  120  120  120  120  120  120
120  120  120  120  120  120  120  120  120  120  120  120  120  120  120  120  120  120  120  120  120
120  120  120  120  120  120  120  120  120  120  120  120  120  120  120  120  120  120  120  120  120
120  120  120  120  120  120  120  120  120  120  120  120  120  120  120  120  120  120  120  120  120
120  120  120  120  120  120  120  120  120  120  120  120  120  120  120  120  120  120  120  120  120
120  120  120  120  120  120  120  120  120  120  120  120  120  120  120  120  120  120  120  120  120
120  120  120  120  120  120  120  120  120  120  120  120  120  120  120  120  120  120  120  120  120
120  120  120  120  120  120  120  120  120  120  120  120  120  120  120  120  120  120  120  120  120
120  120  120  120  120  120  120  120  120  120  120  120  120  120  120  120  120  120  120  120  120
120  120  120  120  120  120  120  120  120  120  120  120  120  120  120  120  120  120  120  120  120
# histogram of cluster sizes:
#         210          1
#         231          1
### Smax=          231
###   nc=            2
###   no=            2
```

Listing A5. $\alpha = 4$, $T = 0.5$, $n_o = 2$, $n_c = 2$, $S_{\max}/L^2 \approx 67\%$.

```
# irun=          92
# it=     100001, lambda:
 96   96   96   96   96   96   96   96   96   96   96   96   96   96    2    2    2    2    2    2    2
 96   96   96   96   96   96   96   96   96   96   96   96   96   96    2    2    2    2    2    2    2
 96   96   96   96   96   96   96   96   96   96   96   96   96   96    2    2    2    2    2    2    2
 96   96   96   96   96   96   96   96   96   96   96   96   96   96    2    2    2    2    2    2    2
 96   96   96   96   96   96   96   96   96   96   96   96   96   96    2    2    2    2    2    2    2
 96   96   96   96   96   96   96   96   96   96   96   96   96   96    2    2    2    2    2    2    2
 96   96   96   96   96   96   96   96   96   96   96   96   96   96    2    2    2    2    2    2    2
 96   96   96   96   96   96   96   96   96   96   96   96   96   96    2    2    2    2    2    2    2
 96   96   96   96   96   96   96   96   96   96   96   96   96   96    2    2    2    2    2    2    2
 96   96   96   96   96   96   96   96   96   96   96   96   96   96    2    2    2    2    2    2    2
 96   96   96   96   96   96   96   96   96   96   96   96   96   96    2    2    2    2    2    2    2
 96   96   96   96   96   96   96   96   96   96   96   96   96   96    2    2    2    2    2    2    2
 96   96   96   96   96   96   96   96   96   96   96   96   96   96    2    2    2    2    2    2    2
 96   96   96   96   96   96   96   96   96   96   96   96   96   96    2    2    2    2    2    2    2
 96   96   96   96   96   96   96   96   96   96   96   96   96   96    2    2    2    2    2    2    2
 96   96   96   96   96   96   96   96   96   96   96   96   96   96    2    2    2    2    2    2    2
 96   96   96   96   96   96   96   96   96   96   96   96   96   96    2    2    2    2    2    2    2
 96   96   96   96   96   96   96   96   96   96   96   96   96   96    2    2    2    2    2    2    2
 96   96   96   96   96   96   96   96   96   96   96   96   96   96    2    2    2    2    2    2    2
 96   96   96   96   96   96   96   96   96   96   96   96   96   96    2    2    2    2    2    2    2
 96   96   96   96   96   96   96   96   96   96   96   96   96   96    2    2    2    2    2    2    2
# histogram of cluster sizes:
#         147          1
#         294          1
### Smax=          294
###   nc=            2
###   no=            2
```

Listing A6. $\alpha = 3$, $T = 1.5$, $n_o = 30$, $n_c = 399$, $S_{\max} = 4$.

```
# irun=          23
# t=      100001, lambda:
245   40  126  419   51  440   90  245  142   22  274  103  441  245  419  274   22  187  406  147   91
245  194  187  441   40  142  280   22  419  280  406  142  316   40  441  126  194  187  194   51  334
441  316   22   90   97  147   40  280  169  103   43  245   91  280  194  194  194   43   90  316  187
147   90   90   43  126  226  142  194   40  419  440   91  173  185  185  419  419  406  211  327   43
 40   97   90  419  419   51  274  334   97  245   91  147  147   43  274  173  185  194  334  126  334
316   22  173  327  316  419  440  194   51  187  419  327   43  419  406  245   90   40  280  280  194
245   22  187  226  194  327  187  441   90   97  280  440  334  334  173   40  327  226  185  316   90
173  173  316  173  235  406  316  406  185   22  142  147   97  406  327   22  327   22  334  245   22
 43  327  327   43  126   43  173  147   91  406  274   40  441  103   22  126  245  316   43  419  440
280  441   40  173  440   90  440  316  327  406   97  406  185  294  406  294   90  316  173  406  280
406  441  280  440   40   91   40  187  294  235  316   51   22   51   22  142  419   22   22  142  103
 51   22  294   51  126  294  187  440  226  187  169  280  406  441  327  316  185   22  406   40   43
211  294  334   97   22  294  173   40   91   51  235   51  441  316  142  245  274  211  147  235   43
 40   97   40  226  327  327  294  274  226  334   43  294  327  235   40   40   22  142  194   51  169
440   22   22  334  211  440   40  334  245  126  147  316  187   91  280  280  316  441  211  245  441
173  142  294  142  103   91  126  245  173   51  280  211  187  173  441   40  274   97  440   51  103
226  406  194  185   51   51  274  173  147  419   40  185  103  211  194  406  334  211  274  274  441
 43   40  294  185  327   90  327  334  419  316  245  103  419  211  185   40  226  280  235  280  103
142  294  194  280  142  142  274  334   43   51  187  185  334  327  406  126  103  226  142  419  441
187  294  334  441  441  211   97  211  226  334  294  173  147  235  406   43  103  406  142   51  103
194   51  169  327  327  126  245   90   97   22  142  169  441  185  185  142  440  185  245  142  194
# histogram of cluster sizes:
#           1          361
#           2           35
#           3            2
#           4            1
### Smax=           4
###   nc=         399
###   no=          30
```

Listing A7. $\alpha = 6$, $T = 2.5$, $n_o = 32$, $n_c = 415$, $S_{\max} = 3$.

```
# irun=          10
# t=      100001, lambda:
121  191   91  372   52  120  421  191  424  238  421  361  327  128  330  330  198  294  238  193  286
343  421   52  424  233  193  424  273  330  233  198  421  330  327  286  304  416  286  360  294  416
361  134  276  204  204  330  284  361  198  198  343  327  330  258  273  330  273  120  343  191    9
128  204  204  233  286  360  154  238  193  317  327  121  120  424  284  197  361  120  421  121  284
330  343  294  330  191  121  372    9  361  284  421  198  233  360  191  330  330  204  121  154  317
317  304  258  258  134  273  421  233  424  361  286  286   52  273  238  304  360  424  193  317  238
317  233  360  121   91  204   91  284  284  121  284  276  154  154  273    9  327  128  120  360  421
120  416  286  233  330  198  372  284  273  121   52  294  258  372  193  191  372  360  258  330  238
360  294  121   91  233  286  361  276   52  121  317  154  286  286  134    9  121  193  361  198  273
198  360  198  121  421  276  120  233  424  416  193  424  330  330  343  424  198  372  284  304  198
191  286  416  330  361  258   52  128  121  421  424  121  134  233  284  421   91   52  304  372  304
258    9  193  360  258  327  276  204  286  154  273  198  197  204  154  134  317  424  193  284  128
317  421  154  372  330  193   52  238  304  294  191  361  198  360  204  273  198  424  191  276  191
372  258    9  424  154  204   91  317   52  128  372  284  361  343  154  343  327  286  360  424  197
191   91  197  276  121  197  294  193    9  154  304    9  304  330  317  233  191  204  154  327  233
284  421  330  154  317  416  294  330  361  343    9  372  204  421  421  360  424  360  421  416  343
198  134  128  372  317  304    9  343  154  154   52  128  421  343  421  233  286  424  121  327  258
421   52  121  191   52  360  284  421  204  360  191  360  154  258  360  276  294  421  134  360  238
120  198  294  304  294  294  128  286  317  361  294  421  361  286   52  424  330  317  294  424  286
304  134  204  191  360  276  128  154  121  372  416  204  330  317  198  154  233  121  284  120  134
128  191   91  361   52    9  120  193  276  286  197  360  421  286  198  120  421  134  416  361  286
# histogram of cluster sizes:
#           1          392
#           2           20
#           3            3
### Smax=           3
###   nc=         415
###   no=          32
```

References

1. Castellano, C.; Fortunato, S.; Loreto, V. Statistical physics of social dynamics. *Rev. Mod. Phys.* **2009**, *81*, 591–646. [CrossRef]
2. Stauffer, D. A biased review of sociophysics. *J. Stat. Phys.* **2013**, *151*, 9–20. [CrossRef]
3. Sen, P.; Chakrabarti, B.K. *Sociophysics: An Introduction*; Oxford Univeristy Press: Oxford, UK, 2014. Available online: https://archive.org/details/sociophysicsintr0000senp/ (accessed on 13 October 2023).
4. Schweitzer, F. Sociophysics. *Phys. Today* **2018**, *71*, 40–46. [CrossRef]
5. Sobkowicz, P. Social simulation models at the ethical crossroads. *Sci. Eng. Ethics* **2019**, *25*, 143–157. [CrossRef] [PubMed]
6. Jusup, M.; Holme, P.; Kanazawa, K.; Takayasu, M.; Romić, I.; Wang, Z.; Geček, S.; Lipić, T.; Podobnik, B.; Wang, L.; et al. Social physics. *Phys. Rep.* **2022**, *948*, 1–148. [CrossRef]
7. Galam, S.; Chopard, B.; Masselot, A.; Droz, M. Competing species dynamics: Qualitative advantage versus geography. *Eur. Phys. J. B* **1998**, *4*, 529–531. [CrossRef]
8. Galam, S. Application of statistical physics to politics. *Phys. A Stat. Mech. Its Appl.* **1999**, *274*, 132–139. [CrossRef]
9. Chopard, B.; Droz, M.; Galam, S. An evolution theory in finite size systems. *Eur. Phys. J. B* **2000**, *16*, 575–578. [CrossRef]
10. Galam, S. Minority opinion spreading in random geometry. *Eur. Phys. J. B* **2002**, *25*, 403–406. [CrossRef]
11. Galam, S.; Chopard, B.; Droz, M. Killer geometries in competing species dynamics. *Physica A* **2002**, *314*, 256–263. [CrossRef]
12. Galam, S. Modelling rumors: The no plane Pentagon French hoax case. *Physica A* **2003**, *320*, 571–580. [CrossRef]
13. Galam, S. Contrarian deterministic effects on opinion dynamics: 'The hung elections scenario'. *Physica A* **2004**, *333*, 453–460. [CrossRef]
14. Galam, S. The dynamics of minority opinions in democratic debate. *Physica A* **2004**, *336*, 56–62. [CrossRef]
15. Galam, S.; Vignes, A. Fashion, novelty and optimality: An application from Physics. *Physica A* **2005**, *351*, 605–619. [CrossRef]
16. Gekle, S.; Peliti, L.; Galam, S. Opinion dynamics in a three-choice system. *Eur. Phys. J. B* **2005**, *45*, 569–575. [CrossRef]
17. Galam, S. Heterogeneous beliefs, segregation, and extremism in the making of public opinions. *Phys. Rev. E* **2005**, *71*, 046123. [CrossRef]
18. Borghesi, C.; Galam, S. Chaotic, staggered, and polarized dynamics in opinion forming: The contrarian effect. *Phys. Rev. E* **2006**, *73*, 066118. [CrossRef]
19. Galam, S. From 2000 Bush–Gore to 2006 Italian elections: Voting at fifty-fifty and the contrarian effect. *Qual. Quant.* **2007**, *41*, 579–589. [CrossRef]
20. Galam, S.; Jacobs, F. The role of inflexible minorities in the breaking of democratic opinion dynamics. *Physica A* **2007**, *381*, 366–376. [CrossRef]
21. Galam, S. Sociophysics: A review of Galam models. *Int. J. Mod. Phys. C* **2008**, *19*, 409–440. [CrossRef]
22. Chen, P.; Redner, S. Majority rule dynamics in finite dimensions. *Phys. Rev. E* **2005**, *71*, 036101. [CrossRef]
23. Oliveira, L.S.; Rodrigues, A.C.; Forgerini, F.L. Reputation in Majority Rule Model leading to democratic states. *J. Phys. Conf. Ser.* **2019**, *1391*, 012042. [CrossRef]
24. Holley, R.A.; Liggett, T.M. Ergodic theorems for weakly interacting infinite systems and voter model. *Ann. Probab.* **1975**, *3*, 643–663. [CrossRef]
25. Lima, F.W.S.; Malarz, K. Majority-vote model on $(3, 4, 6, 4)$ and $(3^4, 6)$ Archimedean lattices. *Int. J. Mod. Phys. C* **2006**, *17*, 1273–1283. [CrossRef]
26. Lambiotte, R.; Redner, S. Dynamics of vacillating voters. *J. Stat. Mech. Theory Exp.* **2007**, *2007*, L10001. [CrossRef]
27. Fernandez-Gracia, J.; Suchecki, K.; Ramasco, J.J.; San Miguel, M.; Eguiluz, V.M. Is the voter model a model for voters? *Phys. Rev. Lett.* **2014**, *112*, 158701. [CrossRef] [PubMed]
28. Sznajd-Weron, K.; Sznajd, J. Opinion evolution in closed community. *Int. J. Mod. Phys. C* **2000**, *11*, 1157–1165. [CrossRef]
29. Sznajd-Weron, K. Sznajd model and its applications. *Acta Phys. Pol. B* **2005**, *36*, 2537–2547. Available online: https://www.actaphys.uj.edu.pl/fulltext?series=Reg&vol=36&page=2537 (accessed on 13 October 2023).
30. Sznajd-Weron, K.; Sznajd, J. Who is left, who is right? *Physica A* **2005**, *351*, 593–604. [CrossRef]
31. Malarz, K.; Kułakowski, K. The Sznajd dynamics on a directed clustered network. *Acta Phys. Pol. A* **2008**, *114*, 581–588. [CrossRef]
32. Sznajd-Weron, K.; Sznajd, J.; Weron, T. A review on the Sznajd model—20 years after. *Physica A* **2021**, *565*, 125537. [CrossRef]
33. Malarz, K.; Kułakowski, K. Indifferents as an interface between Contra and Pro. *Acta Phys. Pol. A* **2010**, *117*, 695–699. [CrossRef]
34. Öztürk, M.K. Dynamics of discrete opinions without compromise. *Adv. Complex Syst.* **2013**, *16*, 1350010. [CrossRef]
35. Bańcerowski, P.; Malarz, K. Multi-choice opinion dynamics model based on Latané theory. *Eur. Phys. J. B* **2019**, *92*, 219. [CrossRef]
36. Kowalska-Styczeń, A.; Malarz, K. Noise induced unanimity and disorder in opinion formation. *PLoS ONE* **2020**, *15*, e0235313. [CrossRef]
37. Dworak, M.; Malarz, K. Vanishing opinions in Latané model of opinion formation. *Entropy* **2023**, *25*, 58. [CrossRef] [PubMed]
38. Martins, A.C.R. Discrete opinion dynamics with M choices. *Eur. Phys. J. B* **2020**, *93*, 1. [CrossRef]
39. Zubillaga, B.; Vilela, A.; Wang, M.; Du, R.; Dong, G.; Stanley, H. Three-state majority-vote model on small-world networks. *Sci. Rep.* **2022**, *12*, 282. [CrossRef]
40. Li, L.; Zeng, A.; Fan, Y.; Di, Z. Modeling multi-opinion propagation in complex systems with heterogeneous relationships via Potts model on signed networks. *Chaos* **2022**, *32*, 083101. [CrossRef]
41. Doniec, M.; Lipiecki, A.; Sznajd-Weron, K. Consensus, polarization and hysteresis in the three-state noisy q-voter model with bounded confidence. *Entropy* **2022**, *24*, 983. [CrossRef]

42. Xiong, F.; Liu, Y.; Wang, L.; Wang, X. Analysis and application of opinion model with multiple topic interactions. *Chaos* **2017**, *27*, 083113. [CrossRef] [PubMed]
43. Galam, S. The drastic outcomes from voting alliances in three-party democratic voting (1990–2013). *J. Stat. Phys.* **2013**, *151*, 46–68. [CrossRef]
44. Wu, D.; Szeto, K.Y. Analysis of timescale to consensus in voting dynamics with more than two options. *Phys. Rev. E* **2018**, *97*, 042320. [CrossRef]
45. Mobilia, M. Polarization and consensus in a voter model under time-fluctuating influences. *Physics* **2023**, *5*, 517–536. [CrossRef]
46. Nowak, A.; Szamrej, J.; Latané, B. From private attitude to public opinion: A dynamic theory of social impact. *Psychol. Rev.* **1990**, *97*, 362–376. [CrossRef]
47. Darley, J.M.; Latané, B. Bystander intervention in emergencies—Diffusion of responsibility. *J. Personal. Soc. Psychol.* **1968**, *8*, 377–383. [CrossRef] [PubMed]
48. Latané, B.; Harkins, S. Cross-modality matches suggest anticipated stage fright a multiplicative power function of audience size and status. *Percept. Psychophys.* **1976**, *20*, 482–488. [CrossRef]
49. Latané, B. The psychology of social impact. *Am. Psychol.* **1981**, *36*, 343–356. [CrossRef]
50. Kacperski, K.; Hołyst, J.A. Phase transitions as a persistent feature of groups with leaders in models of opinion formation. *Physica A* **2000**, *287*, 631–643. [CrossRef]
51. Hołyst, J.A.; Kacperski, K.; Schweitzer, F. Phase transitions in social impact models of opinion formation. *Physica A* **2000**, *285*, 199–210. [CrossRef]
52. Hołyst, J.A.; Kacperski, K.; Schweitzer, F. Social impact models of opinion dynamics. In *Proceedings of the Annual Reviews of Computational Physics IX*; Stauffer, D., Ed.; World Scientific: Singapore, 2011; pp. 253–273. [CrossRef]
53. Asch, S.E. Studies of independence and conformity: I. A minority of one against a unanimous majority. *Psychol. Monogr.* **1956**, *70*, 1–70. [CrossRef]
54. Bandura, A. *Social Learning Theory*; General Learning Press: New York, NY, USA, 1971. Available online: https://archive.org/details/BanduraSocialLearningTheory/ (accessed on 13 October 2023).
55. Milgram, S. *Obedience to Authority: An Experimental View*; Harper & Row, Publishers: New York, NY, USA, 1974. Available online: https://archive.org/details/obediencetoautho0000milg/ (accessed on 13 October 2023).
56. Cialdini, R.B. *Influence: The Psychology of Persuation*; HarperCollins e-books: New York, NY, USA, 1983. Available online: https://archive.org/details/influence_202107/ ((accessed on 13 October 2023).
57. Abelson, R.P.; Aronson, E.; McGuire, W.J.; Newcomb, T.M.; Rosenberg, M.J.; Tannenbaum, P.H. (Eds.) *Theories of Cognitive Consistency: A Sourcebook*; Rand McNally and Company: Chicago, IL, USA, 1968. Available online: https://archive.org/details/theoriesofcognit0000unse_m8y1/ (accessed on 13 October 2023).
58. Fesinger, L. *A Theory of Cognitive Dissonance*; Stanford University Press: Stanford, CA, USA, 1968. Available online: https://archive.org/details/FestingerLeonATheoryOfCognitiveDissonance1968StanfordUniversityPress/ (accessed on 13 October 2023).
59. Bahr, D.B.; Passerini, E. Statistical mechanics of opinion formation and collective behavior: Micro-sociology. *J. Math. Sociol.* **1998**, *23*, 1–27. [CrossRef]
60. Bahr, D.B.; Passerini, E. Statistical mechanics of collective behavior: Macro-sociology. *J. Math. Sociol.* **1998**, *23*, 29–49. [CrossRef]
61. Hoshen, J.; Kopelman, R. Percolation and cluster distribution. 1. Cluster multiple labeling technique and critical concentration algorithm. *Phys. Rev. B* **1976**, *14*, 3438–3445. [CrossRef]
62. Landau, D.P.; Binder, K. *A Guide to Monte Carlo Simulations in Statistical Physics* ; Cambridge University Press: New York, NY, USA, 2009. [CrossRef]
63. Frijters, S.; Krüger, T.; Harting, J. Parallelised Hoshen–Kopelman algorithm for lattice-Boltzmann simulations. *Comput. Phys. Commun.* **2015**, *189*, 92–98. [CrossRef]
64. Kotwica, M.; Gronek, P.; Malarz, K. Efficient space virtualisation for Hoshen–Kopelman algorithm. *Int. J. Mod. Phys. C* **2019**, *30*, 1950055. [CrossRef]
65. Muller, H. Some genetic aspects of sex. *Am. Nat.* **1932**, *66*, 118–138. [CrossRef]
66. Malarz, K.; Tiggemann, D. Dynamics in Eigen quasispecies model. *Int. J. Mod. Phys. C* **1998**, *9*, 481–490. [CrossRef]
67. Lin, N. *Social Capital: A Theory of Social Structure and Action*; Cambridge University Press: Cambridge, UK, 2001. [CrossRef]
68. Lin, N. Building a network theory of social capital. *Connections* **1999**, *22*, 28–51. Available online: https://faculty.washington.edu/matsueda/courses/590/Readings/ (accessed on 13 October 2023).
69. Valori, L.; Picciolo, F.; Allansdottir, A.; Garlaschelli, D. Reconciling long-term cultural diversity and short-term collective social behavior. *Proc. Natl. Acad. Sci. USA* **2012**, *109*, 1068–1073. [CrossRef]
70. Katz, E.; Lazarsfeld, P.F. *Personal Influence: The Part Played by People in the Flow of Mass Communications*; The Free Press/Mcmillan Publishers Co., Inc.: New York, NY, USA, 1955. Available online: https://archive.org/details/personalinfluenc0000katz/ (accessed on 13 October 2023).
71. Bineham, J.L. A historical account of the hypodermic model in mass communication. *Commun. Monogr.* **1988**, *55*, 230–246. [CrossRef]
72. Putnam, R.D. Bowling alone: The collapse and revival of American community; In *CSCW'00: Proceedings of the 2000 ACM Conference on Computer Supported Cooperative Work, Philadelphia, PA, USA, 2–6 December 2000*; Kellog, W., Whittaker, S., Eds.; Association for Computing Machinery: New York, NY, USA, 2000; p. 357. [CrossRef]

73. Burt, R.S. Social contagion and innovation: Cohesion versus structural equivalence. *Am. J. Sociol.* **1987**, *92*, 1287–1335. [CrossRef]
74. Burt, R.S. The social capital of opinion leaders. *Ann. Am. Acad. Polit. Soc. Sci.* **1999**, *566*, 37–54. [CrossRef]
75. Mavridis, C.; Tsakas, N. Social capital, communication channels and opinion formation. *Soc. Choice Welf.* **2021**, *56*, 635–678. [CrossRef]
76. Iijima, R.; Kamada, Y. Social distance and network structures. *Theor. Econ.* **2017**, *12*, 655–689. [CrossRef]
77. Malarz, K.; Galam, S. Square-lattice site percolation at increasing ranges of neighbor bonds. *Phys. Rev. E* **2005**, *71*, 016125. [CrossRef] [PubMed]
78. Galam, S.; Malarz, K. Restoring site percolation on damaged square lattices. *Phys. Rev. E* **2005**, *72*, 027103. [CrossRef]

physics

MDPI

Article

Mathematical Programming for the Dynamics of Opinion Diffusion

Andrea Ellero *, Giovanni Fasano and Daniela Favaretto

Department of Management, Ca' Foscari University of Venice, 30123 Venice, Italy; fasano@unive.it (G.F.); favaret@unive.it (D.F.)

* Correspondence: ellero@unive.it

Abstract: The focus of this paper is on analyzing the role and the choice of parameters in sociophysics diffusion models by leveraging the potentialities of sociophysics from a mathematical programming perspective. We first present a generalised version of Galam's opinion diffusion model. For a given selection of the coefficients in our model, this proposal yields the original Galam's model. The generalised model suggests guidelines for possible alternative selection of its parameters that allow it to foster diffusion. Examples of the parameters selection process as steered by numerical optimisation, taking into account various objectives, are provided.

Keywords: sociophysics; stochastic models; Galam's diffusion model; linear programming; non-linear programming; diffusion control

1. Introduction

Opinion diffusion is a fundamental aspect of human interaction, playing a pivotal role in the spread of innovations, and the emergence of consensus. Understanding how opinions spread and evolve within social networks is crucial for comprehending the emergence of collective behavior, the formation of public opinion, and the polarisation of societies.

Social sciences provide plenty of applications, where the importance of diffusion dynamics is studied and witnessed, such as marketing (see, e.g., [1–3]), agent-based modelling (see, e.g., [4]), and sociophysics, where the papers of Serge Galam serve as a guide for researchers (see, e.g., [5]).

Galam's opinion diffusion model (see [6]) provides a framework for investigating and understanding the dynamics of opinion formation in social systems. Galam's model sheds light on the mechanisms that drive opinion dynamics. In particular, Galam's model describes the formation of collective opinion convergence processes using concepts from statistical physics. To analyse them, along with several generalisations of the model, the authors of the current paper have proposed the alternatives in Ref. [7] (showing some drawbacks when applying Galam's model with a relatively small number of agents), and in Ref. [8] (where a novel subset of agents, namely the opinion leaders, are introduced, to duly represent a possible communication bias). Furthermore, additional literature covering a number of keynote issues can be found in Refs. [5,9,10] and the references therein.

The standard Galam's model in Ref. [6] assumes that individuals in a group influence and are influenced by their peers. After several iterations within repeated group discussions, each agent may consequently influence the opinion diffusion. Galam's stylised model simplifies the complexity of opinions by considering a binary state, where individuals can adopt either of two possible opinions, thus allowing a straightforward mathematical analysis of the diffusion process.

We suggest a generalisation of Galam's model and then offer evidence in an optimisation framework that some of the diffusion parameters can be successfully adjusted in order to steer the diffusion process. Combining Galam's model with specific Linear Programming (LP) formulations, we take into account both theoretical and numerical results

Citation: Ellero, A.; Fasano, G.; Favaretto, D. Mathematical Programming for the Dynamics of Opinion Diffusion. *Physics* **2023**, *5*, 936–951. https://doi.org/10.3390/physics5030061

Received: 29 June 2023
Accepted: 10 August 2023
Published: 11 September 2023

concerning the role of the model's parameters in order to better control the dynamics of the diffusion. Namely, we aim at maximising the probability that an opinion can spread among agents.

This paper is organised as follows. The essential characteristics of Galam's model [6] are recapped in Section 2. Then, in Section 3, we present and discuss a generalisation of Galam's model. The optimisation of the parameters of the new model via linear programming allows one to speed up opinion diffusion at a fixed step of the process: the optimisation problem is discussed in Section 4, while Section 5 provides numerical examples of its resolution. Section 6 discusses a dynamic programming extension of the optimisation problem that optimises the values of the parameters along the whole diffusion process, till convergence. Section 7 contains final remarks and concludes the paper.

2. Recap of Galam's Model

This section reviews the foundational elements of Galam's model [6]. Think of a group of N people (agents), each of whom may hold one of two opposing views (let us say, '+' or '−'), regarding a specific issue. These agents meet, for several repeated rounds within subgroups of smaller cardinality, to converse and possibly change one anothers' minds.

Each agent at round (time) t has a probability a_k of being a member of a subgroup of size k, and L is the maximum allowed size of groups. In the original formulation of Galam's model, the values $a_1, \ldots, a_L$ are exogenous parameters such that

$$\sum_{k=1}^{L} a_k = 1, \qquad a_k \geq 0, \quad k = 1, \ldots, L.$$

Joining a group discussion at time t, any agent may, at the beginning of the subsequent time $t+1$, modify their position (e.g., '+' becomes '−' or vice versa) in accordance with a unique law: the majority rule. Indeed, all agents in a subgroup take the position of the majority in that group in the end of time t. The rule for reversing opinion is slightly biased in favour of the negative opinion '−', since tie breaks are in favour of '−'. This may lead to a substantial bias in the case where the subgroup has an even number of members.

Calling $P_+(t)$ the probability that an agent opinion is '+' at time t, the probability of opinion '−' is then $P_-(t) = 1 - P_+(t)$. The probability $P_+(t+1)$ is estimated in Ref. [6] as

$$P_+(t+1) = \sum_{k=1}^{L} a_k \sum_{j=\lfloor \frac{k}{2}+1 \rfloor}^{k} C_j^k P_+(t)^j \{1 - P_+(t)\}^{k-j}, \tag{1}$$

where $\lfloor z \rfloor$ is the largest integer less or equal to z, and C_j^k is the binomial coefficient, $\binom{k}{j}$.

Let $N_+(t)$ ($N_-(t) = N - N_+(t)$) be the number of agents who at time step t have opinion '+' ('−'). Observe that for small values of N, setting $P_+(0) = N_+(0)/N$, where $N_+(0)$ is the number of agents thinking '+' at time $t = 0$, for any $t \geq 1$ the quantity $P_+(t)$ may possibly differ from the 'actual' frequency of '+', i.e., $N_+(t)/N$ (see [7]). Nevertheless, the estimated probability $P_+(t+1)$ always lies in the interval $[0, 1]$ being

$$\sum_{j=\lfloor \frac{k}{2}+1 \rfloor}^{k} C_j^k P_+(t)^j \{1 - P_+(t)\}^{k-j} < \sum_{j=0}^{k} C_j^k P_+(t)^j \{1 - P_+(t)\}^{k-j} = 1.$$

A relevant role in the model is taken by the value $\hat{P}_+$ such that

$$\begin{aligned}
&\text{when } P_+(0) > \hat{P}_+ \text{ then } \lim_{t\to\infty} P_+(t) = 1, \\
&\text{when } P_+(0) < \hat{P}_+ \text{ then } \lim_{t\to\infty} P_+(t) = 0, \\
&\text{when } P_+(0) = \hat{P}_+ \text{ then } P_+(t) = P_+(0), \qquad \forall\, t > 0.
\end{aligned}$$

$\hat{P}_+$ constitutes the threshold value such that when $N_+(0)/N$ is greater than $\hat{P}_+$, then all agents will eventually have opinion '+'. Conversely, when $t \to \infty$ all the agents will definitely think '−' if $N_+(0)/N < \hat{P}_+$. The value $\hat{P}_+$ is the fixed point of Equation (1) and is called the killing point or tipping point of the model. In Figure 1, one can observe the dynamics described by relation (1), when the largest cardinality L of the subgroups takes integer values in the interval $[3, 20]$ and $a_k = 1/L$, for any k: for each of the 18 curves, the killing point is identified by the intersection with the bisector line other than the extreme points $(0, 0)$ and $(1, 1)$. Observe that for small values of L no killing point appears (or equivalently, it coincides with the extreme point $(1, 1)$). Figure 2 shows a zoomed in perspective of Figure 1.

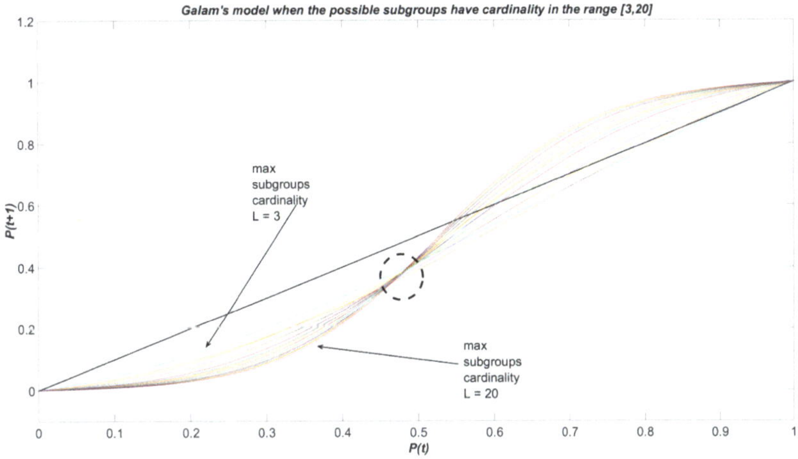

Figure 1. Dynamic curves described by relation (1), when the largest cardinality L of the subgroups takes integer values in the interval $[3, 20]$ and $a_k = 1/L$, for any k.

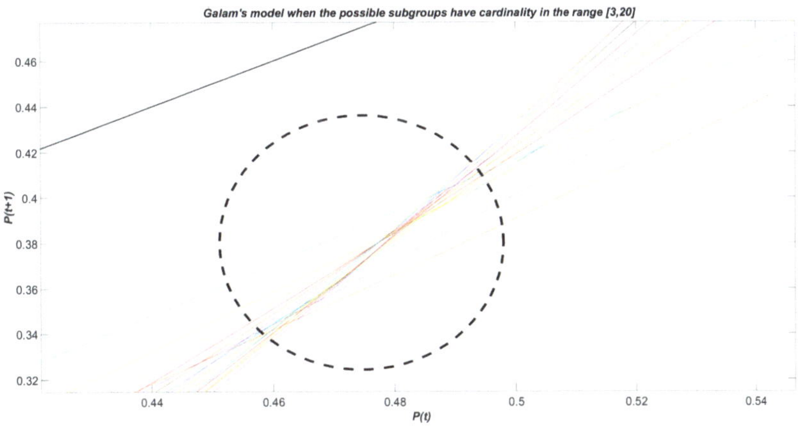

Figure 2. Dynamic curves described by relation (1), when the largest cardinality L of the subgroups takes integer values in the interval $[3, 20]$ and $a_k = 1/L$, for any k: magnification of the area of intersections among curves.

3. Capturing Additional Dynamics Using a Generalised Galam's Model

According to model (1), a strict majority of members with positive opinion is a necessary and sufficient condition to have opinion '+' for all the subgroup members.

Since the ultimate choice in a group is frequently the product of a discussion, rather than a simple tally of the two opposing viewpoints in the group, a rule that is so rigorous yet so straightforward could not correctly reflect the actual opinion dynamics in practice. As an illustration, think about the propagation of political opinions. The probability that everyone in the group would share the same view (e.g., +) at the close of a conversation may rise smoothly with the number of members who had that opinion prior to the meeting. As a second application, the reader may consider the audience of social networks, where the dynamics of information spreading was widely investigated in the dedicated literature (see, e.g., [11]).

Accordingly, we suggest to generalise Galam's model (1) taking into account a softer dynamics of opinion change, i.e., without using rigid majority as a necessary and sufficient condition to shift the entire group to the same view. In particular, we assume that after the conversations, all agents in a group will have opinion '+' with a probability α_j^k that is dependent on the size k of the group and the number j of agents that had opinion '+' before the discussion. We use the following definition to model the relationship between the probabilities of having an opinion '+' before (i.e., $P_+(t)$) and after (i.e., $P_+(t+1)$) discussion (see [12] for a preliminary version of the model):

$$P_+(t+1) = \sum_{k=1}^{L} a_k \sum_{j=0}^{k} \alpha_j^k C_j^k P_+(t)^j \{1 - P_+(t)\}^{k-j}. \tag{2}$$

Model (2) generalises model (1), since it is possible to select indeed the probabilities $\{\alpha_j^k\}$ that exactly replicate model (1) (see also Figure 3):

$$\alpha_j^k = \begin{cases} 0 & \text{if } j < \left\lfloor \frac{k}{2} + 1 \right\rfloor, \\ 1 & \text{if } j \geq \left\lfloor \frac{k}{2} + 1 \right\rfloor. \end{cases} \tag{3}$$

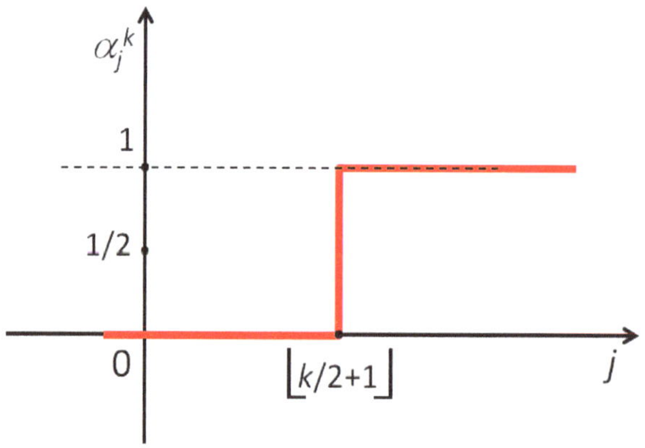

Figure 3. For a given $1 \leq k \leq L$, the model (3) considers the above step-shaped choice for the coefficients $\{\alpha_j^k\}$. In particular, if $j < \lfloor k/2+1 \rfloor$ then $\alpha_j^k = 0$, otherwise $\alpha_j^k = 1$, so that this choice corresponds exactly to obtain the original Galam's model (1).

As mentioned, probabilities $\{\alpha_j^k\}$ should reasonably increase with the number j of '+' in a subgroup, and decrease with respect to the size of the group k. In this regard, it is reasonable to expect for the probability α_j^k to become negligible when the number j of '+' in a subgroup is relatively small (i.e., much lower than the strict majority), and to become close to 1 as the number of '+' noticeably increases (i.e., higher than strict majority). In general, the probability that all agents in the subgroup will adopt the opinion '+' is conceivably a function rising from zero to one as j increases. If one assumes that this function of j is linear, one may precisely define the probabilities α_j^k as (see Figure 4)

$$\alpha_j^k = \begin{cases} 0\,, & \text{if } j < \left\lfloor \frac{k}{2} + 1 \right\rfloor + z_j^k - h, \\ \frac{1}{2} + \frac{\left[j - \left(\left\lfloor \frac{k}{2} + 1 \right\rfloor + z_j^k \right) \right]}{2h}\,, & \text{if } \left\lfloor \frac{k}{2} + 1 \right\rfloor + z_j^k - h \leq j < \left\lfloor \frac{k}{2} + 1 \right\rfloor + z_j^k + h, \\ 1\,, & \text{if } j \geq \left\lfloor \frac{k}{2} + 1 \right\rfloor + z_j^k + h. \end{cases} \tag{4}$$

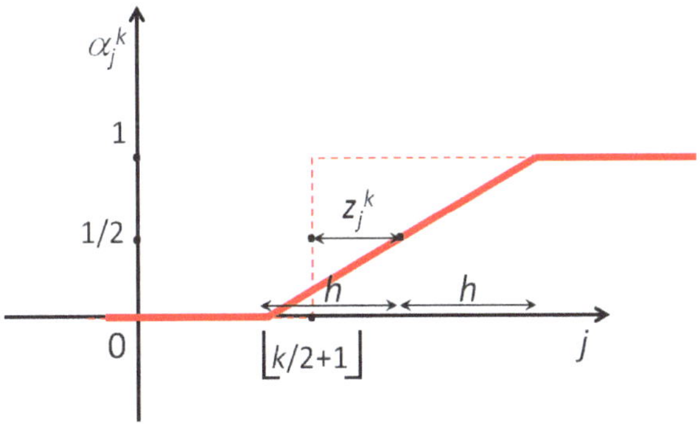

Figure 4. For a given $1 \leq k \leq L$, the model (4) considers the above *piecewise-linear* choice for the coefficients $\{\alpha_j^k\}$, where z_j^k represents a shift with respect to the abscissa $\lfloor k/2 + 1 \rfloor$, and $1/(2h)$ is the slope of the ramp.

When $z_j^k = h - 0$, the probabilities defined in Equation (4) are exactly the same as in Equation (3). The value of z_j^k in (4) represents a shift (either positive or negative) from the value $\lfloor k/2 + 1 \rfloor$, while the slope of the ramp in Figure 4 is $1/(2h)$.

To better grasp the geometric insight behind the choice of the parameters in (4), let us consider the simple numerical example reported in Ref. [6], where $L = 4$, $a_1 = 0$, and $a_2 = a_3 = a_4 = 1/3$ (and $h = z_j^k = 0$ for all j and k). As reported in Ref. [6], the killing point (KP) lies in between 0.84 and 0.87 (see Figure 5). Then, in Figure 6, there is also the model (2) with the choice (4), after setting $h = z_j^k = 0.5$ for any j and k. One can immediately infer that the dynamics of Galam's model [6] is definitely spoiled when the choice for the coefficients $\{\alpha_j^k\}$ in Equation (3) is replaced by Equation (4). Equivalently, even in case $P_+(t) = 1$ (complete consensus among agents to think $+$) then the choice (4) imposes a regression towards the stable point given by the origin.

Hence, the presence of a killing point for Galam's model seems to be strictly related to the expression (1), and introducing the coefficients $\{\alpha_j^k\}$ definitely also spoils the killing point. Moreover, one observes that, in case the majority rule is not explicitly applied in Equation (1), then the dynamics of the model is strongly altered.

Nevertheless, the model (2) retains a number of properties and can suggest fruitful results, following both of the next guidelines:

- it provides clues on the perspective for studying the process of diffusion of information;
- it may suggest proper ways to control information spreading among agents.

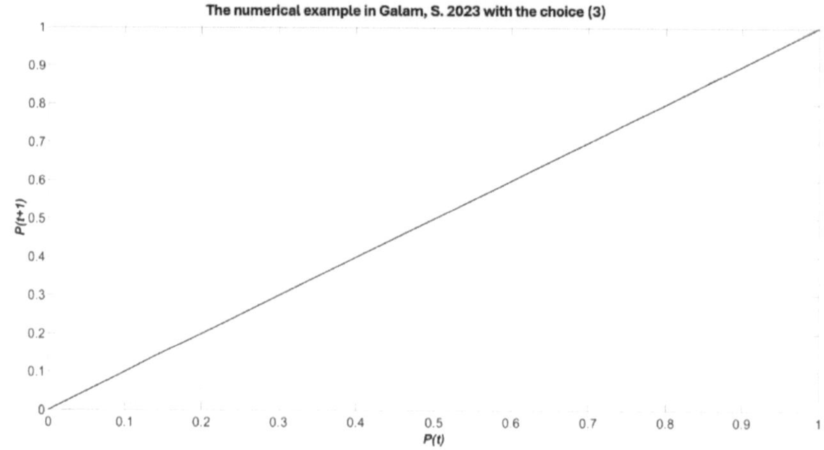

Figure 5. The numerical example from Ref. [6], with $L = 4$, $a_1 = 0$, $a_2 = a_3 = a_4 = 1/3$ and probabilities (3).

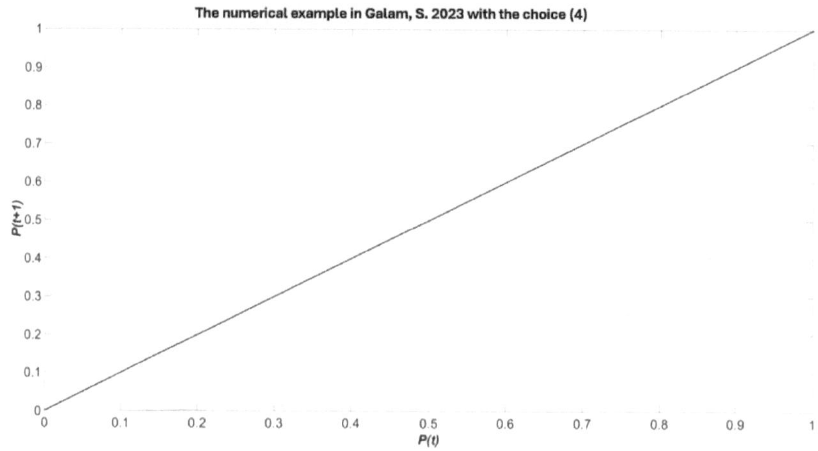

Figure 6. The numerical example from Ref. [6], with $L = 4$, $a_1 = 0$, $a_2 = a_3 = a_4 = 1/3$ but probabilities (4).

At least a couple of intriguing issues in applications can steer the choice of the coefficients $\{\alpha_j^k\}$, ranging from small–medium- to large-scale (number of agents) applications:

- One Period Analysis. One may assess values for $\{\alpha_j^k\}$ so that, considering at time t the probability $P_+(t)$, one miximizes the probability $P_+(t+1)$ at time $t+1$. In this regard, the values $\{\alpha_j^k\}$ represent costs for steering the information among agents and fostering the positive opinion: the larger $\{\alpha_j^k\}$ (i.e., higher costs) the larger the probability to convey a group to opinion '+'. Certainly, the combination of the proper values for each of the coefficients $\{\alpha_j^k\}$ is strongly related to the value $P_+(t)$, the probabilities $\{a_k\}$, and the binomial coefficients C_j^k;
- Multi-Period Analysis. Similarly to the previous item, one may decide to select $\{\alpha_j^k\}$ such that, considering again at time t the probability $P_+(t)$, one miximizes the probability $P_+(T)$ at time $T > t$, and compute the entire sequence $P_+(t+1), P_+(t+2), \ldots, P_+(T)$ of intermediate probabilities.

The last two scenarios can be declined in a variety of contexts, including production and marketing, where consumers play the role of agents and the aforementioned perspectives may be the answer to the question: what is (if any) the necessary extra effort in terms of advertising campaign, in order to promote a certain spread of the products or customers' expectations on items? In this regard, the quantities $\{\alpha_j^k\}$ summarise the resulting effort, while from a mathematical programming perspective they represent unknowns to be assessed. This also suggests that a number of additional questions may arise adopting the model (2), mapping relevant practical instances where a tentative forecast is sought in order to plan the use of (possibly scarce) resources to control the process of diffusion of the information. For the sake of brevity, we itemise below some proposals that are representative but not exhaustive:

- Feasibility Problem. Given the value of $P_+(t)$ and the target value $\bar{P}$, one may determine a set of values for the coefficients $\{\alpha_j^k\}$ such that, starting at t from $P_+(t)$, at time $T > t$, $P_+(T) \geq \bar{P}$;
- Focus Group Problem. Given the value of $P_+(t)$ and the target value $\bar{P}$ and the time $T > t$, one may be interested to determine a set of values for the coefficients $\{\alpha_j^k\}$ such that $\alpha_m^{(k)} \leq \alpha_j^k \leq \alpha_M^{(k)}$, where α_m, α_M are possible bounds, and $P_+(T) \geq \bar{P}$. This scenario models a specific interest on those subgroups of cardinality k.

4. Single Period Maximisation of Diffusion

As from the analysis in Section 3, one of the possible proposals to couple the dynamics described through the model (2) with a single period optimisation framework, is given by the following LP scheme:

$$\max_{\alpha} \quad \sum_{k=1}^{L} a_k \sum_{j=0}^{k} \alpha_j^k C_j^k P_+(t)^j \{1 - P_+(t)\}^{k-j}, \tag{5}$$

$$\alpha_j^k \leq \alpha_{j+1}^k \qquad j = 0, 1, \ldots, k-1; \; k = 1, \ldots, L, \tag{6}$$

$$\alpha_j^k \geq \alpha_j^{k+1} \qquad j = 0, 1, \ldots, k; \; k = 1, \ldots, L-1, \tag{7}$$

$$\sum_{k=1}^{L} \sum_{j=0}^{k} \alpha_j^k \leq b(t), \tag{8}$$

$$0 \leq \alpha_j^k \leq 1 \qquad j = 0, 1, \ldots, k; \; k = 1, \ldots, L. \tag{9}$$

The last optimisation problem assumes that the value $P_+(t)$ is given, so that both the objective function and the constraints are linear in the set $\{\alpha_j^k\}$. Though the meaning associated with the objective function is relatively clear, for the constraints some clarifications seem necessary:

- the constraints in Equation (6) state that for a given dimension k of a subset, larger values of the number of agents thinking '+' imply a larger probability α_j^k;
- similarly, in the constraints in Equation (7), when the number of agents thinking '+' remains constant, the larger the set cardinality (i.e., k), the smaller the probability α_j^k;
- the constraint in Equation (8) expresses the overall cost (i.e., it is a budget constraint) allowed to assess the probabilities $\{\alpha_j^k\}$, at time t;
- since the quantities $\{\alpha_j^k\}$ need to represent probabilities, they are required to fulfill also the constraint (9).

In order to avoid either infeasible or straightforward solutions for the optimisation Problems (5)–(9), the following lemmas gives some indications on the choice of the parameter $b(t)$, for any t (for a proof, see [12]).

Lemma 1. *Let one be given the linear Programs* (5)–(9) *with the choice* ($1 \le k \le L$ *and* $0 \le j \le k$)

$$\alpha_j^k = \begin{cases} 0, & for\ any\ j < \left\lfloor \frac{k}{2} + 1 \right\rfloor, \\ 1, & for\ any\ j \ge \left\lfloor \frac{k}{2} + 1 \right\rfloor. \end{cases} \tag{10}$$

Then, relations (10) *satisfy the constraints* (6), (7) *and* (9)*. Moreover, if Equation* (10) *also satisfies Equation* (8)*, then the value of the objective function in Equation* (5) *coincides with* $P_+(t+1)$ *in Equation* (1).

Note that Lemma 1 gives a straightforward idea of the fact that the solution of the linear Programs (5)–(9) is substantially nothing else but a generalisation of Equation (1). Furthermore, the next result will give a precise indication on the possible values for the budget $b(t)$ in Equation (8).

Lemma 2. *Let one be given the optimisation problems* (5)–(9) *and let* $\{\alpha_j^k\}$ *be assigned as in Equation* (10)*, for any* $1 \le k \le L$ *and* $0 \le j \le k$*. Then,*

$$\sum_{k=1}^{L} \sum_{j=0}^{k} \alpha_j^k = \begin{cases} \left(\frac{L}{2}\right)^2 + \frac{L}{2}, & when\ L\ is\ even, \\ \left(\frac{L-1}{2}\right)^2 + L, & when\ L\ is\ odd. \end{cases} \tag{11}$$

Note that the right sides of Equation (11) explicitly give some hints for the selection of the parameter $b(t)$ in Equation (8), both whether in case L it is even or it is odd.

Regarding some technicalities associated with the solution of the optimisation problems (5)–(9), one observes that it is a continuously differentiable problem, where all the functions are linear. Hence, it is a convex problem, so that the set of all its solutions (possibly an empty set or a singleton) is a convex set. This implies that for any given pair of its solutions, all the points in the segment joining them will be solutions, too. Moreover, being a linear program, all its (possible) solutions will lie on vertices of the feasible set, and not in the interior of the feasible set. As is known, this makes its solution relatively simple (throughout any solver based on the simplex method) and achievable in polynomial time. Hence, large-scale instances are well tractable and allow for a scalable solution in a number of applications with a great number of agents (e.g., problems involving social media or large social groups).

We complete this section highlighting that one further generalisation of the linear programs (5)–(9) suggests to disaggregate the budget constraint and to replace the inequality (8) by the set of constraints,

$$\sum_{j=0}^{k} \alpha_j^k \le b_k(t), \qquad k = 1, \dots, L.$$

Observe that the value $b_k(t)$, for any k, represents a budget devoted to possibly affect the solution after working uniquely on the subsets of cardinality k. This increases the flexibility of the model without compromising its overall complexity (being the resulting model yet a linear program).

5. Preliminary Numerical Experience

This Section is devoted to provide a numerical experience where the potentialities of the formulations (5)–(9) are exploited. In this regard, the investigation here is limited to a couple of meaningful instances. The first one is a small-scale (number of unknowns) problem, while the second one attempts to scale the first's by 100.

5.1. Small-Scale Instance

Regarding the value of the coefficients in (5)–(9), we consider for the small-scale instance the following setting:

$$\begin{cases} L = 10, \\ \{a_k\} = \{0.1, 0, 0.15, 0, 0.3, 0, 0.3, 0, 0.15, 0\}, \\ P_+(t) = 0.65. \end{cases}$$

The above choice corresponds to possibly allow only subgroups of dimensions 1, 3, 5, 7, and 9, with probabilities that include null values and possibly repeated values. Furthermore, we set the initial percentage of the population thinking '+' to $P_+(t) = 65\%$. Finally, in order to avoid an empty feasible set in Equations (5)–(9) we also selected, respectively, the values $b(t) = 5$ and $b(t) = 10$, based on Lemma 2. The outcomes of our numerical experience are obtained using the solver CPLEX 20.1.0.0 available on the NEOS Server platform [13]. Let us remark that CPLEX is definitely among the best and faster solvers for LP, along with the alternative solver BARON. Our numerical experience on the above two instances is summarised as follows:

- $b(t) = 5$: the problems (5)–(9) is coded using AMPL (A Mathematical Programming Language; see [14,15]). The presolve tool in CPLEX eliminated 65 constraints from the formulation, so that the simplified problem was reduced to 65 variables and 110 linear inequality constraints. Regarding the final solution, CPLEX performed 13 dual simplex iterations (i.e., iterations of a method for LP which represents the counterpart of the simplex method, after exploiting the duality theory), with an overall time for the solution not larger than a couple of seconds. Finally, the value of $P_+(t+1)$ found by the solver is

$$P_+(t+1) = 0.215128625$$

with the set of variables $(\alpha_j^k)^*$ given by

	k = 1	k = 2	k = 3	k = 4	k = 5	k = 6	k = 7	k = 8	k = 9	k = 10
j = 0	0	0	0	0	0	0	0	0	0	0
j = 1	1	0	0	0	0	0	0	0	0	0
j = 2	.	0	0	0	0	0	0	0	0	0
j = 3	.	.	2/3	2/3	2/3	0	0	0	0	0
j = 4	.	.	.	2/3	2/3	0	0	0	0	0
j = 5	.	.	.	.	2/3	0	0	0	0	0
j = 6	.	.	.	.	.	0	0	0	0	0
j = 7	.	.	.	.	.	.	0	0	0	0
j = 8	.	.	.	.	.	.	.	0	0	0
j = 9	.	.	.	.	.	.	.	.	0	0
j = 10	.	.	.	.	.	.	.	.	.	0

- $b(t) = 10$: again the problem (5)–(9) is coded using AMPL and the presolve tool in CPLEX eliminated 65 constraints from the formulation, so that again the simplified problem was reduced to 65 variables and 110 linear inequality constraints. CPLEX performed 18 dual simplex iterations with an overall time for computation similar to the case where it is $b(t) = 5$. Finally, the value of $P_+(t+1)$ found by the solver is

$$P_+(t+1) = 0.4392275888$$

with the set of variables $(\alpha_j^k)^*$ given by the matrix,

	k = 1	k = 2	k = 3	k = 4	k = 5	k = 6	k = 7	k = 8	k = 9	k = 10
j = 0	1	0	0	0	0	0	0	0	0	0
j = 1	1	0	0	0	0	0	0	0	0	0
j = 2	.	0	0	0	0	0	0	0	0	0
j = 3	.	.	1	1	1	0	0	0	0	0
j = 4	.	.	.	1	1	0.285714	0.285714	0	0	0
j = 5	.	.	.	.	1	0.285714	0.285714	0	0	0
j = 6	.	.	.	.	.	0.285714	0.285714	0	0	0
j = 7	.	.	.	.	.	.	0.285714	0	0	0
j = 8	.	.	.	.	.	.	.	0	0	0
j = 9	.	.	.	.	.	.	.	.	0	0
j = 10	.	.	.	.	.	.	.	.	.	0

Note that while allowing a larger value of the budget (for example, $b(t) = 10$), the solver is able to select a larger number of nonzero unknowns, so that the final value of $P_+(t+1)$ is almost doubled with respect to the case $b(t) = 5$. In actual applications, this implies that, in order to obtain almost twice the effect of the previous case, on this instance twice the value of the budget must be allowed.

5.2. Large-Scale Instance

Here, we focus on the same instance of Section 5.1, where conversely, a larger number of unknowns is considered. The novel setting of the parameters for the solver are given as follows:

$$\begin{cases} L = 1000 \\ a_k = 1/L, \qquad k = 1, \dots, L, \\ P_+(t) = 0.65. \end{cases}$$

The above choice corresponds to possibly allow all the subgroups with the same probability, and again the initial percentage of the population thinking '+' is equal to $P_+(t) = 65\%$. Similarly to the small-scale instance, we consider two values for the budget, $b(t) = 50$ and $b(t) = 100$. The computation is again performed adopting the solver CPLEX 20.1.0.0 on the NEOS Server [13]. The outcomes of the resulting two instances are summarised as follows:

- $b(t) = 50$: the problems (5)–(9) is coded using AMPL and the presolve tool in CPLEX eliminated 501,500 constraints from the formulation. The overall simplified problem contained 501,500 variables and 1,001,000 linear inequality constraints. CPLEX performed 192 dual simplex iterations with a time of computation smaller than 20 s, showing that the linear model can well scale the time of computation. The final value of $P_+(t+1)$ found by the solver is

$$P_+(t+1) = 0.01061571566$$

with the set of variables $(\alpha_j^k)^*$ given by

	k = 1	k = 2	k = 3	k = 4	k = 5	k = 6	k = 7	k = 8	k = 9	k = 10	k = 11	k = 12	k ≥ 13
j = 0	1	0	0	0	0	0	0	0	0	0	0	0	0
j = 1	1	1	1	0	0	0	0	0	0	0	0	0	0
j = 2	.	1	1	1	1	0	0	0	0	0	0	0	0
j = 3	.	.	1	1	1	1	1	0	0	0	0	0	0
j = 4	.	.	.	1	1	1	1	1	0	0	0	0	0
j = 5	.	.	.	.	1	1	1	1	1	1	0	0	0
j = 6	.	.	.	.	.	1	1	1	1	1	1	0	0
j = 7	.	.	.	.	.	.	1	1	1	1	1	0.8333	0
j = 8	.	.	.	.	.	.	.	1	1	1	1	0.8333	0
j = 9	.	.	.	.	.	.	.	.	1	1	1	0.8333	0
j = 10	.	.	.	.	.	.	.	.	.	1	1	0.8333	0
j = 11	.	.	.	.	.	.	.	.	.	.	1	0.8333	0
j = 12	.	.	.	.	.	.	.	.	.	.	.	0.8333	0
j = 13	.	.	.	.	.	.	.	.	.	.	.	.	0
j = 14	.	.	.	.	.	.	.	.	.	.	.	.	0
j > 14	.	.	.	.	.	.	.	.	.	.	.	.	0

- $b(t) = 200$: the outcomes are quite similar to the case where $b(t) = 50$, with 501,500 indicating the overall number of variables and 1,001,000 the linear inequality constraints. CPLEX performed 30,854 simplex iterations with a time of computation smaller than 20 s, showing again that the linear model can well scale the time of computation. The final value of $P_+(t+1)$ found by the solver was now

$$P_+(t+1) = 0.0242061448$$

and we do not report the value of the unknowns for the sake of simplicity.

Observe that in both the last two examples, given that subgroups are allowed to be quite large, despite relatively high budget values, the levels reached by the probabilities $P_+(t+1)$ are quite low.

As a general achievement, we can immediately describe the pattern of the nonzero unknowns which was respected in all four numerical tests reported above. In particular, we can prove that the solution of the formulations (5)–(9) only fills the principal submatrix of unknowns of order $h \times h$, depending h on the value of the budget parameter along with the number of possible subgroups L. Finally, this also allows one to straightforwardly assess a number of null unknowns and possibly reduce the complexity of the overall formulation.

6. A Multi-Period Model

This Section considers a multi-period reformulation of the LP (5)–(9), in order to further generalise the model (2). In particular, in place of the single-period model (5)–(9), where at time step t the quantity $P_+(t+1)$ is maximised starting from $P_+(t)$, we can consider a multi-period approach with the ultimate goal of maximising $P_+(T+1)$, where T is the time horizon. The latter approach is summarised in the following Nonlinear Program (NLP):

$$\max_{\alpha} \quad P_+(T+1), \tag{12}$$

$$P_+(t+1) = \sum_{k=1}^{L} a_k \sum_{j=0}^{k} \alpha_j^k(t) C_j^k P_+(t)^j (1 - P_+(t))^{k-j}, \qquad t = 1, \dots, T, \tag{13}$$

$$\alpha_j^k(t) \le \alpha_{j+1}^k(t) \qquad t = 1, \dots, T; \; j = 0, 1, \dots, k-1; \; k = 1, \dots, L, \tag{14}$$

$$\alpha_j^k(t) \ge \alpha_j^{k+1}(t) \qquad t = 1, \dots, T; \; j = 0, 1, \dots, k; \; k = 1, \dots, L-1, \tag{15}$$

$$\sum_{k=1}^{L} \alpha_j^k(t) \le b_j(t) \qquad t = 1, \dots, T; \; j = 0, 1, \dots, k, \tag{16}$$

$$0 \le \alpha_j^k(t) \le 1 \qquad t = 1, \dots, T; \; j = 0, 1, \dots, k; \; k = 1, \dots, L. \tag{17}$$

The above NLP includes the unknowns

$$\alpha_j^k(t), \qquad t = 1, \dots, T; \; j = 0, 1, \dots, k; \; k = 1, \dots, L,$$

which represent a larger number of variables with respect to Equations (5)–(9). This implies that the larger the time horizon T, the larger the number of unknowns of the resulting problem formulation. Moreover, as already remarked, Equations (12)–(17) represents an NLP, which in general increases the difficulty of its solution. In particular, the nonlinearities in Equation (13) might yield a nonconcave overall maximisation problem, which implies that a certain number of local maxima (which are not global) might arise. In addition, unlike Equations (5)–(9), the feasible set of Equations (12)–(17) is no longer a polyhedron, so that the final solutions are not necessarily located on the boundary of the feasible set. As a consequence, the choice of the starting value $P_+(1)$ in Equation (13) becomes a key factor for at least a couple of reasons:

- the sequence $\{P_+(t)\}$ strongly depends on $P_+(1)$, in a similar fashion of the single-period formulation (5)–(9);
- the choice of $\{P_+(t)\}$ strongly affects the local maximum provided by the solver adopted for Equations (12)–(17).

Regarding the constraints in Equation (13), observe that the constraints just state a recursion for the function in Equation (2), at each time step. Moreover, the constraints in Equations (14), (15), and (17) have a similar meaning of the constraints in Equations (6), (7), and (9), for the single-period formulation. Finally, the budget constraints in Equation (16) have a possible double formulation, due to the multi-period scheme.

Proposition 1. *Let the sequence* $\{\alpha_j^k(t)\}$ *be a global solution of Equations* (5)–(9), *for any* $t = 1, \dots, T$. *Then,* $\{\alpha_j^k(t)\}_{t=1,\dots,T}$ *is a global solution of Equations* (12)–(17).

Proof. Observe that from relation (2), one has the following equalities:

$$\begin{aligned} &P_+(t+1) - P_+(t) = \\ &\sum_{t=2}^{T+1} \left[\sum_{k=1}^{L} a_k \sum_{j=0}^{k} \alpha_j^k(t) C_j^k P_+(t)^j (1 - P_+(t))^{k-j} - \sum_{k=1}^{L} a_k \sum_{j=0}^{k} \alpha_j^k(t-1) C_j^k P_+(t-1)^j (1 - P_+(t-1))^{k-j} \right] = \\ &\sum_{t=2}^{T+1} \Delta P_+(t). \end{aligned}$$

Thus, considering that $P_+(1) \in \Re$ is a constant value, the maximisation in Equation (12) is equivalent to the maximisation,

$$\max \sum_{t=2}^{T+1} \Delta P_+(t). \tag{18}$$

Moreover, observe that if $P_+(t)$ is given, $\Delta P_+(t+1)$ depends only on the unknowns,

$$\alpha_j^k(t), \qquad j = 0, \dots, k;\ k = 1, \dots, L,$$

and the maximisation in Equation (5) is equivalent to

$$\max_{\alpha} \Delta P_+(t+1).$$

Now, suppose the sequence $\{\alpha_j^k(t)\}_{j=0,\dots,k;\ k=1,\dots,L}$ is a solution of Equations (5)–(9); then, it also satisfies Equation (13) for t, as well as Equation (14)–(17). Thus, if the sequences,

$$\left\{\alpha_j^k(t)\right\}_{j=0,\dots,k;\ k=1,\dots,L} \qquad\qquad t = 1, \dots, T \tag{19}$$

solve the problem (5)–(9), for $t = 1, \dots, T$, then by Equation (18), the values in Equation (19) are a feasible point of Equations (12)–(17), satisfying also Equation (18), i.e., Equation (19) is a solution of Equations (12)–(17). □

Lemma 3. *Let* $\left\{\alpha_j^k(t)\right\}$ *be a global solution of Equations* (5)–(9)*, for any* $t \in \{1, 2, \dots, T\}$*, where the inequality* (8) *is replaced by the nonlinear inequality,*

$$\phi_t\left[\sum_{k=1}^{L} \alpha_j^k(t)\right] \le b_j(t), \qquad \phi_t : \Re^{\frac{(L+1)(L+2)}{2}-1} \to \Re.$$

Then, $\left\{\alpha_j^k(t)\right\}_{t=1,\dots,T}$ *is also a global solution of Equations* (12)–(17)*, with Equation* (16) *replaced by*

$$\phi_t\left[\sum_{k=1}^{L} \alpha_j^k(t)\right] \le b_j(t), \qquad t = 1, \dots, T;\ j = 0, \dots, k. \tag{20}$$

Proof. The proof trivially follows the guidelines of the proof for Proposition 1. □

Lemma 4. *Let* $\left\{\alpha_j^k(t)\right\}$ *be a global solution of Equations* (5)–(9)*, for any* $t \in \{1, 2, \dots, T\}$*, where the inequality* (8) *is replaced by the nonlinear inequality,*

$$\Phi\left[\sum_{k=1}^{L} \alpha_j^k(t)\right] \le b_j(t), \qquad \Phi : \Re^{\frac{(L+1)(L+2)}{2}-1} \to \Re,$$

and Φ *is any convex function. Then,* $\left\{\alpha_j^k(t)\right\}_{t=1,\dots,T}$ *is also a feasible solution of Equations* (12)–(17)*, with Equation* (16) *replaced by*

$$\Phi\left[\sum_{t=1}^{T} \lambda_t \sum_{k=1}^{L} \alpha_j^k(t)\right] \le b_j(t), \qquad j = 0, \dots, k;\ 0 \le \lambda_t \le 1;\ \sum_{t=1}^{T} \lambda_t = 1, \tag{21}$$

being $b_j \ge \sum_{t=1}^{T} \lambda_t b_j(t)$.

Proof. The convexity of Φ and the hypotheses imply

$$\Phi\left[\sum_{t=1}^{T}\lambda_t\sum_{k=1}^{L}\alpha_j^k(t)\right] \leq \sum_{t=1}^{T}\lambda_t\Phi\left[\sum_{k=1}^{L}\alpha_j^k(t)\right] \leq \sum_{t=1}^{T}\lambda_t b_j(t) \leq b_j,$$

so that $\left\{\alpha_j^k(t)\right\}_{t=1,\ldots,T}$ also satisfy Equation (21). The rest of the proof follows the guidelines of the proof for Proposition 1. □

Remark 1. *Note that from Lemma 4, if the sequences* $\left\{\alpha_j^k(t)\right\}$, $t = 1,\ldots,T$, *are global solutions of Equations* (5)–(9) *for* $t = 1,\ldots,T$, *then the sequence* $\left\{\alpha_j^k(t)\right\}_{t=1,\ldots,T}$ *is a feasible solution but possibly not a global solution of Equations* (12)–(17), *with Equation* (16) *replaced by*

$$\Phi\left[\sum_{t=1}^{T}\lambda_t\sum_{k=1}^{L}\alpha_j^k(t)\right] \leq b_j(t), \qquad j = 0,\ldots,k;\ 0 \leq \lambda_t \leq 1;\ \sum_{t=1}^{T}\lambda_t = 1,$$

and $b_j \geq \sum_{t=1}^{T}\lambda_t b_j(t)$.

A Numerical Example

We provide here a preliminary numerical example, where the solution of the formulation (12)–(17) is considered, and the budget constraints (16) are replaced by the unique one (which aggregates possible different budgets at the epoches $1,\ldots,T$):

$$\sum_{k=1}^{L}\sum_{j=0}^{k}\sum_{t=1}^{T}\alpha_j^k(t) \leq \text{budget}.$$

Since the multi-period formulation is indeed nonlinear and possibly nonconvex, we preferred to preliminarily investigate its solutions through a renowned nonlinear (NLP) solver from the literature. In particular, we adopted Knitro 13.2.0 from the NEOS Server [13]. Nevertheless, we are aware that since the problem might be nonconvex (unless the last property is explicitly proved to hold for Equations (12)–(17)), the choice of the starting point by the NLP solver may be crucial, to both outreach an accurate final solution and to reduce the overall computational burden. The parameters of the two instances considered are summarised as follows (respectively):

$$\begin{cases} L = 10 \\ a_k = 1/L, \qquad k = 1,\ldots,L, \\ P_+(t) = 0.65, \\ T = 6, \\ \text{budget} = 20, \end{cases}$$

and

$$\begin{cases} L = 10, \\ a_k = 1/L, \qquad k = 1,\ldots,L, \\ P_+(t) = 0.65, \\ T = 6, \\ \text{budget} = 100. \end{cases}$$

In what follows, we briefly show that, due to the smaller budget (i.e., $20 < 100$) in the first scenario, it yields worse results (i.e., a smaller value of $P_+(T)$) with respect to the second scenario. Indeed, adopting the first set of parameters for our multi-period formulation, Knitro presolver was able to eliminate 403 constraints and 1 variable, so that the final formulation included 395 variables and 660 constraints. Moreover, for the probabilities

$\{P_+(t)\}$, we obtain the following nonmonotone sequence (due to the nonlinearity of the multi-period model)

$$\begin{aligned} P_+(1) &= 0.65, \\ P_+(2) &= 9.21406 \times 10^{-5}, \\ P_+(3) &= 1.51607 \times 10^{-5}, \\ P_+(4) &= 2.37203 \times 10^{-5}, \\ P_+(5) &= 2.06585 \times 10^{-5}, \\ P_+(6) &= 0.499547. \end{aligned}$$

Conversely, adopting the second set of parameters for our multi-period formulation, Knitro presolver was again able to eliminate 403 constraints and 1 variable, yielding for the probabilities the values

$$\begin{aligned} P_+(1) &= 0.65, \\ P_+(2) &= 0.140855, \\ P_+(3) &= 0.0613564, \\ P_+(4) &= 0.0521609, \\ P_+(5) &= 0.0573217, \\ P_+(6) &= 1. \end{aligned}$$

Again, one observes a nonmonotone behaviour for the sequence $\{P_+(t)\}$, due to nonlinearities of the multi-period model. For the sake of completeness let us remark that the solution of the last two nonlinear instances by Knitro required no more than 10 s each.

7. Conclusions

In this paper, we studied the problem of possibly enhancing standard dynamics from sociophysics (namely the model (1)), using a mathematical programming perspective. We were interested about following a couple of lines of research. First, we coupled the model (1) with an LP scheme in order to possibly control the probability $P_+(t+1)$ for a given probability $P_+(t)$, after leveraging the unknowns α_j^k in Equation (2). The final value of these variables gives a measure of the effort which is required to steer the diffusion process of the opinion '+' for the subgroups of cardinality k. The dynamics are strongly based on the idea of possibly maximising $P_+(t+1)$. A numerical experience is also reported in this regard, showing that the computational effort for solving LPs associated with this first proposal is definitely modest, even in the case where the solution of large instances is sought.

As a second task, we also extended the (one period) LP formulations to a more general multi-period formulation, and in this case it was necessary to compute the sequence $\{P_+(t+1), P_+(t+2), \ldots, P_+(T)\}$ for a given value of $P_+(t)$ and for a given time horizon, T. Some theoretical properties showing a connection between the solutions of both our proposals have been given, though much work yet is required, including a broad numerical experience also in the multi-period case.

Author Contributions: Conceptualization, A.E., G.F. and D.F.; Methodology, A.E., G.F. and D.F.; Software, A.E., G.F. and D.F.; Validation, A.E., G.F. and D.F.; Writing—original draft, A.E., G.F. and D.F.; Writing—review & editing, A.E., G.F. and D.F. All authors have read and agreed to the published version of the manuscript.

Funding: G.F. wishes to thank the CNR-INM—Institute of Marine Engineering, Italian National Research Council, and the GNCS (Gruppo Nazionale per il Calcolo Scientifico) group within INdAM (Istituto Nazionale di Alta Matematica "Francesco Severi"), Italy, for their support.

Data Availability Statement: Not applicable.

Conflicts of Interest: The authors declare no conflict of interest. The funders had no role in the design of the study; in the collection, analyses, or interpretation of data; in the writing of the manuscript; or in the decision to publish the results.

References

1. Bass, F.M. A new product growth for model consumer durables. *Manag. Sci.* **1969**, *15*, 215–227. [CrossRef]
2. Moore, G.A. *Crossing the Chasm: Marketing and Selling Disruptive Products to Mainstream Customers*; Harper Business: New York, NY, USA, 2014. Available online: https://archive.org/details/crossingthechasm_202002/mode/2up (accessed on 5 August 2023).
3. Rogers, E.M. *Diffusion of Innovations*; The Free Press: New York, NY, USA, 2003.
4. Schelling, T.C. Dynamic models of segregation. *J. Math. Sociol.* **1971**, *1*, 143–186. [CrossRef]
5. Galam, S. Sociophysics: A review of Galam models. *Int. J. Mod. Phys. C* **2008**, *19*, 409–440. [CrossRef]
6. Galam, S. Modelling rumors: The no plane pentagon french hoax case. *Phys. A* **2003**, *320*, 571–580. [CrossRef]
7. Ellero, A.; Fasano, G.; Sorato, A. A modified Galam's model for word-of-mouth information exchange. *Physica A* **2009**, *388*, 3901–3910. [CrossRef]
8. Ellero, A.; Fasano, G.; Sorato, A. Stochastic model for agents interaction with opinion–leaders. *Phys. Rev. E* **2013**, *87*, 042806. [CrossRef] [PubMed]
9. Galam, S. *Sociophysics. A Physicist's Modeling of Psycho-Political Phenomena*; Springer Science + Business Media, LLC: New York, NY, USA, 2012. [CrossRef]
10. Galam, S.; Jacobs, F. The role of inflexible minorities in the breaking of democratic opinion dynamics. *Physica A* **2003**, *381*, 366–376. [CrossRef]
11. Backstrom, L.; Huttenlocher, D.; Kleinberg, J.; Lan, X. Group formation in large social networks: Membership, growth, and evolution. In Proceedings of the 12th ACM SIGKDD International Conference on Knowledge Discovery and Data Mining, Philadelphia, PA, USA, 20–23 August 2006; pp. 44–54. [CrossRef]
12. Ellero, A.; Fasano, G.; Favaretto, D. An application of Linear Programming to sociophysics models. In Proceedings of the Modeling and Analysis of Complex Systems and Processes, Venice, Italy, 22–24 October 2020; pp. 23–33. Available online: https://ceur-ws.org/Vol-2795/ (accessed on 5 August 2023).
13. NEOS Server: State-of-the-Art Solvers for Numerical Optimization. Available online: https://neos-server.org/neos (accessed on 5 August 2023).
14. Fourer, R. Modeling languages versus matrix generators for linear programming. *ACM Trans. Math. Softw.* **1983**, *9*, 143–183. [CrossRef]
15. Fourer, R.; Gay, D.M.; Kernighan, B.W. *AMPL: A Modeling Language for Mathematical Programming*; Duxbury Press/Thomson Co.: Belmont, CA, USA, 2002. Available online: https://vanderbei.princeton.edu/307/textbook/AMPLbook.pdf (accessed on 5 August 2023).

physics

MDPI

Article

Phase Transition in the Galam's Majority-Rule Model with Information-Mediated Independence

André L. Oestereich [1], Marcelo A. Pires [2], Silvio M. Duarte Queirós [3] and Nuno Crokidakis [4,*]

1 Departamento de Economia, Universidade de Brasília, Brasília 70910-900, DF, Brazil; andrelo@id.uff.br
2 Eixo das Tecnologias, Universidade Federal de Alagoas, Delmiro Gouveia 57480-000, AL, Brazil; marcelo.pires@delmiro.ufal.br
3 Centro Brasileiro de Pesquisas Físicas and National Institute of Science and Technology for Complex Systems, Rio de Janeiro 22290-180, RJ, Brazil; sdqueiro@cbpf.br
4 Instituto de Física, Universidade Federal Fluminense, Niterói 24210-346, RJ, Brazil
* Correspondence: nunocrokidakis@id.uff.br

Abstract: We study the Galam's majority-rule model in the presence of an independent behavior that can be driven intrinsically or can be mediated by information regarding the collective opinion of the whole population. We first apply the mean-field approach where we obtained an explicit time-dependent solution for the order parameter of the model. We complement our results with Monte Carlo simulations where our findings indicate that independent opinion leads to order–disorder continuous nonequilibrium phase transitions. Finite-size scaling analysis show that the model belongs to the mean-field Ising model universality class. Moreover, results from an approach with the Kramers–Moyal coefficients provide insights about the social volatility.

Keywords: social dynamics; Galam model; collective phenomena in social systems; nonequilibrium phase transitions; order-disorder Monte Carlo simulation

1. Introduction

Opinion dynamics is one of the most attractive topics in sociophysics. This recent research area uses tools and concepts of statistical physics to describe some aspects of social and political behavior [1–4]. From the theoretical point of view, opinion models are interesting to physicists because they can present order–disorder transitions, hysteresis, scaling, and universality, among other typical features of physical systems, which have attracted much of attention [5–14]. Concerning sociologists, these methods are useful to improve forecasting by means of controlled toy models that can be run multiple times and help fine-tune field studies as well [15]. In addition to the interesting properties of opinion dynamics models, per se, such dynamics have also been applied in various fields such as finance and business [16], and epidemic dynamics with the presence of conflicting opinions [17–23], among others [3].

Among the most studied models, one can highlight the voter model [24,25], the Sznajd model [26], the Deffuant model [27], the kinetic exchange opinion models [28], and the majority rule model [1,29–31]. All the mentioned models are built based on distinct microscopic rules that control the dynamics of interactions among agents. The Sznajd model considers a two-state (up/down spins) outflow dynamics, where a group of agents sharing a common opinion influence the group's neighbors to follow the group's opinion. The model presents a phase transition between the positive and the negative consensus: initial densities of spins up smaller than $1/2$ lead eventually to all spins down, and densities greater than $1/2$ to all spins up, i.e., consensus absorbing states where the system cannot escape [26]. On the other hand, the Deffuant model considers the opinions as continuous variables, and the interactions depend on the "distance" among pairs of opinions, which defines the concept of bounded confidence. Depending on the value of such bounded

Citation: Oestereich, A.L.; Pires, M.A.; Duarte Queirós, S.M.; Crokidakis, N. Phase Transition in the Galam's Majority-Rule Model with Information-Mediated Independence. *Physics* **2023**, *5*, 911–922. https://doi.org/10.3390/physics5030059

Received: 6 June 2023
Revised: 14 July 2023
Accepted: 25 July 2023
Published: 31 August 2023

confidence, the population can evolve to consensus (all equal opinions) or to polarization (population divided into two distinct opinions). No phase transition is observed [27]. The majority rule model considers groups of g agents that interact through a quite simple rule: all agents in the group follow the local majority. In case of even values of g, a probability k defines which opinion will win the debate inside the group. The results of the model, regarding consensus and phase transitions, are similar to those observed in the Sznajd model [29]. Some applications of the majority rule model are mentioned in the following. Finally, the kinetic exchange opinion models are based on dynamics of wealth exchange. Interactions are pairwise and consider continuous opinions originally, or discrete three-state opinions ($+1$, -1 or 0 states) [28]. Both formulations lead the population to undergo order–disorder phase transitions, similar to what occurs in spin models. Absorbing states, where all agents are in the neutral state (all opinions 0 in the population), are observed. One can find that such absorbing states are distinct from the ones observed in the previous models, where all agents share opinion $+1$ or -1. Such kinds of consensus states are observed in kinetic exchange opinion models only in particular situations [28].

We are especially interested in the majority rule model, proposed by Serge Galam [1]. In this model, random groups of agents are chosen, and after the interaction of such agents, all of them assume the initial majority opinion. The model was studied elsewhere [32–39], and it was applied to a series of practical problems, like antivax movement [40], USA [41] and French [42] presidential elections, and terrorism [43], among many others.

Independence in opinion making and the failure of group influence was considered in several opinion dynamics models [44–52]. A recent extension of Galam's model in Ref. [32] considered the impact of independence in social dynamics. In that case, with probability q, an individual acts independently of the majority opinion of their group and chooses at random one of the two possible opinions. The introduction of that condition, quantified by the parameter, paves the way to the occurrence of an order–disorder nonequilibrium phase transition that does not occur in the original majority-rule model [1].

In the present study, we move farther afield than the independence mechanism considered in Ref. [32], and we take into account the overall global opinion of the population when an agent decides to act independently of the group's opinion. With this, one can find a more detailed picture of the process of independence, since agents can now take global opinion into account when they ignore in-group majority. This change manages to incorporate the concept of "impersonal influence" [53] established within political science, the goal of which is to quantify the influence of the anonymous mass of individuals outside one's small world composed of family, (close) friends, and acquaintances. That impersonal influence encompasses polls, reader's comments on news on digital media, and the individual's general perception by consulting social networks that can have an effect on their decision-making process.

The strength of the information-mediated independence can be controlled by a new parameter, g, that gauges the impact of the global population opinion, which is the macrostate, on the individuals. This impact can be of a contrarian nature, for negative values of g, where agents tend to take the opposite opinion to the population or it can be positive and reinforce the predominant opinion, thus helping the building of consensus.

In this paper, we move along the lines of canonical considerations over complex systems for which microscopic and macroscopic features influence one another. We develop an analytical framework in order to understand the results from numerical simulations. All results suggest the occurrence of order–disorder transitions, and the estimates of the critical exponents indicate that the model is in the mean-field Ising model universality class.

2. Model and Methods

Herein, we analyze a majority-rule model with independence; however, differently to Ref. [32] we assume a density-dependent probability, f_t, where t denotes the time, for changing the current opinion independently of the interacting group.

2.1. Model

Let us consider a population of N individuals, $i = 1, ..., N$, with opinions A or B, with respect to a given issue, that map into a stochastic variable, o_i, such that $o_i(A,t) = +1$ and $o_i(B,t) = -1$. Macroscopically, we compute the density of agents with opinion A,

$$\eta_A(t) \equiv \frac{1}{N}\sum_{i=1}^{N} \delta_{o_i(t),+1}, \tag{1}$$

with $\delta_{a,b}$ the Kronecker delta, and the density of agents with opinion B,

$$\eta_B(t) \equiv \frac{1}{N}\sum_{i=1}^{N} \delta_{o_i(t),-1} = 1 - \eta_A(t). \tag{2}$$

The mean opinion, from which we establish the macroscopic state of the system, reads:

$$m(t) \equiv \frac{1}{N}\sum_{i=1}^{N} o_i(t) = \eta_A(t) - \eta_B(t). \tag{3}$$

The dynamics of each individual is governed at each time step, t, by the following set of rules:

- an individual with opinion A can change to opinion B through two mechanisms:
  - with probability q, the individuals act independently of their group. In that case, the individuals change their opinion with probability $f^{AB}(t) = f(1 - g\,m(t))$;
  - otherwise, if the individuals do not act on their own, then there is a probability $1 - q$ that they change their opinion according to a local majority-rule, $A + 2B \to 3B$.
- on the other hand, an individual with opinion B can flip to opinion A through two mechanisms:
  - with probability q, the individuals decide to whether act independently of their group or not. In that case, the agent will change their opinion with probability $f^{BA}(t) = f(1 + g\,m(t))$;
  - otherwise, if the the individuals do not act on their own, then there is a probability $1 - q$ that those individuals change their opinion according to a local majority-rule, $B + 2A \to 3A$.

The rules listed above are translated into the transition matrix,

$$W(t) \equiv \begin{bmatrix} w_1 & w_2 \\ w_3 & w_4 \end{bmatrix} = \begin{bmatrix} q\,f^{AB}(t) & 1-q \\ 1-q & q\,f^{BA}(t) \end{bmatrix}. \tag{4}$$

Note that the definitions $f(t)^{AB} = f(1 - g\,m(t))$ and $f(t)^{BA} = f(1 + g\,m(t))$ imply that

$$\eta_A(t) > \eta_B(t) \Rightarrow m(t) > 0 \Rightarrow f(t)^{BA} > f(t)^{AB},$$

as expected.

Let us have a closer look at the parameters involved in the model: the parameter q is related to the backbone of our approach establishing the relative weight of the local peer-pressure, $p = 1 - q$, leading to a decision-making process wherein the individual either submits to the local majority (a conformist behavior) or the decision-making dynamics is carried out on its own. The probability $f^{XY}(t)$—related to the latter case—is naturally shaped by the assessment of the state of affairs provided by the global state, $m(t)$, so that a standard propensity to change opinion through reflection, f, is either boosted or mitigated. Epistemologically, the shaping of the probability is equivalent to the process of risk taking versus risk aversion, described within prospect theory [54]. Herein, we assume a linearized form $f^{XY}(m(t)) = f + v\,m(t) + \mathcal{O}(m(t)^2)$, so that, depending on the sign of v, one has either a follower or a contrarian impact. If $g = 0$, then $f^{XY}(t) = f$ and one recovers the results of Ref. [32].

2.2. Simulation Details

Our Monte Carlo simulations are structured within an agent-based framework, as individuals constitute the underlying object of study in social theories [55]. In the algorithm here, we consider a computational array of size N to store the opinion of each agent. In each time t, a Monte Carlo step (MCS) that represents a complete iteration through all agents is applied. During each interaction, the simulation chooses a group of 3 agents at random, considering their current opinion and applying specific rules. These rules are summarized in Table 1 and define how an agent's opinion may change based on various conditions and probabilities. After each MCS, we implement a simultaneous-parallel updating. This means that the updated opinions are applied to all agents at the same time, ensuring that the changes in opinions are synchronized across the entire population.

Table 1. Agent-based rules of the model.

Three-Agent Interaction	Transition Probability
Each agent with opinion A can flip to opinion B through two mechanisms:	
1. $A \to B$	$p^{(1)}_{A\to B} = q\, f^{AB}(t)$
2. $A + 2B \to 3B$	$p^{(2)}_{A\to B} = (1-q)\, \eta_B(t)^2$
Each agent with opinion B can flip to opinion A through two mechanisms:	
1. $B \to A$	$p^{(3)}_{B\to A} = q\, f^{BA}(t)$
2. $B + 2A \to 3A$	$p^{(4)}_{B\to A} = (1-q)\, \eta_A(t)^2$

3. Results and Discussion

3.1. Analytical Results

Using the mean-field approach, one can obtain a set of ordinary differential equations that describes the time evolution of the competing opinions in a population. To derive the rate of change of opinions A and B at time t, one needs to consider that each opinion is influenced by the intrinsic independent behavior (controlled by the parameter f), information-driven independence (modulated by the parameter g), and local interactions. Thus, based on the rules summarized in Table 1, one obtains the following mean-field equations:

$$\frac{d\eta_A(t)}{dt} = q\, f^{BA}(t)\, \eta_B(t) + (1-q)\, \eta_A(t)^2\, \eta_B(t) - q\, f^{AB}(t)\, \eta_A(t) - (1-q)\, \eta_A(t)\, \eta_B(t)^2, \tag{5}$$

$$\frac{d\eta_B(t)}{dt} = q\, f^{AB}(t)\, \eta_A(t) + (1-q)\, \eta_B(t)^2\, \eta_A(t) - q\, f^{BA}(t)\, \eta_B(t) - (1-q)\, \eta_B(l)\, \eta_A(t)^2, \tag{6}$$

$$f^{AB}(t) = f(1 - g\, m(t)), \tag{7}$$

$$f^{BA}(t) = f(1 + g\, m(t)). \tag{8}$$

From Equations (1)–(3), namely, that

$$\eta_A(t) = \frac{1}{2}(1 + m(t)), \quad \eta_B(t) = \frac{1}{2}(1 - m(t)), \quad \eta_A(t)\, \eta_B(t) = \frac{1}{4}(1 - m(t))^2, \tag{9}$$

the set of differential equations yields the ordinary differential equation for the global state, $m(t)$:

$$\frac{dm(t)}{dt} = -2q\, f\, (1-g) m(t) + (1-q)\, m(t) \frac{1 - m(t)^2}{2}. \tag{10}$$

In other words, starting from a given initial condition, $m(0) = m_0$, the macroscopic state evolves and eventually reaches a stationary state, $dm/dt = 0$; that state is lower-bounded by the maximal state of disagreement than $m = 0$, whereas when the population presents unanimity, $|m| = 1$. Thus, we expect that for certain conditions dictated by the parameters of the problem, the system can evade the final stationary state of disagreement and end up in a situation for which $|m| \neq 0$, i.e., a majority of individuals favoring opinion $A(B)$. Physically, $m(t)$ is thus defined as an order parameter. That turns out clearer when we consider that the population adjust is macrostate m aiming at minimizing its so-called Hamiltonian function, $\mathcal{H}$.

That is best understood when one recasts the previous equation into

$$\frac{dm(t)}{dt} = -\frac{\partial \mathcal{H}}{\partial m} = r\, m(t) + u\, m(t)^3, \tag{11}$$

where

$$r = \frac{1}{2}\{q[4f(1-g) + 1] - 1\}, \qquad u = \frac{1}{2}(q-1). \tag{12}$$

Therefore, the analytical form of $\mathcal{H}$,

$$\mathcal{H}(m) = -\frac{1}{2} r\, m^2 - \frac{1}{4} u\, m^4, \tag{13}$$

dictates not only the dynamics of the parameter m, but its stable outcome. First, since $q \leq 1$ and $u < 0$, the stability of the process is assured, as the fourth-order term is positive. In the limit $q \to 1$, the agents act independently from the local group, and totally rely on their assessment of the position of the whole population, and one has: $\lim_{q \to 1} u = 0^-$ and $\lim_{q \to 1} r > 0$, which leads to nontrivial minima of $\mathcal{H}$ at $m_c \neq 0$. By $u < 0$, the emergence of those $m \neq 0$ minima are related to the change of convexity of $\mathcal{H}$ at $m = 0$ from $\frac{d^2\mathcal{H}}{dm^2}|_{m=0} > 0$ to a concave profile $\frac{d^2\mathcal{H}}{dm^2}|_{m=0} < 0$. The fulfillment of the concave condition implies

$$|m| = m_c = \sqrt{-\frac{r}{u}} = \sqrt{1 - \frac{4fq(1-g)}{1-q}}, \qquad |m| \leq 1, \tag{14}$$

the graphical representation of which can be seen in Figure 1. Using relations (12) one finds:

$$m \sim (q_c - q)^{\beta}, \tag{15}$$

where $\beta = 1/2$ and

$$q_c = \frac{1}{1 + 4f(1-g)} \tag{16}$$

defines the critical peer-pressure relative weight, $p_c \equiv 1 - q_c$. Let us note that the instances with $g < 0$ imply a smaller value of q_c and, then a larger p_c in what we assume as a freethinker-prone behavior; on the other hand, when $g > 0$ we regard it as a conformist-prone case.

Equation (15) with $\beta = 1/2$ suggests a phase transition in the same universality class of the mean-field Ising model. We discuss this point in more detail in Section 3.3, where we exhibit the results of Monte Carlo simulations of the model.

Equation (16) corresponds to the limit $t \to \infty$ of the solution to Equation (17):

$$m(t) = \left[\exp(-2rt)\left(\frac{1}{m_0^2} + \frac{u}{r}\right) - \frac{u}{r}\right]^{-1/2} = m_c\left[\exp(-2rt)\left(\frac{m_c^2}{m_0^2} - 1\right) + 1\right]^{-1/2}, \tag{17}$$

where m_0 is the initial condition of the system.

One can further explore the dynamical behavior of the system, especially when the parameters are set at their critical values and one lets the system evolve. In that case, two situations deserve a particular attention: (i) when the initial state corresponds to an unanimity, $m_0 = 1$, the factor given by $\exp(-2rt)\left(m_c^2/m_0^2 - 1\right)$ in Equation (17) can be seen as perturbation, (ii) whereas for the same factor, this factor dominates Equation (17) when the initial condition is that of full disagreement ($m_0 \to 0$). These two possibilities result in two quite different behaviors of $m(t)$ in the short-term.

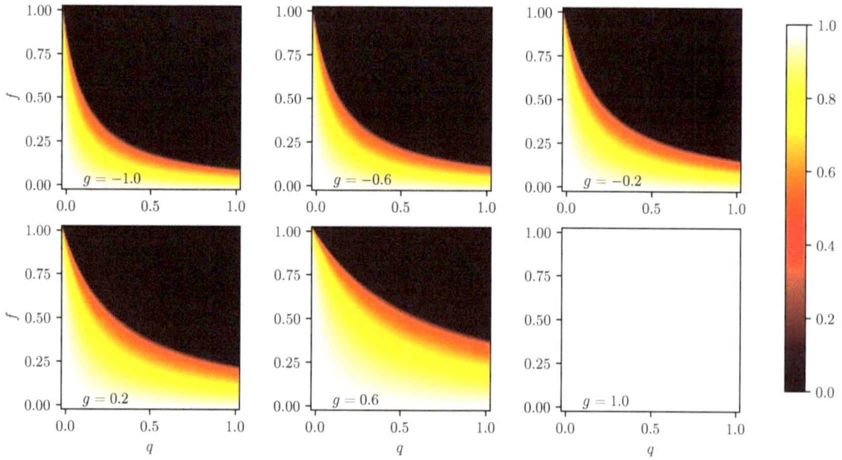

Figure 1. Graphical representations of Equation (14) for the stationary state solution, $|m|$, in the plane f versus q for typical values of g. As negative values of g imply a contrarian effect and positive a follower, the ordered region increases with an increase of g. Note that for $g = 1.0$, $|m| = 1$ for all values of f and q, as predicted by Equation (14).

3.2. Probabilistic Approach

The previous deterministic approach can be further seasoned when fluctuations are taken into account. Recalling that, for a population of N individuals the macroscopic state m, changes by $\mu = \pm 2/N$ every time the individuals switch their opinion with each opinion fraction varying by $1/N$, one finds:

$$\eta_A(t+1) - \eta_A(t) = \frac{1}{N} p^{\dagger}(t) - \frac{1}{N} p(t), \tag{18}$$

as soon as the focus is on the time evolution of the fraction of individuals with opinion A at time $t+1$, with

$$p^{\dagger}(m,t) = w_1\, \eta_A + w_2\, \eta_A\, \eta_B^2, \tag{19}$$

corresponding to the probability that the number of people with opinion A increases by one individual, and

$$p(m,t) = w_3\, \eta_B\, \eta_A^2 + w_4\, \eta_B, \tag{20}$$

giving the probability that the number of people with opinion A diminishes by one individual. The quantities (19) and (20) are identified as the operators of, respectively, creation and destruction in the probability space [56].

Taking into consideration that $p^{\dagger}$ and p correspond to an increment and a reduction of the macroscopic state by $\mu = 2/N$, respectively, one can establish the following master equation for the evolution of m for a time step, $\epsilon = 1/N$:

$$\eta(m, t+\epsilon) = p^{\dagger}(m-\mu, t)\, \eta(m-\mu, t) + p(m+\mu, t)\, \eta(m+\mu, t) + \bar{p}(m,t)\eta(m,t), \tag{21}$$

with $\bar{p} \equiv 1 - p^{\dagger} - p$ quantifying the maintenance of the macroscopic state. Formally, Equation (21) fits within the (normalized) one-step class of stochastic processes and, thus,

$$\eta(m,t) = \exp[\mathbf{L}_{KM}(m,t)]\eta(m_0, 0) \qquad \text{with} \qquad \eta(m,0) = \delta(m - m_0), \tag{22}$$

where, bearing in mind that a normalized quantity is computed and not just $N_A - N_B$, the Kramers–Moyal operator reads:

$$\mathbf{L}_{KM}(m,t) = \sum_{n=1}^{\infty} \frac{(-\mu)^{-n}}{n!} \frac{\partial^n}{\partial m^n} \left[p_{m,t} + (-1)^n \, p^{\dagger}_{m,t} \right]. \tag{23}$$

In considering $\mu \to 0$ so that the variance of m_t is kept fixed and equal to $\sigma^2_m(t)$, we neglect the terms of order $n > 2$, and the formal solution obtains the form of a Fokker–Planck equation:

$$\frac{\partial \eta(m,t)}{\partial t} = -\mu^{-1} \frac{\partial}{\partial m} [D_1(m,t)\, \eta(m,t)] + \frac{\mu^{-2}}{2} \frac{\partial^2}{\partial m^2} [D_2(m,t)\, \eta(m,t)]. \tag{24}$$

Therefrom we identify the Kramers–Moyal coefficients,

$$D_1(m,t) \propto p(m,t) - p^{\dagger}(m,t), \tag{25}$$

that defines the shape of the effective potential wherein the macroscopic dynamics of the order parameter evolves in time; on the other hand,

$$D_2(m,t) \propto p(m,t) + p^{\dagger}(m,t) \tag{26}$$

characterizes the magnitude of the fluctuations, which in the present social system we associate with the concept of social volatility [57].

Plugging the previous relations (19)/(20) for the probability creation/annihilation operators into Equations (25) and (26), one finally obtains:

$$D_1(m,t) \propto r\, m(t) + u\, m(t)^3, \tag{27}$$

as given by the effective-Hamiltonian, Equation (11) Landau approach. The second-order coefficient,

$$D_2(m,t) \propto 1 + q\,(4f - 1) - [1 + q(4\,f\,g - 1)]\, m(t)^2, \tag{28}$$

indicates a macroscopic feature of this model that is worth noting: the magnitude of the fluctuations—i.e., the social volatility—exhibited by the system depends on its state in such a form that as m increases and approaches $m = 1$ (unanimity), the volatility decreases. On the contrary, when the group shows strong disagreement, $m \approx 0$, the fluctuations approach the sate of maximal volatility. Such a behavior contrasts with what is measured in quantitative finance where the realized volatility is directly proportional to price variations [58]. If one considers that, within a physical context, D_2 is related to the (local) temperature of a physical system, one can assert that our model is able to capture the cooling down and the heating up of a social system as the system approaches or departs from consensus.

Alternatively, the fluctuations given by $D_2(m,t)$ can be understood from another perspective: given that $p^{\dagger}$ and p, respectively, correspond to an increment and a reduction of the macroscopic state by $\mu = 2/N$, one can then interpret D_1 as the imbalance between the likelihood of increment and reduction of $p(m)$, whereas D_2 to be related to the average over increment and reduction, which is nonvanishing. Thisis related to the microscopic change of opinion that each individual can make and which corresponds to a source of the macroscopic fluctuations that end up being expressed by the social volatility. Complementarily, those fluctuations yield an entropy production that can be associated with the total information output due to the microscopic interaction between agents. Therefore, around a consensus, one measures a less volatile state as that state is less entropic and vice versa.

3.3. Monte Carlo Simulation and Finite-Size Scaling

Figure 2a shows a comparison between the analytical solution, as elucidated in Section 3.1, and the Monte Carlo simulation for the Galam model with information-mediated independence. We plot the stationary values of the macrostate obtained from simulations for a population size $N = 10^4$ and from Equation (14) for typical values of g and fixed $f = 0.5$. One can see a good agreement between the results obtained by both methods of calculations. One as well observes the order–disorder phase transitions, which mark a collective change in the population behavior. These transitions indicate a macroscopic change from the so-called ordered state, characterized by the presence of a well-defined majority ($|m| > 0$), to the disordered-state, characterized by the absence of a clear majority ($|m| \sim 0$). When $q = 0$, the result of the original Galam model, i.e., a consensus in the population (all agents sharing opinion A or B) is recovered.

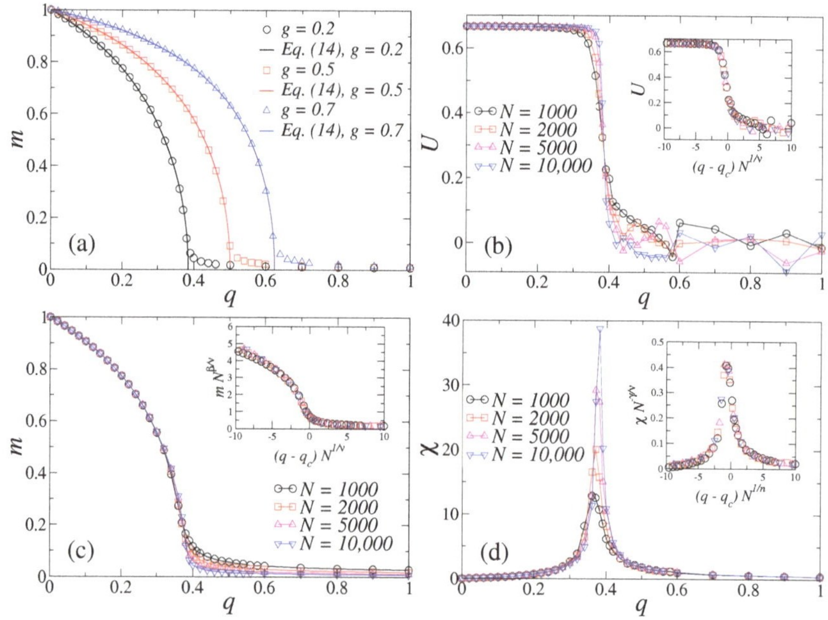

Figure 2. The model Monte Carlo results for $f = 0.5$ averaged over 100 simulations. (**a**) Stationary order parameter, m (collective opinion), as a function of q. The symbols represent the simulations for $N = 10^4$ and typical values of g, and the lines show the analytical results obtained from Equation (14). The results at fixed $g = 0.2$ and distinct population sizes N for, and the corresponding finite-size scaling analysis for (**b**) the Binder cumulant, U (30), (**c**) the order parameter, m, and (**d**) the susceptibility, χ (29) are also shown. The following critical probability, $q_c \approx 0.385$ and the critical exponents, $\beta \approx 0.50$, $\gamma \approx 1.00$, and $\nu \approx 2.00$, are obtained; see text for details.

In order to verify the universality class of the model, we performed numerical simulations for distinct population sizes and applied a so-called scaling analysis. In addition to the order parameter, m we also computed the fluctuations χ of the order parameter (or "susceptibility"), defined as

$$\chi \equiv N\left(\langle m^2 \rangle - \langle m \rangle^2\right) \tag{29}$$

and the Binder cumulant U, defined as [59]

$$U \equiv 1 - \frac{\langle m^4 \rangle}{3\,\langle m^2 \rangle^2}. \tag{30}$$

The brackets here denote the averaging over distinct realizations of the dynamics.

As an example, we exhibit in Figure 2 the finite-size scaling (FSS) analysis of the order parameter, the susceptibility, and the Binder cumulant for four lattice sizes, for $f = 0.5$ and $g = 0.2$. We obtain the critical value, q_c, by the crossing of the Binder cumulant curves, as seen in Figure 2b, to be $q_c \approx 0.385$, in quite a good agreement with the analytical result of Equation (16), $q_c \approx 0.3846$. The critical exponents, β, γ, and ν, are found by the best collapse of data. The FSS analysis was based on the standard relations,

$$\begin{aligned} m(q,N) &\sim N^{-\beta/\nu}, &(31)\\ \chi(q,N) &\sim N^{\gamma/\nu}, &(32)\\ U(q,N) &\sim \text{constant}, &(33)\\ q_c(N)-q_c &\sim N^{-1/\nu} &(34) \end{aligned}$$

Considering Equations (31)–(34), the values of the critical exponents, $\beta \approx 0.5$, $\gamma \approx 1.0$, and $\nu \approx 2.0$, are obtained. The data collapses are exhibited in Figure 2b–d. We have also verified that for other values of f and g, the same exponents are obtained. The results suggest that the model belongs to the mean-field Ising model universality class, and it is also in the same universality class of the Sznajd model and kinetic exchange opinion models in the presence of independence [44,47,50,60].

The above results, namely, Equation (34), bridge with the dynamical analysis as Equation (17) sets up a relaxation timescale, τ, of the macroscopic parameter that is inversely proportional to r. Comparing the exponential factor in Equation (17) with the typical term related to the relaxation, $e^{t/\tau}$, one obtains the relaxation time, $\tau = -1/(2r)$. Then, plugging Equation (16) into the definition of r, one finds: $r = (q - q_c)/(2q_c)$, and, finally,

$$\tau \sim (q_c - q)^{-1}. \tag{35}$$

Explicitly, at criticality, one considers a relaxation timescale of the order parameter, $m(t)$, that diverges with the same scale-invariant functional form as the correlation length. Since the propagator given by the Fokker–Planck Equation (24) rules all relaxation quantities of $m(t)$, the same slowing down near the transition is found for the self-correlation function of $m(t)$, $\langle m(t')\, m(t)\rangle \sim \exp[-|t' - t|/\tau]$.

4. Conclusions

In this paper, we studied an extension of the Galam's majority-rule model. For this purpose, we introduced the mechanism of independence, considering that individuals can act independently of their interaction groups with a given probability q that is complementary to the peer-pressure weight, $p = 1 - q$. In addition, the individuals inspect the global population opinion and such opinion affects their independent probability. When an individual does not act independently of the group, the individual follows the local majority opinion, as in the original Galam model.

We observed that the independence mechanism leads the population to undergo a critical change of behavior at $q = q_c$ in which a minimal consensus $m \neq 0$—where m is the order parameter of the model—optimizes the overall state of the population better than the case of complete disagreement. Within that phase transition context, we derived an expression for order parameter, $m(t)$. From its stationary solution, we obtained the critical behavior, $m \sim (q_c - q)^{\beta}$ with $\beta = 1/2$. We found that as one approached the critical transition, the relaxation of the overall state is ever slower with its typical timescale, $\tau \sim (q_c - q)^{-1}$. The other canonical critical exponents, γ and ν, were obtained through Monte Carlo simulations. From the set of critical exponents, we verified that the model is in the ubiquitous universality class of the mean-field Ising model. This result is expected since, as opinions are mapped into random variables, $o_i = \pm 1$, the phase transition corresponds to a group $\mathbb{Z}_2$ symmetry breaking of which the Ising model is the quintessential case.

Let us mention that, while our model is not defined by a physical Hamiltonian, the identification with the Ising universality class arises from a series of results coming from three methods: mean-field approach, Monte Carlo simulations, and finite-size scaling analysis.

From the microscopic dynamics, we derived the probabilistic evolution of m. The results obtained allowed us to confirm the critical behavior of m from the first Kramers–

Moyal coefficient, and from the second, the nature of the fluctuations that can be coined as social volatility. In respect of the latter, we learned that the magnitude of the volatility depends on the state of the population in a inverse proportion relation way, such that, in this case, herding in opinion tends to induce less agitation in the population. Further insights into this subject matter to be discussed in our future studies.

Author Contributions: Conceptualization, A.L.O., M.A.P., S.M.D.Q. and N.C.; methodology, A.L.O., M.A.P., S.M.D.Q. and N.C.; software, A.L.O., M.A.P., S. M.D.Q. and N.C.; validation, A.L.O., M.A.P., S.M.D.Q. and N.C.; formal analysis, A.L.O., M.A.P., S. M.D.Q. and N.C.; investigation, A.L.O., M.A.P., S.M.D.Q. and N.C. resources, A.L.O., M.A.P., S.M.D.Q. and N.C.; data curation, A.L.O., M.A.P., S.M.D.Q. and N.C.; writing—original draft preparation, A.L.O., M.A.P., S.M.D.Q. and N.C.; writing—review and editing, A.L.O., M.A.P., S.M.D.Q. and N.C.; visualization, A.L.O., M.A.P., S.M.D.Q. and N.C.; supervision, S.M.D.Q. and N.C.; project administration, S.M.D.Q. and N.C.; funding acquisition, S.M.D.Q. and N.C. All authors have read and agreed to the published version of the manuscript.

Funding: This research was funded by National Council for Scientific and Technological Development (grant numbers: 302348/2022-0 and 310893/2020-8) and by Fundação Carlos Chagas Filho de Amparo à Pesquisa do Estado do Rio de Janeiro (grant number: 203.217/2017).

Data Availability Statement: The data can be obtained from the corresponding author upon a reasonable request.

Conflicts of Interest: The authors declare no conflict of interest.

References

1. Galam, S. Sociophysics: A review of Galam models. *Int. J. Mod. Phys. C* **2008**, *19*, 409–440. [CrossRef]
2. Sen, P.; Chakrabarti, B.K. *Sociophysics: An Introduction*; Oxford University Press: Oxford, UK, 2014.
3. Castellano, C.; Fortunato, S.; Loreto, V. Statistical physics of social dynamics. *Rev. Mod. Phys.* **2009**, *81*, 591–646. [CrossRef]
4. da Luz, M.G.; Anteneodo, C.; Crokidakis, N.; Perc, M. Sociophysics: Social collective behavior from the physics point of view. *Chaos Solit. Fract.* **2023**, *170*, 113379. [CrossRef]
5. Dinkelberg, A.; O'Sullivan, D.J.; Quayle, M.; Maccarron, P. Detecting opinion-based groups and polarization in survey-based attitude networks and estimating question relevance. *Adv. Complex Syst.* **2021**, *24*, 2150006. [CrossRef]
6. Sobkowicz, P. Whither now, opinion modelers? *Front. Phys.* **2020**, *8*, 587009. [CrossRef]
7. Martins, A.C. Discrete opinion dynamics with M choices. *Eur. Phys. J. B* **2020**, *93*, 1. [CrossRef]
8. Baumann, F.; Lorenz-Spreen, P.; Sokolov, I.M.; Starnini, M. Modeling echo chambers and polarization dynamics in social networks. *Phys. Rev. Lett.* **2020**, *124*, 048301. [CrossRef] [PubMed]
9. Anagnostopoulos, A.; Becchetti, L.; Cruciani, E.; Pasquale, F.; Rizzo, S. Biased opinion dynamics: When the devil is in the details. *Inf. Sci.* **2022**, *593*, 49–63. [CrossRef]
10. Zubillaga, B.J.; Vilela, A.L.M.; Wang, M.; Du, R.; Dong, G.; Stanley, H.E. Three-state majority-vote model on small-world networks. *Sci. Rep.* **2022**, *12*, 282. [CrossRef] [PubMed]
11. Oestereich, A.L.; Pires, M.A.; Crokidakis, N. Three-state opinion dynamics in modular networks. *Phys. Rev. E* **2019**, *100*, 032312. [CrossRef]
12. Oestereich, A.L.; Pires, M.A.; Queirós, S.D.; Crokidakis, N. Hysteresis and disorder-induced order in continuous kinetic-like opinion dynamics in complex networks. *Chaos Solit. Fract.* **2020**, *137*, 109893. [CrossRef]
13. Vazquez, F.; Saintier, N.; Pinasco, J.P. Role of voting intention in public opinion polarization. *Phys. Rev. E* **2020**, *101*, 012101. [CrossRef] [PubMed]
14. Oestereich, A.L.; Crokidakis, N.; Cajueiro, D.O. Impact of memory and bias in kinetic exchange opinion models on random networks. *Phys. A Stat. Mech. Its Appl.* **2022**, *607*, 128199. [CrossRef]
15. Imai, K. *Quantitative Social Science. An Introduction*; Princetion University Press: Princeton, NJ, USA, 2018.
16. Zha, Q.; Kou, G.; Zhang, H.; Liang, H.; Chen, X.; Li, C.C.; Dong, Y. Opinion dynamics in finance and business: A literature review and research opportunities. *Financ. Innov.* **2020**, *6*, 44. [CrossRef]
17. Fang, F.; Ma, J.; Li, Y. The coevolution of the spread of a disease and competing opinions in multiplex networks. *Chaos Solit. Fract.* **2023**, *170*, 113376. [CrossRef] [PubMed]
18. Pires, M.A.; Oestereich, A.L.; Crokidakis, N. Sudden transitions in coupled opinion and epidemic dynamics with vaccination. *J. Stat. Mech. Theory Exp.* **2018**, *2018*, 053407. [CrossRef]
19. Pires, M.A.; Oestereich, A.L.; Crokidakis, N.; Queirós, S.M.D. Antivax movement and epidemic spreading in the era of social networks: Nonmonotonic effects, bistability, and network segregation. *Phys. Rev. E* **2021**, *104*, 034302. [CrossRef]
20. Oestereich, A.; Pires, M.; Crokidakis, N.; Cajueiro, D. Optimal rewiring in coupled opinion and epidemic dynamics with vaccination. *arXiv* **2023**, arXiv:2305.02488. [CrossRef]

21. Franceschi, J.; Pareschi, L.; Bellodi, E.; Gavanelli, M.; Bresadola, M. Modeling opinion polarization on social media: Application to Covid-19 vaccination hesitancy in Italy. *arXiv* **2023**, arXiv:2302.01028. [CrossRef]
22. Harari, G.S.; Monteiro, L.H.A. An epidemic model with pro and anti-vaccine groups. *Acta Biotheor.* **2022**, *70*, 20. [CrossRef]
23. Wang, J.; Xu, W.; Chen, W.; Yu, F.; He, J. Information sharing can suppress the spread of epidemics: Voluntary vaccination game on two-layer networks. *Phys. A Stat. Mech. Its Appl.* **2021**, *583*, 126281. [CrossRef]
24. Clifford, P.; Sudbury, A. A model for spatial conflict. *Biometrika* **1973**, *60*, 581–588. [CrossRef]
25. Holley, R.A.; Liggett, T.M. Ergodic theorems for weakly interacting infinite systems and the voter model. *Ann. Probab.* **1975**, *3*, 643–663. Available online: https://www.jstor.org/stable/2959329 (accessed on 20 July 2023). [CrossRef]
26. Sznajd-Weron, K.; Sznajd, J.; Weron, T. A review on the Sznajd model—20 years after. *Phys. A Stat. Mech. Its Appl.* **2021**, *565*, 125537. [CrossRef]
27. Deffuant, G.; Neau, D.; Amblard, F.; Weisbuch, G. Mixing beliefs among interacting agents. *Adv. Complex Syst.* **2000**, *3*, 87–98. [CrossRef]
28. Biswas, S.; Chatterjee, A.; Sen, P.; Mukherjee, S.; Chakrabarti, B.K. Social dynamics through kinetic exchange: The BChS model. *Front. Phys.* **2023**, *11*, 1196745. [CrossRef]
29. Galam, S. Application of statistical physics to politics. *Phys. A Stat. Mech. Its Appl.* **1999**, *274*, 132–139. [CrossRef]
30. Galam, S. Minority opinion spreading in random geometry. *Eur. Phys. J. B-Condens. Matter Complex Syst.* **2002**, *25*, 403–406. [CrossRef]
31. Krapivsky, P.L.; Redner, S. Dynamics of Majority Rule in Two-State Interacting Spin Systems. *Phys. Rev. Lett.* **2003**, *90*, 238701. [CrossRef]
32. Crokidakis, N.; de Oliveira, P.M.C. Inflexibility and independence: Phase transitions in the majority-rule model. *Phys. Rev. E* **2015**, *92*, 062122. [CrossRef]
33. Crokidakis, N.; de Oliveira, P.M.C. The first shall be last: Selection-driven minority becomes majority. *Phys. A Stat. Mech. Its Appl.* **2014**, *409*, 48–52. [CrossRef]
34. Abilhoa, W.D.; de Oliveira, P.P.B. Density classification based on agents under majority rule: Connectivity influence on performance. In *Distributed Computing and Artificial Intelligence, Proceedings of the 16th International Conference, Ávila, Spain, 26–28 June 2019;* Herrera, F., Matsui , K., Rodríguez-González, S., Eds.; Springer: Cham, Switzerland, 2020; pp. 163–170. [CrossRef]
35. Galam, S. *Sociophysics. A Physicist's Modeling of Psycho-Political Phenomena;* Springer Science+Business Media, LLC: New York, NY, USA, 2012; Ch. 7. [CrossRef]
36. Gimenez, M.C.; Reinaudi, L.; Paz-García, A.P.; Centres, P.M.; Ramirez-Pastor, A.J. Opinion evolution in the presence of constant propaganda: Homogeneous and localized cases. *Eur. Phys. J. B* **2021**, *94*, 35. [CrossRef]
37. Krapivsky, P.; Redner, S. Divergence and consensus in majority rule. *Phys. Rev. E* **2021**, *103*, L060301. [CrossRef]
38. Muslim, R.; Anugraha, R.; Sholihun, S.; Rosyid, M.F. Phase transition and universality of the three-one spin interaction based on the majority-rule model. *Int. J. Mod. Phys. C* **2021**, *32*, 2150115. [CrossRef]
39. Azhari; Muslim, R. The external field effect on the opinion formation based on the majority rule and the *q*-voter models on the complete graph. *Int. J. Mod. Phys. C* **2023**, *34*, 2350088. [CrossRef]
40. Pires, M.A.; Crokidakis, N. Dynamics of epidemic spreading with vaccination: Impact of social pressure and engagement. *Phys. A Stat. Mech. Its Appl.* **2017**, *467*, 167–179. [CrossRef]
41. Galam, S. The Trump phenomenon: An explanation from sociophysics. *Int. J. Mod. Phys. B* **2017**, *31*, 1742015. [CrossRef]
42. Galam, S. Unavowed abstention can overturn poll predictions. *Front. Phys.* **2018**, *6*, 24. [CrossRef]
43. Galam, S. Identifying a would-be terrorist: An ineradicable error in the data processing? *Chaos Solit. Fract.* **2023**, *168*, 113119. [CrossRef]
44. Sznajd-Weron, K.; Tabiszewski, M.; Timpanaro, A.M. Phase transition in the Sznajd model with independence. *Europhys. Lett.* **2011**, *96*, 48002. [CrossRef]
45. Nyczka, P.; Sznajd-Weron, K. Anticonformity or independence?—Insights from statistical physics. *J. Stat. Phys.* **2013**, *151*, 174–202. [CrossRef]
46. Nyczka, P.; Sznajd-Weron, K.; Cisło, J. Phase transitions in the q-voter model with two types of stochastic driving. *Phys. Rev. E* **2012**, *86*, 011105. [CrossRef] [PubMed]
47. Muslim, R.; Kholili, M.J.; Nugraha, A.R. Opinion dynamics involving contrarian and independence behaviors based on the Sznajd model with two-two and three-one agent interactions. *Phys. D Nonlinear Phenom.* **2022**, *439*, 133379. [CrossRef]
48. Vieira, A.R.; Crokidakis, N. Phase transitions in the majority-vote model with two types of noises. *Phys. A Stat. Mech. Its Appl.* **2016**, *450*, 30–36. [CrossRef]
49. Muslim, R.; Anugraha, R.; Sholihun, S.; Rosyid, M.F. Phase transition of the Sznajd model with anticonformity for two different agent configurations. *Int. J. Mod. Phys. C* **2020**, *31*, 2050052. [CrossRef]
50. Crokidakis, N. Phase transition in kinetic exchange opinion models with independence. *Phys. Lett. A* **2014**, *378*, 1683–1686. [CrossRef]
51. Abramiuk, A.; Sznajd-Weron, K. Generalized independence in the q-voter model: How do parameters influence the phase transition? *Entropy* **2020**, *22*, 120. [CrossRef]
52. Vieira, A.R.; Anteneodo, C.; Crokidakis, N. Consequences of nonconformist behaviors in a continuous opinion model. *J. Stat. Mech. Theory Exp.* **2016**, *2016*, 023204. [CrossRef]

53. Mutz, D.C. Impersonal influence: Effects of representations of public opinion on political attitudes. *Political Behav.* **1992**, *14*, 89–122. [CrossRef]
54. Kahneman, D.; Tversky, A. Prospect theory: An analysis of decision under risk. *Econometrica* **1979**, *47*, 263–291. [CrossRef]
55. Conte, R.; Gilbert, N.; Bonelli, G.; Cioffi-Revilla, C.; Deffuant, G.; Kertesz, J.; Loreto, V.; Moat, S.; Nadal, J.P.; Sanchez, A.; et al. Manifesto of computational social science. *Eur. Phys. J. Spec. Top.* **2012**, *214*, 325–346. [CrossRef]
56. van Kampen, N.G. *Stochastic Processes in Physics and Chemistry*; North Holland: Amsterdam, The Netherlands, 2007.
57. Behrens, T.E.J.; Hunt, L.T.; Woolrich, M.W.; Rushworth, M.F.S. Associative learning of social value. *Nature* **2008**, *456*, 245–249. [CrossRef]
58. Queirós, S.M.D. On non-Gaussianity and dependence in financial time series: A nonextensive approach. *Quant. Financ.* **2005**, *5*, 475–487. [CrossRef]
59. Binder, K. Finite size scaling analysis of Ising model block distribution functions. *Z. Phys. B.* **1981**, *43*, 119–140. [CrossRef]
60. Calvelli, M.; Crokidakis, N.; Penna, T.J. Phase transitions and universality in the Sznajd model with anticonformity. *Phys. A Stat. Mech. Its Appl.* **2019**, *513*, 518–523. [CrossRef]

physics

MDPI

Article

Opinion Dynamics Systems via Biswas–Chatterjee–Sen Model on Solomon Networks

Edmundo Alves Filho [1], Francisco Welington Lima [1,*], Tayroni Francisco Alencar Alves [1], Gladstone de Alencar Alves [2] and Joao Antonio Plascak [3,4,5]

1 Dietrich Stauffer Computational Physics Laboratory, Departamento de Física, Universidade Federal do Piauí, Teresina 64049-550, PI, Brazil; edmundoalves12@gmail.com (E.A.F.); tay@ufpi.edu.br (T.F.A.A.)
2 Departamento de Física, Universidade Estadual do Piauí, Teresina 64002-150, PI, Brazil; alves.gladstone@gmail.com
3 Departamento de Física, Centro de Ciências Exatas e da Natureza (CCEN), Universidade Federal da Paraíba, Cidade Universitária, João Pessoa 58051-970, PB, Brazil; pla@uga.edu
4 Departamento de Física, Universidade Federal de Minas Gerais, C. P. 702, Belo Horizonte 30123-970, MG, Brazil
5 Department of Physics and Astronomy, University of Georgia, Athens, GA 30602, USA
* Correspondence: wel@ufpi.edu.br

Abstract: The critical properties of a discrete version of opinion dynamics systems, based on the Biswas–Chatterjee–Sen model defined on Solomon networks with both nearest and random neighbors, are investigated through extensive computer simulations. By employing Monte Carlo algorithms on SNs of different sizes, the magnetic-like variables of the model are computed as a function of the noise parameter. Using the finite-size scaling hypothesis, it is observed that the model undergoes a second-order phase transition. The critical transition noise and the respective ratios of the usual critical exponents are computed in the limit of infinite-size networks. The results strongly indicate that the discrete Biswas–Chatterjee–Sen model is in a different universality class from the other lattices and networks, but in the same universality class as the Ising and majority-vote models on the same Solomon networks.

Keywords: Biswas–Chatterjee–Sen model; Solomon networks; finite-size scaling; universality class

Citation: Filho, E.A.; Lima, F.W.; Alves, T.F.A.; Alves, G.d.A.; Plascak. J.A. Opinion Dynamics Systems via Biswas–Chatterjee–Sen Model on Solomon Networks. *Physics* **2023**, *5*, 873–882. https://doi.org/10.3390/physics5030056

Received: 12 June 2023
Accepted: 3 August 2023
Published: 17 August 2023

1. Introduction

Although a great deal of attention has been given to sociophysics in the last forty years, it is over the past two decades that the interest in this area has been indeed enhanced, mainly due to the dynamics properties that are present in social systems or networks (see, e.g., Refs. [1–9]). The early models in this new scenario have been proposed by Stauffer [6] and Galam [9] using local majority rule arguments, which are also called dynamics of opinions.

There are also other models that are useful for simulating the behavior of people in a community where each person can either be influenced by the neighbors or be a means to influence these neighbors. Just to cite a few among the great variety of dynamical systems, which are be of particular interest to the present study, let us mention the Ising model (IM), originally proposed to treat the thermal properties of magnetic systems [10–12], the majority-vote model (MVM) [13], and the Biswas, Chatterjee, and Sen (BChS) model [5].

The IM, MVM, and BChS model have been treated by using Monte Carlo simulations on different lattices and networks. It has been noticed that the existence of a transition–whether of first or second order–and, in the latter case, the corresponding critical exponents, strongly depend on the type of the network. Magnetic models of the Ising kind have been reviewed in Ref. [14] and, more recently, the universality class of the BChS model, in comparison with the MVM, has been treated in Ref. [15]. In Ref. [15], besides regular lattices, topologies like Apollonian networks, regular and directed Barabási-Albert networks, regular and directed Erdös-Rènyi random graphs, among others, have been considered. In particular, a more recent review on the BChS model [16] has more details of what has been determined so far in this dynamical system.

From the results collected in the literature, mainly regarding the dynamics of the above models, it has been noticed that: (i) one cannot say, using just basic arguments, whether a known dynamical model, on a given network, could or could not undergo a specific phase transition and (ii) in the case the system does present a second-order phase transition, what should its universality class be.

Thus, it is worth investigating what kind of dynamic behavior the discrete BChS model presents when defined on Solomon networks (SNs). In addition, since the BChS model on SNs, to our best knowledge, has not been treated so far in the literature, it is also beneficial to compare its new behavior with the previous results obtained using different networks (and also to compare it with the IM and MVM). Actually, one can further implement Table 3 of Ref. [15]. It should be stressed that the novelty of the SNs over other networks is that the SNs have a more realistic approach to the behavior of a community of people as soon as SNs take into account the influence of different types of neighbors one faces, for example, at home and at work. This is indeed a new ingredient that has not been included in the lattices, networks, and random graphs studied so far.

Thus, in the present study, the BChS dynamical system on SNs is studied through the Monte Carlo simulations allied with the finite-size-scaling techniques. The paper is organized as follows. In Section 2, we briefly present the SNs and the BChS model in its discrete version, together with the magnetic-like variables considered here, such as the magnetization, susceptibility and the reduced fourth-order Binder cumulant. Section 2 also provides with the main ingredients of the Monte Carlo simulations employed to compute the evolution of the physical variables, which allow understanding the transition type. The results obtained are presented in Section 3. Section 4 addresses the conclusions and gives some remarks.

2. Solomon Networks, Biswas–Chatterjee–Sen Model and Monte Carlo Simulation Details

2.1. Solomon Networks

As mentioned in Section 1, the behavior of people in a community is such that each person can be influenced by the neighbors and can, at the same time, influence the neighbors. However, in reality, the neighbors at home differ from the neighbors at the workplace except, certainly, when everybody works at home. This more general situation can be modeled by using two different lattices, one representing the home lattice and the other the workplace lattice. From this general point of view, one can consider linear chains of length L and $N = L$ sites in one dimension (1D), and square lattices with $N = L^2$ sites in 2D (can be naturally extended to higher dimensions). If one labels by a site i ($1 \leq i \leq N$) the sites in the workplace lattice, the home lattice to be labeled $P(i)$, to recognize $P(i)$ is a random permutation of the order established in the workplace lattice. Thus, each person occupies two entirely different sites, i and $P(i)$, in the workplace lattice and home lattice, respectively. A sketch of a part of a Solomon network in two-dimensions is shown in Figure 1. From Figure 1, one can see that the neighborhood of a site on one lattice is different from the neighborhood of the corresponding permuted site on the other lattice. The one-dimensional case can be straightforwardly obtained from Figure 1.

Such a network composed of two chains or two square lattices has been suggested in Refs. [17,18]. These types of networks are called Solomon networks [19], mainly because each person is equally shared by two lattices, just as in the King Solomon's biblical story. These SNs are also close to small-world networks [20,21]. In this way, within each lattice we have the proper type of interaction defined by the corresponding model (e.g., IM, MVM, BChS and others). As a result, the net interaction of the variables defined at site i is a sum of the corresponding interactions of the site i with its neighboring sites on the workplace lattice, added to the interactions with the neighbor sites of $P(i)$ on the home lattice.

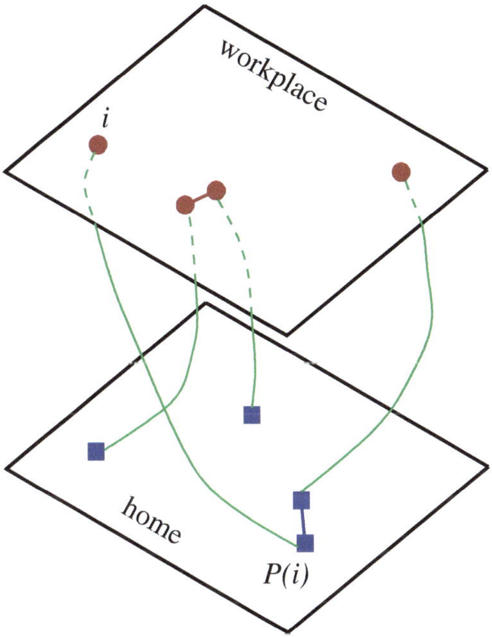

Figure 1. Sketch of a part of Solomon network (SN) in two dimensions. The upper layer represents the workplace lattice, and the corresponding sites, labeled by i, are given by the full circles. The bottom layer represents the home lattice, with the corresponding sites labeled by the permutation $P(i)$ and given by the full squares. The thick (straight) lines represent nearest-neighbor interactions on each layer. The curved lines specify the resulting random permutation in the home lattice of the order established in the workplace lattice.

It is worth mentioning here that extensions can be made on the networks as well as on specific interactions on each lattice. For example, one may introduce some correlation between residence and workplace lattices, making $P(i)$ not completely random, reflecting the situation when people select workplaces closer to their homes. More generally, instead of just representing the workplace and home, the lattices can represent any two types of groups of people, or even of more than two groups. Different topologies for the two groups could also be considered. Although all these generalizations indeed warrant further investigations, in what follows, we consider only random permutations $P(i)$ in order not to have additional theoretical parameter.

2.2. Biswas–Chatterjee–Sen Model

The BChS model has already been treated on different regular lattices and networks [5,22–24]. Below, we list just the main ingredients of the model on SNs.

Each site i of SNs (having N sites) has an individual. These individuals have opinion variables defined by $\sigma_i(t)$, at a given simulation time step t, that can take only three different values, e.g., -1, 0, or $+1$. The individual opinions $\sigma_i(t)$ are then updated at the following simulation time $t+1$ according to the BChS rules, as follows. (i) First, one initial state is constructed by randomly choosing one of the three opinion values -1, 0, or $+1$ for each site i of the SNs. (ii) Second, a random site i is selected in order to be updated. (iii) Next, a site j (which has a bond to the previously chosen site i) is selected at random and an affinity, μ_{ij}, is ascribed to this ij bond. Note that j, the corresponding site sharing the bond with site i, can be located either on the workplace lattice or home lattice. This affinity parameter, μ_{ij}, is viewed as another discrete variable taking a positive value $+1$, with a probability q to be turned negative to -1. In this way, as has been shown in earlier studies [5,22–24], the probability q acts, in the end, as an external noise, modeling local discordances. (iv) The opinion variables are now updated as

$$\sigma_i(t+1) = \sigma_i(t) + \mu_{ij}\sigma_j(t)\,, \tag{1}$$
$$\sigma_j(t+1) = \sigma_j(t) + \mu_{ij}\sigma_i(t)\,, \tag{2}$$

where $\sigma_i(t+1)$ and $\sigma_j(t+1)$ are the updated opinion states of the two sites i and j, respectively, at the simulation time $t+1$. Note that in these SNs, one sometimes updates sites that are not necessarily neighbors and can even be far apart. (v) Finally, when the opinion state $\sigma_i(t+1)$ of any site happens to be out of the interval $[-1,+1]$, this site to be automatically equal to either $+1$ or -1 depending on the case of $\sigma_i(t+1) > +1$ or $\sigma_i(t+1) < -1$, respectively.

2.3. Magnetic-like Variables of Interest

Averaging the opinion variables $\sigma_i(t)$ over all individuals constitutes an order parameter M, given by

$$M = \left|\sum_{i=1}^{N}\sigma_i(t)\right|/N, \tag{3}$$

where t is chosen to be large enough for the system having reached the stationary state. It has been shown that the system undergoes a phase transition, which is driven by the random configurations, in contrast to the the usual magnetic transitions driven by temperature. This means that for $q < q_c$, where q denotes the noise probability and q_c is its critical value, $M \neq 0$ and an ordered phase is present, while for $q > q_c$, $M = 0$ and the system is in a disordered phase instead. $q = q_c$ indicates a second-order phase transition, in which both ordered and disordered phases become identical and highly correlated (For more details, see [12,13,19,25]).

To fully characterize this phase transition, besides the above order parameter, one can also analyse its magnetic-like quantities, such as the order parameter fluctuation or susceptibility, $\chi(q)$, and the reduced fourth-order Binder cumulant, $U_4(q)$, which are given, respectively, by

$$M(q) = \Big[\langle M\rangle_t\Big]_c\,, \tag{4}$$
$$\chi(q) = N\Big[\langle M^2\rangle_t - \langle M\rangle_t^2\Big]_c\,, \tag{5}$$
$$U_4(q) = 1 - \left[\frac{\langle M^4\rangle_t}{3\langle M^2\rangle_t^2}\right]_c. \tag{6}$$

The expressions (4), (5) and (6), where the brackets $\langle\cdots\rangle_t$ stand for time averages computed after the system has reached the stationary state, and $[\cdots]_c$ represents averages over different initial configurations, to be used in Monte Carlo simulations.

2.4. Monte Carlo Simulation Details

The quantities (4)–(6) were computed as a function of q by using extensive Monte Carlo simulations on SNs of finite sizes ranging from $N = 1000,\ 5000,\ 10{,}000,\ 30{,}000,\ 50{,}000,\ 70{,}000$, up to 100,000 for linear chains (1D) and $L = 16,\ 32,\ 64,\ 128,\ 256$, up to 512 for square lattices (2D) where in the latter case $N = L^2$. The initial 10^5 Monte Carlo steps (MCSs) were discarded where, as described above in this Section, one MCS consists of randomly choosing N sites of the workplace network. The number of MCSs was found to be large enough for the systems to reach their stationary state. Next, the following 2×10^5 MCSs were taken to compute the corresponding time averages. On the other hand, the configurational averages have been obtained by considering 10^3 (for the larger lattices) to 10^4 (for the smaller lattices), different initial configurations for each set of network size N and parameter q.

Close to the critical region, where $q \sim q_c$, and for large networks size N, one can assume the following finite-size scaling (FSS) relations for the quantities defined in (4)–(6) (see Ref. [25]),

$$M(q) = N^{-\beta/\nu} f_M(N^{1/\nu}(q - q_c)), \tag{7}$$

$$\chi(q) = N^{\gamma/\nu} f_\chi(N^{1/\nu}(q - q_c)), \tag{8}$$

$$U_4(q) = f_{U_4}(N^{1/\nu}(q - q_c)), \tag{9}$$

$$q_c(N) = q_c + bN^{-1/\nu}, \tag{10}$$

where β, ν, and γ are the critical exponents of the order parameter, the fluctuation of the order parameter (susceptibility, χ)) and the correlation length, respectively. The functions, $f_k(x)$, with $x = N^{1/\nu}(q - q_c)$ and $k = \{M, \chi, U_4\}$, are the corresponding scaling functions, and b is a non-universal constant. The value q_c is the corresponding critical noise for an infinite-size network. Equation (10) is the scaling behavior of the pseudo-critical noise parameter as a function of the network size N, and $q_c(N)$ is estimated here as the q-value at the peak of the susceptibility.

In Section 3, the scaling laws (7)–(10) are used based on the computed values of the proper variables from the Monte Carlo simulations in order to obtain a description of the critical behavior of the model. It should be stressed here that, while for 1D networks, N is just the number of sites, for 2D networks, N is convenient to replaced with the linear size L.

The scaling equations (7)–(10) are obtained assuming a second-order phase transition, while, however, stay the same in the case of the first-order phase transition, but with the lattice dimension replacing the exponents ratios.

3. Results and Discussion

The dependence of the reduced Binder cumulant, U_4, on the noise parameter, q, for several finite size lattices, is shown in Figure 2a, for the (1D) case and in Figure 2b, for the (2D) case. From Figure 2, one can see that the system indeed undergoes the second-order phase transition as soon as Equation (9) does not depend on the size of the network and U_4 functions cross at the same point for different values of N apart from some finite-size effects, resulting in $q_c = 0.215(2)$ for (1D) and $q_c = 0.216(2)$ for 2D. Interestingly, the critical noise, is almost the same, within the error bars, for both types of lattices. This observation is in contrast with what one obtains by studying the IM and the MVM on the same networks. For the IM, one finds: $T_c = 2.995(3)$ on (1D) [26] and $T_c = 6.985(4)$ on 2D [27], where T_c is the usual critical temperature (here, the number in the parentheses is the statistical error to the last digit). For the MVM, one finds: $T_c = 1.165(4)$ on (1D) [26] and $T_c = 1.915(5)$ on 2D [27], where T_c is now the critical social temperature [27]. The above results are also shown in Table 1.

Equations (7), (8), and (10) can now be exploited to compute the ratios of the critical exponents using the critical noise, q_c, obtained. By computing $M(q_c)$, $\chi(q_c)$, and the difference, $q_c(N) - q_c$, at q_c for different values of N, while plotting the values in log scales as a function of N, a straight line is expected. What distinguishes a second-order phase transition from a first-order phase transition is the value of the exponents being different from the dimension of a lattice. The pseudo-transition noise, $q_c(N)$, is obtained from the maximum value of the susceptibility and is discussed in more details just below. The desired critical exponents ratios are thus straigtforwardly given by the slope obtained from a linear fit to the data. Typical plots of the above quantities are shown in Figure 3 as a function of the sizes N in (1D) (Figure 3a–c) and L in 2D (Figure 3d,e) on SNs. From the alignment of the data in Figure 3, one obtains an additional evidence that corroborates the transition as indeed to be in the second order case. The critical exponent ratios obtained are indicated in Figure 3 and listed in Table 2.

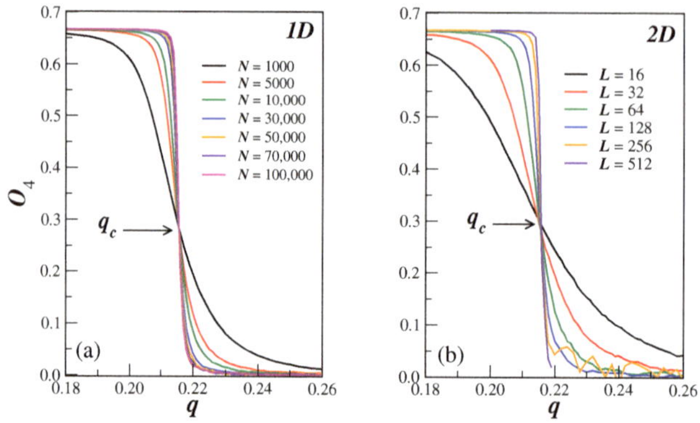

Figure 2. Reduced Binder cumulant, U_4, plotted as function of the noise probability, q, for one dimension (1D) (**a**), and 2D (**b**) for different values of the sizes, N (1D) and L (2D), as indicated. The crossings occur at $q_c = 0.215(2)$ (1D) and 0.216(5) (2D), where the numbers in the parentheses show the statistical errors in the last digit..

Table 1. Critical points (critical value, q_c, of the noise probability, q) of the IM (thermodynamical critical temperature), MVM (critical social temperature) and (BChS) model (critical noise) in (1D) and 2D SNs. The number in the parentheses shows the statistical error in the last digit.

IM		MVM		BChS	
1D	2D	1D	2D	1D	2D
2.995(3)	6.985(4)	1.165(4)	1.915(5)	0.215(2)	0.216(2)

Table 2. Critical exponents ratios, $1/\nu$, β/ν, γ/ν of the BChS model on SNs. The results obtained for the IM and the MVM on the same networks from Refs. [26,27] are included for comparison. The exponents ratio, $\gamma/\nu(q_c^{\max})$, represents the results from the maximum of the susceptibility (see Figure 4d,e). See text for details. The number in the parentheses shows the statistical error in the last digit.

Model	$1/\nu$	β/ν	$\gamma/\nu(q_c)$	$\gamma/\nu(q_c^{\max})$
		Linear Chain (1D)		
IM [26]	0.52(3)	0.237(4)	0.512(4)	0.515(2)
MVM [26]	0.55(5)	0.223(8)	0.534(3)	0.512(3)
BChS	0.52(5)	0.238(7)	0.511(5)	0.524(6)
		Square Lattice (2D)		
IM [27]	0.91(3)	0.52(3)	0.97(4)	0.99(3)
MVM [27]	0.91(2)	0.50(2)	1.00(2)	0.99(2)
BChS	0.92(4)	0.53(4)	1.02(4)	1.06(7)

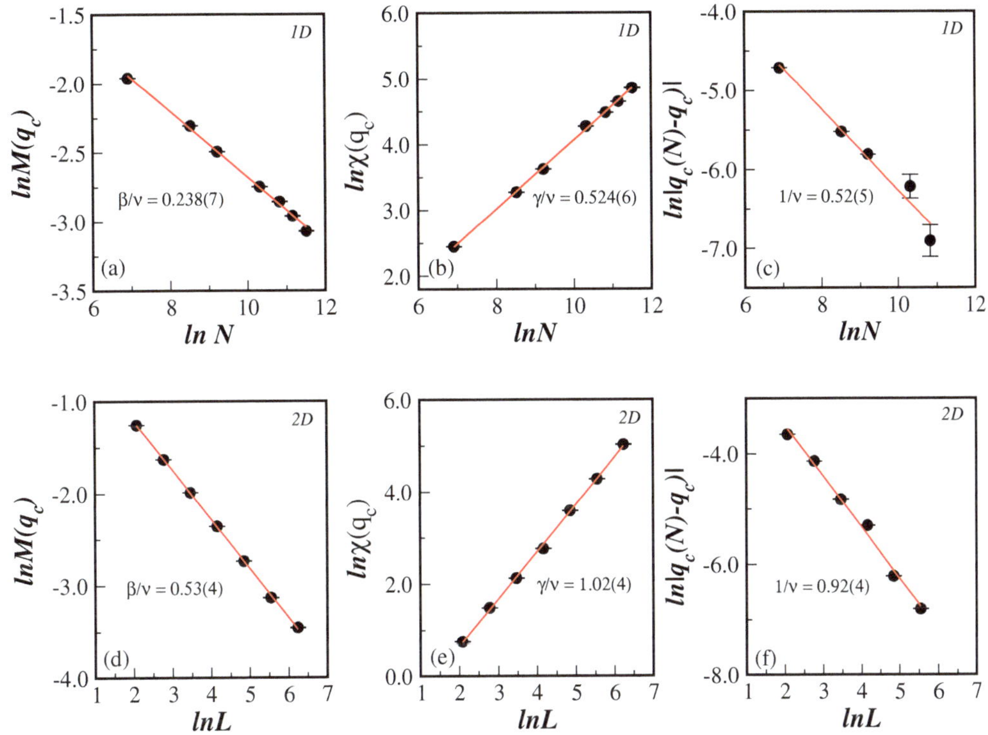

Figure 3. Log-log plot of the order parameter, $M(q_c)$ (**a**,**d**), the susceptibility, $\chi(q_c)$ (**b**,**e**), and the magnitude of the displacement, $q_c(N) - q_c$ (**c**,**f**), as a function of N (1D) and L (2D), on SNs. All quantities are computed at the extrapolated critical noise parameter, q_c, for the corresponding dimension. The ratios of the critical exponents, β, γ and ν, are shown being obtained from the slopes of the linear fits (red lines) to the data.

So far, Equations (7)–(9) have been used only to obtain the critical noise and the critical exponents. However, the equatons can be further exploited to obtain the corresponding universal scaling functions, $f_k(x)$, by analyzing the Monte Carlo data in a still wide-enough range of q. This means that when the scaled-order parameter, $MN^{\beta/\nu}$, the scaled susceptibility, $\chi N^{-\gamma/\nu}$, and the reduced Binder cumulant, U_4, are plotted as a function of the scaled noise probability displacement, $(q - q_c)N^{1/\nu}$, the data collapse to a corresponding universal scaling function each. Figure 4 shows that indeed the collapse is observed for the listed scaling quantities in (1D), see Figure 4a–c, and in 2D, see Figure 4d,e, (recall that for the 2D case, N is replaced by L). The collapse shown in Figure 4 indicates the transition is indeed of a second order. Moreover, the computed critical exponents seem to be accurate, based on the present simulations.

Interestingly, the susceptibility, $\chi(q)$, presents a pronounced peak for q close to the noise transition value q_c, as is depicted in Figure 4b for (1D) and Figure 4e for 2D. This means that one has, at $q_c^{\max}$, a peak with value $\chi(q_c^{\max})$ for any of the network sizes, N or L. Thus, $q_c^{\max}$ is assumed to be the pseudo-critical noise used in Figure 3c,f, and $\chi(q_c^{\max})$ can also be used in Equation (8) to obtain another estimate of the critical exponents ratio, γ/ν. Comparing the results in the fourth and fifth columns of Table 2, one can see that the collapse estimate of γ/ν is well comparable with the critical exponents finding (fourth column in Table 2).

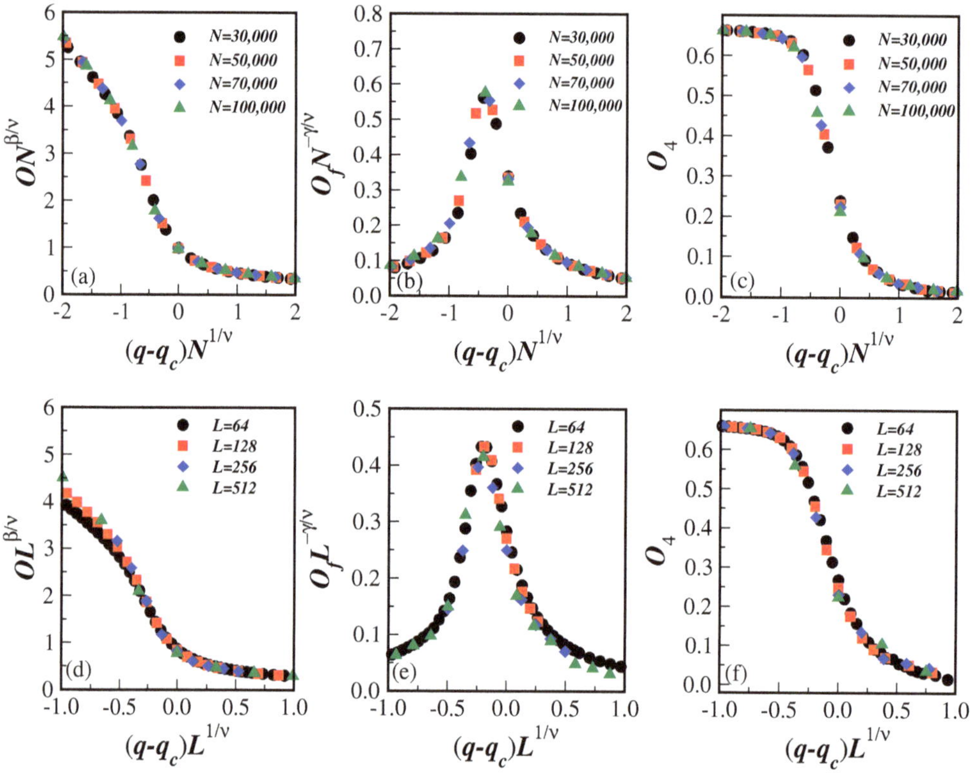

Figure 4. Collapse of the data for the BChS model on SNs of the scaled-order parameter, $MN^{\beta/\nu}$ (**a**,**d**), the scaled susceptibility, $\chi N^{-\gamma/\nu}$ (**b**,**e**), and the scaled reduced Binder cumulant, U_4 (**c**,**f**) as functions of the scaled displacements, $(q-q_c)N^{1/\nu}$, in (1D) (**a**–**c**) and $(q-q_c)L^{1/\nu}$ in 2D (**d**–**f**) for different values of the network sizes, N and L, as indicated.

4. Concluding Remarks

In this paper, the discrete version of the BChS model, defined on one- and two-dimensional SNs, has been studied through extensive Monte Carlo simulations. The variables of the model, coming from the order parameter, have been computed for several values of the local consensus controlling parameter and different network sizes in both dimensions. The finite-size-scaling approach indicates that the system undergoes a second-order phase transition, as soon as the critical exponents and critical ratios certainly differ from the dimension of the network (which would imply a first-order transition instead). However, contrary to the findings in IM and MVM on the same networks, the critical noise values are found to be the same, within the statistical error, for both dimensions.

The independence of the critical point on spatial dimension is quite striking. Certainly, longer simulations would allow to positively detect the obtained equality with smaller statistical errors. Let us note that the critical noise in 2D is slightly larger than in (1D). We consider the simulations in 3D Solomon lattices to better clarify on this intriguing finding.

The BChS model in regular 1D lattices, like the IM, does not have a finite critical point. This is an expected result in classical models with short-range interactions. However, in SNs, these models can sustain an order for finite noise or finite temperatures. This transition could be associated with the fact that despite being in one dimension, due to the permutation of the sites in one of the lattices, the interactions turn out to be longer ranged, in a way that can induce a finite noise or finite temperature phase transition (recall that the 1D IM, where all spins interact with each other, has a phase transition with mean-field critical exponents). From Figure 1, one can realize that the SNs present the effective dimensionality which is higher than the corresponding embedding dimensionality.

The computed critical exponents of the BChS model on SNs are different from those obtained on regular 2D and 3D lattices [28], and are also different from those on other networks and random graphs [15]. As discussed in Section 1, this looks a general trend in these topologies, because even the existence of a phase transition is not straight enough to predict using just physical grounds. Nevertheless, the results conveyed in Table 2 strongly suggest that the BChS model, the IM and MVM fall all in the same universality class when defined on the SNs. The longer-ranged interactions may be a reason that all three models belong the same universality class. Recall that the mean-field approximation is obtained when long-range interactions are present and the exponents are independent of the model type. In addition, this may be the universality class of other similar systems in sociophysics. For completeness, we extend Table 3 of Ref. [15] by including the new universality class of the BChS model on SNs; see Table 3.

Table 3. Extension of Table 3 of Ref. [15] by including the universality class of the BChS model, IM and MVM on SNs, as obtained in the present study. The "proper class" indicates that the exponents are different from any known model. z represents the connectivity of the network or random graph.

Lattice or Network	Universality Class	Ref.
Discrete Biswas–Chatterjee–Sen Model		
Fully connected	Mean field	[5]
Regular dimension-D	d-Dimensional IM	[28]
Apollonian	proper class	[22]
Barabási-Albert	Proper class z-dependent exponents	[15]
Directed Barabási-Albert	MVM z-dependent exponents	[23]
Erdös-Rènyi	Proper class z-dependent exponents	[24]
Directed Erdös-Rènyi	Proper class z-dependent exponents	[24]
Small world	$z \to \infty$ either of Erdös-Rènyi graphs	[24]
Solomon	Proper class including IM and MVM	This paper
Continuum Biswas–Chatterjee–Sen model		
Fully connected	Mean field	[5]
Regular dimension-D	d-Dimensional IM	[28]

Author Contributions: Conceptualization, T.F.A.A.; Methodology and F.W.L.; Validation, E.A.F.; Formal analysis, G.d.A.A., F.W.L. and J.A.P.; Investigation, T.F.A.A.; Writing–original draft, F.W.L.; Writing–review and editing, T.F.A.A., G.d.A.A. and J.A.P.; Supervision, G.d.A.A. and J.A.P. All authors have read and agreed to the published version of the manuscript.

Funding: This research received no external funding.

Data Availability Statement: The data can be obtained from the correspondence upon a reasonable request.

Acknowledgments: Computational support provided by the system SGI Altix 1350 from the CENAPAD.UNICAMP-USP computational park, Campinas, São Paulo-Brazil, is gratefully acknowledged. The supports from Conselho Nacional de Desenvolvimento Científico e Tecnológico (CNPq), Coordenação de Aperfeiçoamento de Pessoal de Nível Superior (CAPES), A Fundação de Amparo à Pesquisa do Estado de Minas Gerais (FAPEMIG),and Fundação Cearense de Apoio ao Desenvolvimento Científico e Tecnológico (FUNCAP) are also gratefully acknowledged.

Conflicts of Interest: The authors declare no conflict of interest.

References

1. Chakrabarti, B.K.; Chakraborti, A.; Chatterjee, A. (Eds.). *Econophysics and Sociophysics: Trends and Perspectives*; Wiley-VCH Verlag GmbH & Co. KGaA: Weinheim, Germany, 2006. [CrossRef]
2. Castellano, C.; Fortunato, S.; Loreto, V. Statistical physics of social dynamics. *Rev. Mod. Phys.* **2009**, *81*, 591–646. [CrossRef]
3. Helbing, D. *Quantitative Sociodynamics. Stochastic Methods and Models of Social Interaction Processes*; Springer: Berlin/Heidelberg, Germany, 2010. [CrossRef]
4. Galam, S. *Sociophysics. A Physicist's Modeling of Psycho-Political Phenomena*; Springer Science+Business Media, LLC: New York, NY, USA, 2012. [CrossRef]
5. Biswas, S.; Chatterjee, A.; Sen, P. Disorder induced phase transition in kinetic models of opinion dynamics. *Physica A* **2012**, *391*, 3257–3265. [CrossRef]
6. Stauffer, D. A biased review of sociophysics. *J. Stat. Phys.* **2013**, *151*, 9–20. [CrossRef]
7. Sen, P.; Chakrabarti, B.K. *Sociophysics: An Introduction*; Oxford University Press: New York, NY, USA, 2013.
8. Noorazar, H. Recent advances in opinion propagation dynamics: A 2020 survey. *Eur. Phys. J. Plus* **2020**, *135*, 521. [CrossRef]
9. Galam, S.; Mod, I.J. The Trump phenomenon: An explanation from sociophysics. *Int. J. Mod. Phys. B* **2017**, *31*, 1742015. [CrossRef]
10. Ising, E. Beitrag zur Theorie des Ferromagnetismus. *Z. Phys.* **1925**, *31*, 235–258. [CrossRef]
11. Ising, E. Contribution to the Theory of Ferromagnetism. Available online: https://www.hs-augsburg.de/~harsch/anglica/Chronology/20thC/Ising/isi_intr.html (accessed on 29 July 2023).
12. Baxter, R.J. *Exactly Solved Models in Statistical Mechanics*; Academic Press: London, UK, 1982.
13. de Oliveira, M.J. Isotropic majority-vote model on a square lattice. *J. Stat. Phys.* **1992**, *66*, 273–281. [CrossRef]
14. Lima, F.W.S.; Plascak, J.A. Magnetic models on various topologies. *J. Phys. Conf. Ser.* **2014**, *487*, 012011. [CrossRef]
15. Alencar, D.S.M.; Alves, T.F.A.; Alves, G.A.; Macedo-Filho, A.; Ferreira, R.S.; Lima, F.W.S.; Plascak, J.A. Opinion dynamics systems on Barabási-Albert networks: Biswas–Chatterjee–Sen model. *Entropy* **2023**, *25*, 183. [CrossRef] [PubMed]
16. Biswas, S.; Chatterjee, A.; Sen, P.; Mukherjee, S.; Chakrabarti, B.K. Social dynamics through kinetic exchange: The BChS model. *Front. Phys.* **2023**, *11*, 1196745. [CrossRef]
17. Erez, T.; Hohnisch, M.; Solomon, S. Statistical economics on multi-variable layered networks. In *Economics: Complex Windows*; Salzano, M., Kirman, A., Eds.; Springer: Milan, Italy, 2005; pp. 201–217. [CrossRef]
18. Goldenberg, J.; Shavitt, Y.; Shir, E.; Solomon, S. Distributive immunization of networks against viruses using the 'honey-pot' architecture. *Nat. Phys.* **2005**, *1*, 184–188. [CrossRef]
19. Malarz, K. Social phase transition in Solomon network. *Int. J. Mod. Phys. C* **2003**, *14*, 561–565. [CrossRef]
20. Pękalski, A. Ising model on a small world network. *Phys. Rev. E* **2001**, *64*, 057104. [CrossRef] [PubMed]
21. Herrero, C.P. Ising model in small-world networks. *Phys. Rev. E* **2002**, *65*, 066110. [CrossRef] [PubMed]
22. Lima, F.W.S.; Sumour, M.A.; Moreira, A.A.; Araújo, A.D. majority-vote and BCS model on Complex Networks. *Phys. A* **2021**, *571*, 125834. [CrossRef]
23. Lima, F.W.S.; Plascak, J.A. Kinetic models of discrete opinion dynamics on directed Barabási–Albert networks. *Entropy* **2019**, *21*, 942. [CrossRef]
24. Raquel, M.T.S.A.; Lima, F.W.S.; Alves, T.F.A.; Alves, G.A.; Macedo-Filho, A.; Plascak, J.A. Non-equilibrium kinetic Biswas–Chatterjee–Sen model on complex networks. *Physcial A* **2022**, *603*, 127825. [CrossRef]
25. Binder, K.; Heermann, D.W. *Monte Carlo Simulation in Statistical Phyics: An Introduction*; Springer: Berlin/Heidelberg, Germany, 2010. [CrossRef]
26. Lima, F.W.S. Equilibrium and non-equilibrium models on Solomon networks. *Int. J. Mod. Phys. C* **2016**, *28*, 1650134. [CrossRef]
27. Lima, F.W.S. Equilibrium and nonequilibrium models on Solomon networks with two square lattices. *Int. J. Mod. Phys. C* **2017**, *28*, 1750099. [CrossRef]
28. Mukherjee, S.; Chatterjee, A. Disorder-induced phase transition in an opinion dynamics model: Results in two and three dimensions. *Phys. Rev. E* **2016**, *94*, 062317. [CrossRef] [PubMed]

physics MDPI

Article

Faculty Hiring Network Reveals Possible Decision-Making Mechanism

Sunjing Zheng [1], Nan Jiang [1,2], Xiaomeng Li [1], Mingzhong Xiao [3,*] and Qinghua Chen [1,*]

1 School of Systems Science, Beijing Normal University, Beijing 100875, China; sunjingzheng@mail.bnu.edu.cn (S.Z.); nanjiang@mail.bnu.edu.cn (N.J.); lixiaomeng@bnu.edu.cn (X.L.)
2 School of Science, Wuhan University of Science and Technology, Wuhan 430081, China
3 School of Artificial Intelligence, Beijing Normal University, Beijing 100875, China
* Correspondence: xmz@bnu.edu.cn (M.X.); qinghuachen@bnu.edu.cn (Q.C.); Tel.: +86-10-58807943 (M.X.); +86-10-58807085 (Q.C.)

Abstract: Social physics (or sociophysics) offers new research perspectives for addressing social issues in various domains. In this study, we explore the decision-making process of doctoral graduates during their transition from graduation to employment, drawing on the ideas of sociophysics. We divide the process into two decision steps and propose a generative model based on appropriate assumptions. This model effectively reproduces empirical data, allowing us to derive essential parameters that influence the decision-making process from empirical observations. Through a comparison of the best-fit parameters, we discover that doctoral graduates in business disciplines tend to exhibit more concentrated employment choices, while those in computer science and history disciplines demonstrate a greater diversity of options. Furthermore, we observe that universities consider factors beyond rankings when selecting doctoral graduates.

Keywords: sociophysics; faculty hiring network; decision-making behavior; generative model

Citation: Zheng, S.; Jiang, N.; Li, X.; Xiao, M.; Chen, Q. Faculty Hiring Network Reveals Possible Decision-Making Mechanism. *Physics* **2023**, 5, 851–861. https://doi.org/10.3390/physics5030054

Received: 22 June 2023
Accepted: 3 August 2023
Published: 11 August 2023

1. Introduction

The exploration of the laws governing social dynamics is a complex and diverse long-term process for humanity, involving numerous disciplines and fields. Contributions to the study of social dynamics have been made by various disciplines, including economics, political science, sociology, psychology, physics, and other disciplines. Despite many disciplines attempting to understand the dynamics of social systems, progress in understanding society has been slow for a considerable period. The main reason for this is the presence of human factors within social systems. The decision-making behavior of individuals and their interactions with others contribute to the complexity of social systems. In situations where traditional data are difficult to obtain, many analyses remain qualitative in nature [1,2].

There is a new idea utilizing statistical physics that considers individuals as entities and accounts for human behavior to understand the underlying mechanisms of human social behavior [3,4]. Human activities are a simple extension of the "natural order", and applying natural laws can uncover the intrinsic rules followed by human activities. Building upon the groundbreaking contributions of scientists such as Auguste Comte, George Kingsley Zipf, and John Quincy Stewart, Serge Galam introduced the concept of "sociophysics" [5]. Galam proposed a theoretical framework called "social models" to explain the formation and evolution of social phenomena and collective behavior [6]. He used physics to study hierarchical systems, voting in group decisions, stability and fragmentation of alliances between nations, minority opinion propagation, and rumor phenomena. Over the past few years, sociophysics has evolved into a multidisciplinary field. It provides a relatively objective and normative epistemological path and principles in studying social science issues such as political science, economics, and communication, revealing the structure and behavioral patterns of social systems at various scales (from individuals to the societal level) and capturing the regularities of long-term societal evolution [7,8].

The complexity of social systems arises from the decision-making behaviors of individuals and the interactions among people. Understanding the underlying mechanisms of human decision-making behavior becomes a powerful tool for expanding thinking and guiding practice. We focus on the social issue of employment decisions for doctoral graduates. Scientific discovery requires the capacity to seek, nurture, and combine internal and external sources of knowledge. Universities, in particular, serve as vital "containers" for the advancement and integration of this knowledge [9], the employment of doctoral graduates as faculty members is a crucial aspect of this continuous cycle. Faculty hiring decisions determine the composition of the academic workforce and directly impact educational outcomes, career trajectories, and the development and dissemination of ideas [10]. Based on principles from sociophysics, we propose a generative model to simulate the decision-making process of doctoral graduates' employment in higher education institutions. We investigate this decision-making process by dividing it into two stages: the application process, where graduates submit resumes to universities, and the recruitment process, where universities consider and hire applicants. Since individual behaviors are mostly determined by their local contexts and internal psychology [11], we analyze the model's parameters to understand the psychological characteristics of doctoral graduates' job-seeking and universities' recruitment processes. The remaining sections of this paper are organized as follows. In Section 2, we first introduce the data source and distribution characteristics, followed by the construction of the generative model. In Section 3, we estimate and compare the parameters of the generative model. Finally, we summarize and discuss the paper in Section 4.

2. Materials and Methods

2.1. Data Sources

In this paper, we use the university faculty hiring data [12] of PhD graduates collected in Ref. [13] covering the three disciplines of computer science, business, and history. The relevant data information of various disciplines is shown in Table 1, including hundreds of universities in each discipline and the flow of tenure or tenure-track faculty between them, as well as the number of universities ranked in U.S. News & World Report and USA National Research Council (NRC) for each discipline. In detail, each tenure-track faculty was composed of a pair of graduating colleges and in-service colleges in the U.S. universities or institutions.

Table 1. Data summary for computer science, business, history and organization rankings [12].

	Computer Science	Business	History
No. of institutions	205	112	144
No. of regular faculty	5032	9336	4556
Mean size of faculty	25	83	32
No. of ranking university by U.S. News & World Report	153	111	143
No. of ranking university by USA National Research Council (NRC)	103	-	122

2.2. Empirical Evidence of Asymmetric Distribution Jump

To obtain a distribution of ranking differences, one first needs a university ranking. Unfortunately, a universally recognized objective and fair ranking does not currently exist. In the present study, we use rankings from some authoritative organizations for the following reasons: Authoritative organizations invest a significant amount of funds and resources in ranking processes to ensure the accuracy and objectivity of the results [14]. The rankings generated by these organizations are widely trusted and used by students and individuals in the education field. Students often rely on these rankings to make decisions regarding their choice of universities, highlighting the influential role that authoritative rankings play in the decision-making process [15]. In this study, we adopted the rankings provided by the U.S. News & World Report and NRC as the reference rankings. In addition, these two organizations' rankings are considered relatively objective, and the Pearson correlation coefficient between the two rankings exceeds 0.8.

Based on the collected authoritative rankings, we analyze the distribution characteristics of ranking differences in the empirical data. u represents the authoritative ranking of the university where the doctoral graduate obtained their doctoral degree, and v represents the authoritative ranking of the university where the doctoral graduate is employed. $r = v - u$ indicates the difference in rankings between the university where the doctoral graduate is employed and the university from which they obtained their doctoral degree. It is worth noting that smaller values of u and v correspond to higher rankings, indicating that the university is better, while larger values correspond to lower rankings, indicating that the university is worse. The function $p(r)$ represents the probability of a rank difference of r.

As shown in Figure 1a–e, the distribution of ranking differences exhibits consistent patterns across different disciplines and authoritative rankings. The ranking differences are concentrated around 0, indicating that doctoral graduates tend to choose institutions for employment that are close in ranking to their own university. This trend is further highlighted by all the data exhibiting a prominent peak at $r = 0$, with 60% of doctoral graduates in the history field, 77% in the business field, and 44% in the computer science field being attributed to them remaining at their alma mater (the university they graduated from) for employment. The distribution also displays asymmetry, with lower frequencies on the left side and higher frequencies on the right side. Positive ranking differences on the right side indicate that doctoral graduates from higher-ranked universities tend to work in lower-ranked universities, while negative ranking differences on the left side represent the opposite situation.

Figure 1f reveals that over 60% of graduates go to universities that are lower in ranking than their alma mater ($r > 0$), while only a small number of graduates are able to secure employment in universities that are higher in ranking ($r < 0$). This empirical evidence is not an isolated finding; similar characteristics have been observed in Ref. [10] across the entire academic community and in eight different domains. The Minimum Violation Rankings (MVR) model developed in Ref. [13] is based on the core assumption that doctoral graduates tend to move more easily from higher-ranked universities to lower-ranked universities, while the reverse movement is less likely. The model introduces the concept of "reverse arcs", which are instances where doctoral graduates move from lower-ranked universities to higher-ranked universities for employment, contrary to the general trend. The goal of the MVR model is to minimize the number of these "reverse arcs" in a stable doctoral graduate transition-to-employment network, thereby providing rankings of university prestige that align with the observed data. We utilize this characteristic as the basis for our subsequent modeling.

2.3. *Generative Model Based on the Process of Doctoral Employment*

We decompose the faculty recruitment process into two consecutive processes: graduates selecting their employer and schools selecting candidates. By setting the selection probabilities for these two processes, we obtain the final joint probability density distribution and generate flow data accordingly.

2.3.1. Two Decision-Making Processes in the Graduation-Employment Process

By observing the distribution of empirical data, one can see that the ranking differences of doctoral graduates' employment institutions exhibit asymmetric distribution characteristics. Upon closer analysis, we determined that the mobility of doctoral graduates is not solely determined by one party, but rather the result of a mutual decision-making process involving both parties. As shown in Figure 2, a transition from u to v indicates the following: First, a doctoral graduate who graduated from university u has applied for a job to university v, and subsequently, university v has approved the student's job application.

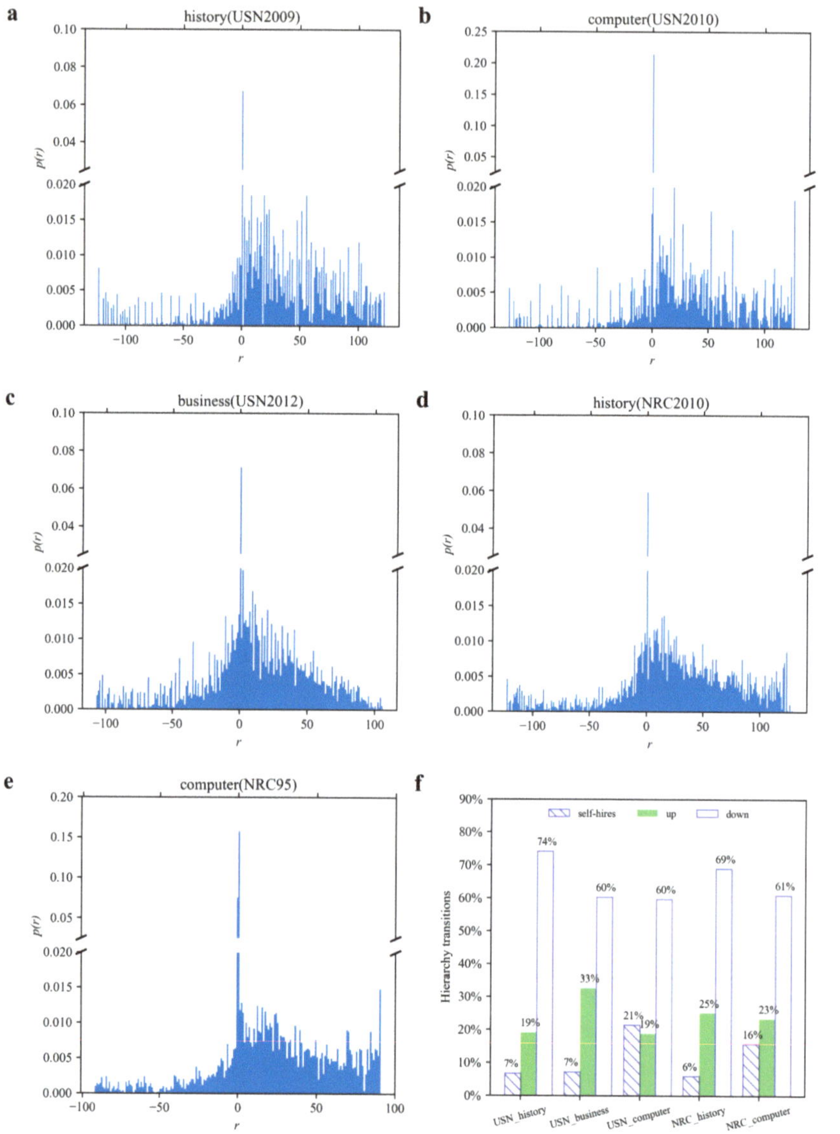

Figure 1. Asymmetry in ranking differences among disciplines in authoritative rankings. $r = v - u$ indicates the difference in rankings between the university where the doctoral graduate is employed and the university from which they obtained their doctoral degree. (**a–c**) The distribution of ranking differences between doctoral graduates' working institutions and their alma maters in the fields of history, computer, and business, respectively, based on the authoritative ranking by U.S. News & World Report. (**d**,**e**) The distribution of ranking differences in the fields of history and computer science, respectively, based on the authoritative ranking by the USA National Research Council (NRC). (**f**) The proportion of hierarchy transitions in real data across different disciplines under various authoritative rankings. The data is obtained from Ref. [12].

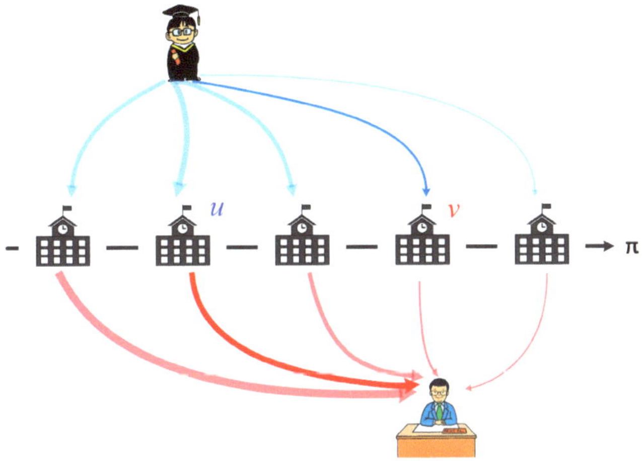

Figure 2. The process of a PhD graduate flowing from u to v can be divided into two subprocesses.

When doctoral graduates choose their place of employment, they pay attention to the differences between their alma mater and the prospective employer university. Due to the broad impact of prestige, better universities have access to more resources, are more likely to gain recognition in the academic community, and often have higher paper acceptance rates [10]. As a result, doctoral graduates tend to prefer employment opportunities at these higher-ranked universities, that is, universities with smaller values of v.

However, doctoral graduates do not rashly apply to top-tier universities considering the costs involved in the application process. When deciding which university to apply to, the graduates speculate on the response they might receive. The graduates are aware that universities generally prefer candidates from higher-ranked institutions. If there is a significant disparity between their alma mater and the target university, there is a higher likelihood of rejection. In light of this, graduates tend to take a conservative approach and apply to universities similar in ranking to their own. The graduates are unwilling to apply to lower ranked institutions, while they recognize the difficulty of securing a position at higher-ranked institutions. Here, we assume that the ranking difference, r, between the graduate's employment institution and their alma mater follows a normal distribution, $N(0, \sigma^2)$, with a probability density given by

$$p_1(r) \propto \frac{1}{\sqrt{2\pi}\sigma} e^{-\frac{r^2}{2\sigma^2}}. \tag{1}$$

To facilitate our analysis, we set the selection standard deviation for all students as σ. Notably, as the selection standard deviation increases, the perceived differences between universities diminish in the eyes of doctoral graduates, and their preferences for specific universities become less pronounced. Conversely, when σ is relatively small, PhD graduates exhibit clear preferences and tend to select universities with rankings similar to their alma mater.

From the perspective of universities, it is indeed true that employers believe that universities with higher prestige have access to better resources to nurture their students. Moreover, doctoral graduates from more prestigious universities are perceived to possess a higher quality. Therefore, universities naturally tend to prioritize applications from students coming from more prestigious institutions. This preference for prestigious universities reinforces their central role in disseminating ideas, shaping academic norms, and influencing broader culture [10]. Therefore, we assume that the selection of doctoral graduates by universities follows an "S"-shaped function, which can be specified as

$$p_2(r) \propto \frac{1}{1+e^{-\frac{r+\Delta}{T}}}. \quad (2)$$

Similarly, r represents the rank difference between the university with recruitment needs and the graduate's graduation university. Δ is an adjustable displacement parameter, and T represents the degree of irrationality. Δ and T are used to adjust the degree of influence of the rank difference on the probability of selection. When $r+\Delta > 0$, the university considers the student to be from a better university and is more willing to offer to them, resulting in a higher selection probability. Conversely, when $r+\Delta < 0$, the university perceives the student to be from a lower-ranked university, leading to a lower selection probability. As soon as T represents the degree of irrationality, a smaller value of T indicates greater rationality of the universities. In this case, universities are almost unwilling to select students from lower-ranked universities. Conversely, a larger value of T suggests lower rationality of the universities. Although universities still tend to hire doctoral graduates from higher-ranked universities, they also consider exceptional doctoral graduates from slightly lower-ranked universities.

2.3.2. Simulation Process of the Generative Model

We decompose the faculty recruitment process into two consecutive processes: student selection of schools for job applications and school selection of whether to give an offer to the student. After both students and universities have gained knowledge of their own and other institutions and have transformed this knowledge into selection probabilities, we generate flow data through a generative mechanism. As shown in Figure 3, the specific generation process is as follows:

(1) We assume there are M schools ranked from 1 to M based on their prestige, with lower ranking values indicating better schools. We generate N graduates for each school.
(2) We randomly select a student who chooses a school to submit their resume using a roulette wheel selection method based on the probability $p_1(r)$, where the standard deviation is a tunable parameter.
(3) After receiving the student's resume, the school decides whether to give an offer to the student based on the probability $p_2(r)$.
(4) If the university extends an offer, we record the ranking of the student's alma mater u and the ranking of the university that extended the offer v. We repeat steps (2) and (3) until all students have made their choices.

The simulated data obtained from the generative model are presented in Section 3.

2.3.3. Theoretical Analysis of the Model

The two decision-making behaviors in the above simulation process come from different entities, assuming that these two decision processes are independent of each other. When there are large values of M and N, one can use mean-field analysis to obtain an expected value of the simulation result, denoted as $p(r)$:

$$p(r) = \frac{p_1(r)\cdot p_2(r)}{\sum_r p_1(r)\cdot p_2(r)} = \frac{\frac{1}{\sqrt{2\pi}\sigma}e^{-\frac{r^2}{2\sigma^2}}\cdot\frac{1}{1+e^{-\frac{r+\Delta}{T}}}}{\sum_r\left(\frac{1}{\sqrt{2\pi}\sigma}e^{-\frac{r^2}{2\sigma^2}}\cdot\frac{1}{1+e^{-\frac{r+\Delta}{T}}}\right)}. \quad (3)$$

When $\Delta = 0$, based on Equation (3) one can see see that as r approaches zero, the joint probability, $p(r)$, increases, indicating that doctoral graduates are more likely to enter schools with rankings similar to their alma mater. Conversely, as r deviates from zero, $p(r)$ decreases, suggesting that the probability of doctoral graduates entering schools with a large ranking difference from their alma mater decreases. The distribution of $p(r)$ is asymmetric and exhibits a characteristic of being high in the middle and low on both sides. The specific impact of the parameters on $p(r)$ are discussed in Section 3.

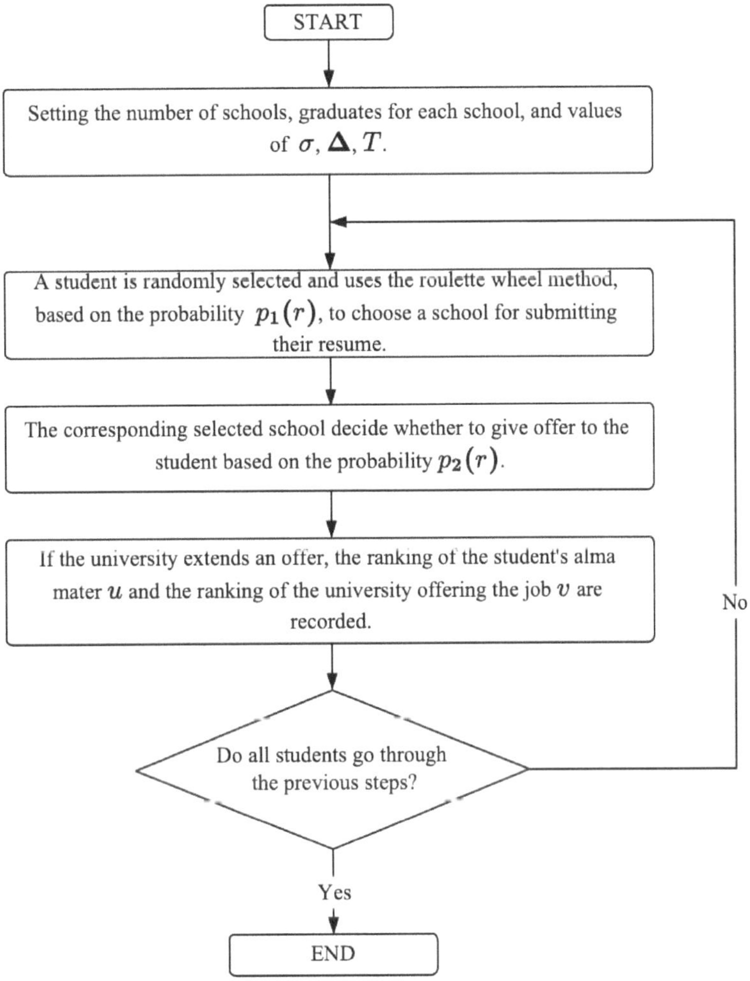

Figure 3. Flowchart of the generation model.

3. Results

3.1. Impact of Parameters on the Generative Model

In this Section, we discuss the effects of three parameters in the model by examining the changes in the joint probability distribution. For the purpose of comparison, $p_1(r)$, $p_2(r)$, and $p(r)$ are plotted together to observe the relationship between the probabilities of individual decisions and the overall process, as shown in Figure 4.

Here, σ represents the concentration level in student choices. When σ is small, students tend to prefer schools with rankings close to their own alma mater. When σ is large, students tend to consider rankings more extensively when selecting the schools to which they want to apply. In Figure 4b, schools dominate the entire recruitment process.

The parameter Δ represents the tolerance level of schools when making offers, which describes the degree to which schools are willing to give offers to doctoral graduates from lower-ranked institutions when selecting students. Smaller values of Δ result in a right-skewed probability density of school choices, leading to a right-skewed joint probability density. As shown in Figure 4c, most students end up in schools with lower rankings than that of their own alma mater. Conversely, when Δ is large, schools have a higher tolerance for doctoral graduates from lower-ranked institutions, as depicted in Figure 4d, and almost give offers to all students. In this case, students become the dominant factor in the entire recruitment process.

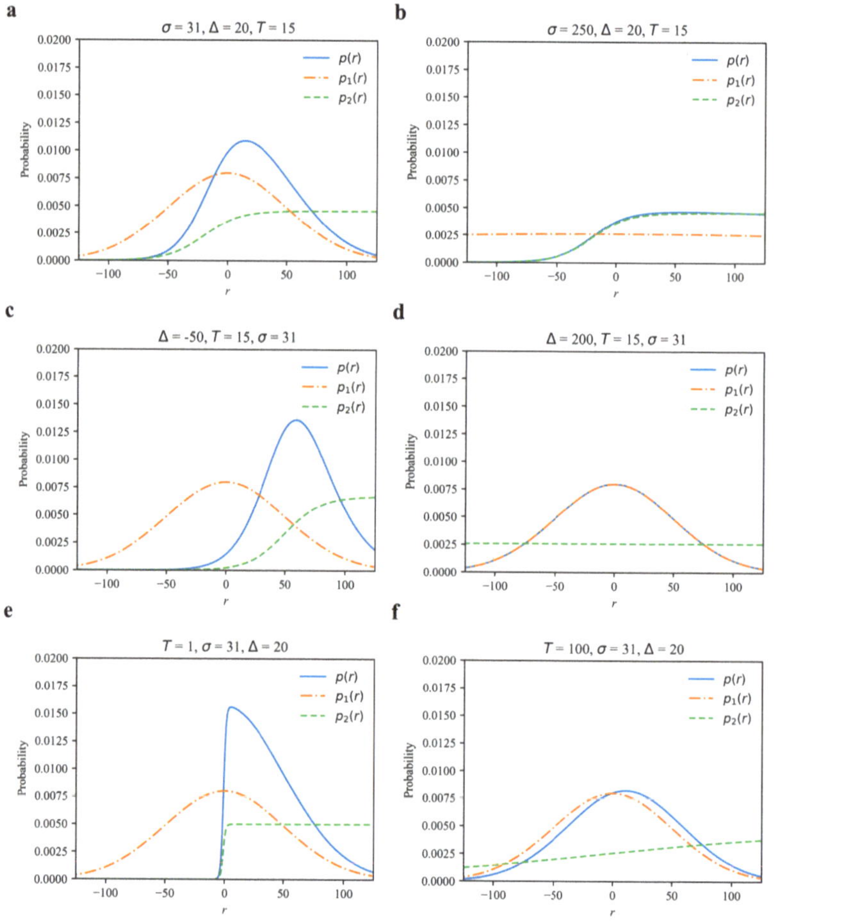

Figure 4. The impact of parameters σ (**a**,**b**), Δ (**c**,**d**), and T (**e**,**f**) on $p_1(r)$, $p_2(r)$, and $p(r)$. See text for details.

The parameter T reflects the degree of importance that universities place on school rankings when making job offers to applicants. As illustrated in Figure 4e, when the irrationality level, T, is small, this indicates a higher level of rationality, with schools attaching more importance to rankings and rarely giving offer to students from lower-ranked institutions. Once the rank difference becomes negative, the selection probability experiences a sharp decline, and schools tend to choose students from higher-ranked alma maters. On the other hand, when the irrationality level T is large, as shown in Figure 4f, schools prioritize other factors beyond rankings, such as students' comprehensive qualities and actual abilities. Importantly, the term "irrationality" here does not imply that schools' decisions are irrational. Rather, it suggests that schools are more likely to consider nonranking factors when making decisions, such as students' overall qualities and practical abilities.

3.2. Parameter Estimation for Empirical Data

We identify the optimal fit parameters for $p(r)$ by adjusting the parameters to minimize the Kolmogorov–Smirnov (KS) statistic between the actual ranking difference distribution and $p(r)$. Then, we use these best-fitting parameters to generate simulation data by using the generative model.

Figure 5 illustrates the actual ranking difference distribution, the joint probability distribution, and the distribution of ranking differences in the simulation data. The blue solid line represents the joint probability density distribution curve, while the red circle represents simulation data generated based on the joint probability density. The joint probability density curve captures the similar distribution characteristics of the actual data rank differences: high probability in the middle and low probability on both sides, with a faster decrease in probability on the left and a slower decrease on the right.

Table 2 displays the optimal fit parameters for the three domains across various ranking authorities. By comparing the best-fit parameters obtained from two authoritative rankings in each domain, we observe a consistent pattern. These parameters can to some extent reflect the decision-making process of students in job seeking and schools in offering admissions.

As shown in Table 2, from the perspective of students, the standard deviation, σ, of the selection distribution is relatively consistent for computer science and history majors, and they are noticeably larger compared to that for business. This indicates that doctoral graduates in the field of business tend to have more concentrated employment choices. In other words, doctoral graduates in the business field prefer to choose universities with smaller ranking differences from their alma mater, while the choices of doctoral graduates in computer science and history are more diverse.

From the perspective of universities, the offset values, Δ, in all three fields are positive. A positive Δ indicates that universities have a certain level of tolerance for graduates from slightly lower-ranked institutions. This means that universities consider the actual abilities of doctoral graduates when selecting them, rather than solely focusing on the ranking of their alma mater. The appropriate value of T also suggests that universities consider factors other than just rankings.

Table 2. The best-fit parameters of a generative model. See text for details.

Authoritative Ranking	Parameter	Computer Science	Business	History
U.S. News & World Report	σ	71	45	69
	Δ	12	12	4
	T	35	51	31
NRC	σ	58	-	65
	Δ	18	-	8
	T	13	-	41

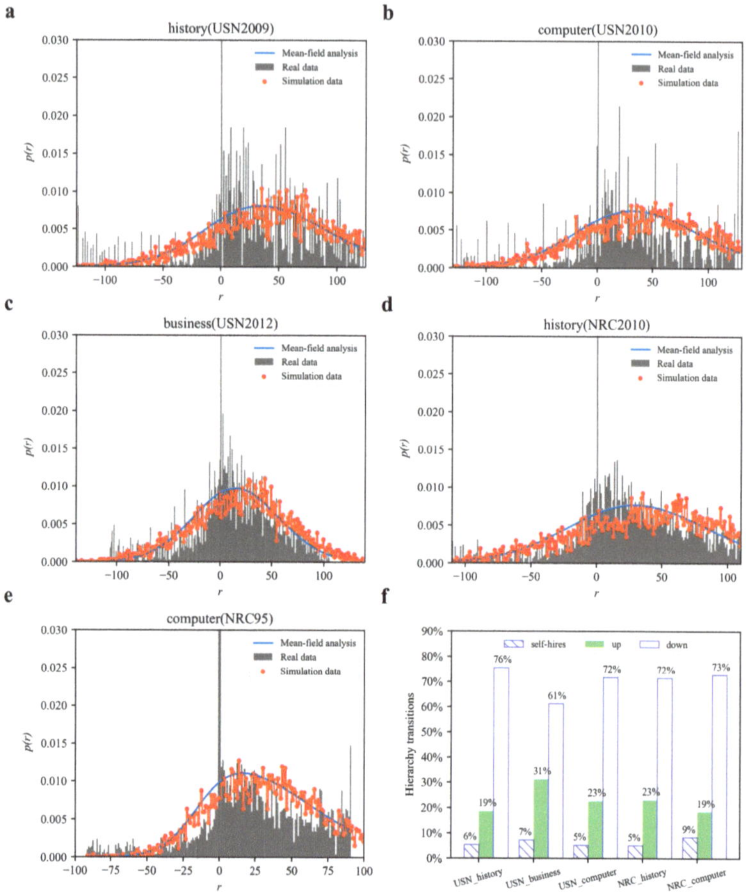

Figure 5. Comparison between the generative model and real data: (**a–c**) the fields of history, computer, and business, when ranked by U.S. News & World Report; (**d**,**e**) the fields of history and computer, when ranked by NRC; and (**f**) the proportion of hierarchy transitions of the simulation data across the various disciplines under different authoritative rankings.

4. Discussion

This study is an attempt to analyze the decision-making process of doctoral graduates in job searching based on the perspective of sociophysics. The process of graduates transitioning from graduation to employment is divided into two steps, and a generative model is proposed based on appropriate assumptions. This model effectively reproduces empirical data and allows for the estimation of important parameters that influence the decision-making process from the empirical data. Among these parameters, σ is related to the diversity of students' resume submissions, Δ reflects the tolerance of universities toward the ranking of graduates' alma mater, and T represents the degree of irrationality in university decision-making. By comparing the best-fitting parameters, we found that doctoral graduates in business fields have more concentrated employment choices, while graduates in computer science and history fields have broader preferences. Furthermore, universities consider factors beyond rankings when selecting doctoral graduates.

Understanding the mechanisms behind human decision-making can broaden our thinking and can guide practical applications. This research provides an initial exploration by proposing a generative model, aiming to inspire further research on decision-making problems in other social systems. Furthermore, with the extended application of this model, it is possible for us to directly obtain the public perception of the excellence levels of various universities from graduation-to-employment data, even in the absence of authoritative ranking information. This approach can also be applied to other domains, such as national residency rates or corporate competitiveness.

The analysis presented here, included faculty members from different ranks, which indeed may have some influence on our conclusions. However, we strongly believe that the choice of doctoral institution reflects graduates' academic achievements and capabilities, playing a significant role in determining their subsequent employment destinations. This belief is further validated by the study in Ref. [13], where the MVR model predicts university rankings and demonstrates a strong correlation with authoritative rankings, highlighting the significance of doctoral graduates' backgrounds in shaping their career paths. Further detailed discussions on this topic remain open for researchers to explore.

Author Contributions: Conceptualization, S.Z.; Methodology, X.L.; Writing—original draft, S.Z.; Writing—review and editing, M.X.; Visualization, N.J.; Supervision, Q.C. All authors have read and agreed to the published version of the manuscript.

Funding: This work is supported by the National Natural Science Foundation of China under Grant 72274020.

Data Availability Statement: The data used in this study can be obtained from the website http://santafe.edu/~aaronc/facultyhiring/ (accessed on 31 July 2023).

Conflicts of Interest: The authors declare no conflicts of interest.

References

1. Pabjan, B. The use of models in sociology. *Phys. A* **2004**, *336*, 146–152. [CrossRef]
2. Weiwei, L. The integration trend and its characteristics of social physics and big data technology. *J. Dialect. Nat.* **2019**, *41*, 80–86. (In Chinese).
3. Bhattacharya, K.; Kaski, K. Social physics: Uncovering human behaviour from communication. *Adv. Phys. X* **2020**, *4*, 96–124. [CrossRef]
4. Jusup, M.; Holme, P.; Kanazawa, K.; Takayasu, M.; Romić, I.; Wang, Z.; Geček, S.; Lipić, T.; Podobnik, B.; Wang, L.; et al. Social physics. *Phys. Rep.* **2022**, *948*, 1–148. [CrossRef]
5. Galam, S. Sociophysics: A review of Galam models. *Int. J. Mod. Phys. C* **2008**, *19*, 409–440. [CrossRef]
6. Galam, S. *Sociophysics. A Physicist's Modeling of Psycho-Political Phenomena*; Springer Science+Business Media, LLC: New York, NY, USA, 2012. [CrossRef]
7. Cheon, T.; Galam, S. Dynamical Galam model. *Phys. Lett. A* **2018**, *382*, 1509–1515. [CrossRef]
8. Ellero, A.; Fasano, G.; Sorato, A. Stochastic model of agent interaction with opinion leaders. *Phys. Rev. E* **2013**, *87*, 042806. [CrossRef] [PubMed]
9. Ferreira, M.; Costas, R.; Servedio, V.; Thurner, S. Scientific mobility, prestige and skill alignment in academic institutions. *arXiv* **2023**, arXiv:2307.06426. [CrossRef]
10. Wapman, K.H.; Zhang, S.; Clauset, A.; Larremore, D.B. Quantifying hierarchy and dynamics in US faculty hiring and retention. *Nature* **2022**, *610*, 120–127. [CrossRef] [PubMed]
11. Chen, Q.; Wang, C.; Wang, Y. Deformed Zipf's law in personal donation. *EPL (Europhys. Lett.)* **2009**, *88*, 38001. [CrossRef]
12. Clauset, A.; Arbesman, S.; Larremore, D.B. Faculty Hiring Networks. Available online: http://santafe.edu/~aaronc/facultyhiring/ (accessed on 31 July 2023).
13. Clauset, A.; Arbesman, S.; Larremore, D.B. Systematic inequality and hierarchy in faculty hiring networks. *Sci. Adv.* **2015**, *1*, e1400005. [CrossRef]
14. Csató, L.; Tóth, C. University rankings from the revealed preferences of the applicants. *Eur. J. Oper. Res.* **2020**, *286*, 309–320. [CrossRef]
15. Johnes, J. University rankings: What do they really show? *Scientometrics* **2018**, *115*, 585–606. [CrossRef]

physics

MDPI

Article

Polarization and Consensus in a Voter Model under Time-Fluctuating Influences

Mauro Mobilia

Department of Applied Mathematics, School of Mathematics, University of Leeds, Leeds LS2 9JT, UK; m.mobilia@leeds.ac.uk

Abstract: We study the effect of time-fluctuating social influences on the formation of polarization and consensus in a three-party community consisting of two types of voters ("leftists" and "rightists") holding extreme opinions, and moderate agents acting as "centrists". The former are incompatible and do not interact, while centrists hold an intermediate opinion and can interact with extreme voters. When a centrist and a leftist/rightist interact, they can become either both centrists or both leftists/rightists. The population eventually either reaches consensus with one of the three opinions, or a polarization state consisting of a frozen mixture of leftists and rightists. As a main novelty, here agents interact subject to time-fluctuating external influences favouring in turn the spread of leftist and rightist opinions, or the rise of centrism. The fate of the population is determined under various scenarios, and it is shown how the rate of change of external influences can drastically affect the polarization and consensus probabilities, as well as the mean time to reach the final state.

Keywords: sociophysics; voter models; stochastic processes; polarization; consensus; mean exit time; fluctuations; noise

Citation: Mobilia, M. Polarization and Consensus in a Voter Model under Time-Fluctuating Influences. *Physics* **2023**, *5*, 517–536. https://doi.org/10.3390/physics5020037

Received: 14 March 2023
Revised: 7 April 2023
Accepted: 13 April 2023
Published: 8 May 2023

1. Introduction

The relevance of parsimonious individual-based models to describe social phenomena at micro and macro levels has a long history [1–3]. In the last few decades, "sociophysics" has grown as a research field that aims at studying collective social behaviour, such as the spread of opinions or the dynamics of cultural diversity, using models and methods from statistical physics [3–14]. Typical questions in sociophysics concern the conditions under which consensus, or long-term opinion diversity, emerges in a population of agents ("voters") whose states ("opinions") change as they interact.

The voter model (VM) [15], closely related to the Ising model [16], has been commonly used to describe how consensus ensues from the interactions between neighbouring voters. In fact, while the classical two-state VM is arguably the simplest and most popular model of opinion dynamics, it rests on a number of oversimplifying assumptions: voters are endowed with only two possible states, they are blind to any external stimuli, have zero self-confidence and are all identical. In reality, members of social communities respond differently to stimuli, as they can interact in groups [1,17–19], and are usually characterised by multiple attributes [20–24]. In light of this, many generalizations of the VM have been proposed: for instance, "zealotry" was introduced in various forms to endow voters with different levels of self-confidence [25–36], while group-size influence is notably captured in the nonlinear q-voter model [37] and its variants [38–46]. It has also been noted that in many cases only some of the attributes characterising agents are actually compatible for social interactions [20,24,47–51]. It was thus suggested that agents whose opinions are too different would not interact, while voters holding close opinions can interact and attain a global consensus. This motivated the study of multi-state VMs, such as the constrained three-state voter model (3CVM) of Refs. [52–54], which is a discrete version of the bounded-compromise model [47–51]. In the 3CVM, incompatible "leftist" and "rightist" voters can only interact with "centrists", and the final outcome is either consensus with one of the three parties, or a polarized state consisting of mixture of leftists and rightists. In the latter case, the population is stuck in a frozen state of "polarization" in which leftists and rightists hold uncompromising opinions.

In addition to interactions among agents, external stimuli or influences, especially social media and news sources, play an increasingly important role in shaping the social environment that in turn affects the collective social behaviour [1,55–61]. Yet, with some recent remarkable exceptions [62–65], the role of social influences is still rarely modelled in sociophysics. Actually, sources of news play both a crucial, yet complex and multifaceted, role in influencing public social behaviour. They are typically characterised by opposing viewpoints, or agenda, that often change over time and can in turn favour consensus or polarization. It is also worth noting that voter-like models subject to a time-periodic external field have recently been studied in the context of economic networks [66,67].

How do the opinions of a social community evolve in a volatile and time-fluctuating environment? Inspired by this question, here, we introduce a generalization of the constrained three-state voter model [52–54] in the presence of binary time-fluctuating external influences. In the same vein as in population dynamics [68–81], for the sake of simplicity, we assume that the media influences endlessly switch from favouring the spread of polarization to promoting centrism, and vice versa. The goal of this work is to determine the fate of the population under various scenarios, and in particular to study how the time variation of the external influences affects the probability to reach polarization or a consensus, as well as the mean time for the population to settle in its final state.

The plan of the paper is as follows: The general formulation of the model is introduced in the next Section. Section 3 is dedicated to a thorough study of the polarization and consensus probabilities. Section 4 focuses on the study of the mean exit time. Section 5 is dedicated to a discussion of the results and to the conclusions. Technical details, including useful results in the absence of external influences, and possible generalizations and applications are discussed in three Appendices.

2. Three-State Constrained Voter Model under Binary Time-Fluctuating Influences

We consider a well-mixed population of N individuals consisting of N_R "rightists", or R-voters, N_L "leftists", or L-voters, and N_C "centrists", or C-voters, with $N = N_L + N_R + N_C$. In the voter model language [7–9,13,15], R and L represent extreme opinions, while C-voters hold an intermediate opinion. In the three-state constrained voter model (3CVM) [53,54], an agent selected at random tries to interact with one of its neighbours, that is any other randomly picked voter, at each microscopic update attempt. When the two agents hold the same opinion or if it is a pair of leftist–rightist (LR or RL), nothing happens. However, if the pair is a centrist C and an L or an R voter (LC, CL, RC or CR), the initial agent adopts the opinion of the neighbour with a probability that depends on the external influences whose effect is encoded into the random variable $\xi(t)$ that fluctuates with the time t; see below in this Section and Figure 1. Hence, while the interactions between L and R voters and centrists C now change in time with ξ, L and R remain incompatible and the system's final state is, as in the static 3CVM, either a consensus state or a state of polarization consisting of a frozen mixture of leftists and rightists [53,54]; see below in this Section.

The main novel ingredient of this study is the modelling of time-fluctuating social influences by means of the coloured dichotomous (telegraph) noise $\xi(t) \in \{-1, 1\}$ [72–74]; see Figure 1. The process $\xi(t)$ simply encodes all the complex effects of social environment and external influences in the endless random switching between two states: here, $\xi = 1$ corresponds to external influences favouring L and R opinions (spread of polarization), whereas $\xi = -1$ favours centrism C (compromise between L and R). Here, the dichotomous noise ξ is always at *stationarity*, and switches from ± 1 to ∓ 1 according to

$$\xi \longrightarrow -\xi, \tag{1}$$

with a rate $(1 - \delta\xi)\nu$, where ν is the *average switching rate* [77], and $1 < \delta < -1$ denotes the switching asymmetry. Accordingly, the average time spent in state $\xi = \pm 1$ before switching to $-\xi$ is $1/(1 \mp \delta)\nu$ (symmetric switching occurs when $\delta = 0$), see Figure 1a–c. At stationarity, $\xi = \pm 1$ with probability $(1 \pm \delta)/2$ [72–74], and its (ensemble-)average is

$$\langle \xi(t) \rangle = \delta, \tag{2}$$

while its autocorrelation is $\langle \xi(t)\xi(t') \rangle - \langle \xi(t) \rangle \langle \xi(t') \rangle = (1 - \delta^2)e^{-2\nu|t-t'|}$.

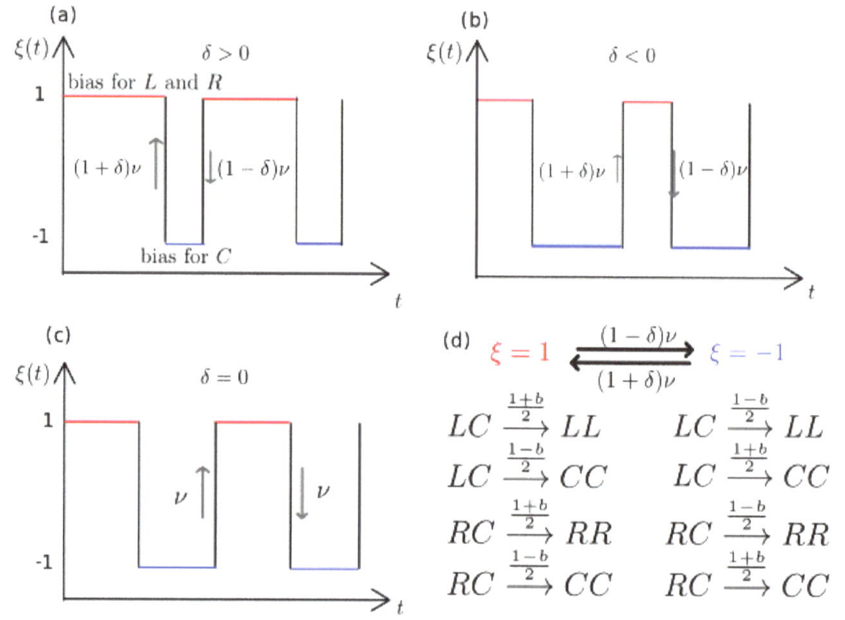

Figure 1. Illustration of the 3CVM under binary social switching $\xi \to -\xi$ at rate $(1 - \delta\xi)\nu$ (where ν is the average switching rate and δ the switching asymmetry). (**a**) $\xi(t)$ versus time t when $\delta > 0$: most time is spent in social state $\xi = 1$ favouring polarization. (**b**) $\xi(t)$ versus t when $\delta < 0$: most time is spent in social state $\xi = -1$ favouring centrism. (**c**) $\xi(t)$ versus t when $\delta = 0$: switching is symmetric and the same average time is spent in $\xi = \pm 1$. (**d**) When $\xi = 1$ there is a bias favouring the spread of L (leftists) and R (rightists): $LC \to LL$ and $RC \to RR$ are the reactions with the highest rate, $(1 + b)/2$ (where b denotes the social influences bias). The social environment $\xi = -1$ favours the spread of centrism: the reactions $LC \to CC$ and $RC \to CC$ have the highest rate, $(1 + b)/2$, under $\xi = -1$. See text for details. In (a)–(c), initially $\xi(0) = 1$.

The 3CVM switching dynamics is therefore defined by the four reactions: $LC \to LL$ and $RC \to RR$, corresponding to the spread of extreme opinions at rate $(1 + b\xi)/2$, and $LC \to CC$ and $RC \to CC$, with centrists replacing L and R voters at rate $(1 - b\xi)/2$. Here, $0 < b < 1$ denotes the social influences bias favouring polarization when $\xi = 1$ and centrism in the social environment $\xi = -1$. The 3CVM switching dynamics can thus be schematically described by the following reactions occurring at each time increment:

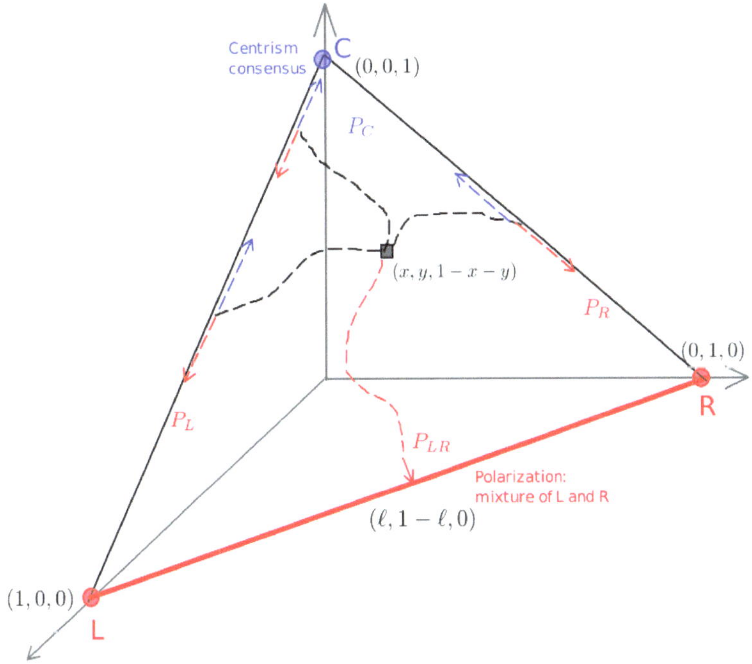

Figure 2. Illustration of the 3CVM dynamics in the simplex $\ell + r + c = (N_L + N_R + N_C)/N = 1$. Circles show the consensus (absorbing) states all-L $(1,0,0)$, all-R $(0,1,0)$, and all-C $(0,0,1)$, and the initial condition is (x, y, z). The thick line indicates polarization state (pol-LR) consisting of a frozen mixture of L and R voters made up of a fraction $\ell = N_L/N$ of L-voters coexisting with an incompatible fraction $1 - \ell$ of R-voters. Dashed lines are typical trajectories: the dynamics ceases when the line pol-LR is reached (polarization with probability P_{LR}), or when there is consensus by reaching one of absorbing states all-L, all-R or all-C (with respective probabilities P_L, P_R and P_C). Once a trajectory reaches the line $\ell = 0$ or $r = 0$, the evolution is restricted on $\ell = 0$, $r = 0$ until there is a consensus. As $\xi(t)$ varies, the dynamics favours in turn the spread of L and R ($\xi = 1$) or that of C ($\xi = -1$). See text for details.

$$
\begin{aligned}
LC &\longrightarrow LL \quad \text{rate}: \frac{1+b\xi}{2}, \qquad LC \longrightarrow CC \quad \text{rate}: \frac{1-b\xi}{2},\\
RC &\longrightarrow RR \quad \text{rate}: \frac{1+b\xi}{2}, \qquad RC \longrightarrow CC \quad \text{rate}: \frac{1-b\xi}{2},
\end{aligned}
$$

where ξ and the rates endlessly fluctuates according to Equation (1). The dynamics of the switching 3CVM is therefore the Markov chain defined by the transition rates (A1) and master Equations (A2) and (A3) [82]; see Appendix A.1. The 3CVM switching dynamics is characterised by three consensus/absorbing states $N_R = N_C = 0$ (all-L), $N_L = N_C = 0$ (all-R), $N_L = N_R = 0$, $N_C > 0$ (all-C), and by the polarization state (pol-LR) where $N_L + N_R = N$, $N_C = 0$; see Figure 2. As in the absence of external influences, the final state of the population is therefore guaranteed to be either one of the consensus/absorbing states or pol-LR [53,54].

It is worth noting that the approach considered here bears some similarities with the two-party model of Refs. [62,63]. However, there are also important differences: first, in the 3CVM, the polarization state corresponds to a frozen mixture (while it is an active state in Refs. [62,63]). Moreover, external influences are here assumed to fluctuate endlessly in time, rather than by establishing a certain number of connections with voters. It is also interesting to notice that the effect of an exogenous time-varying influence has recently been studied in the context of economic networks [66,67]. The ensuing dynamics derived also from an Ising-like (voter-like) model subject to a time-dependent external field [66,67] has been shown to lead to rich dynamics characterized by hysteresis [83], which is not a phenomemon exhibited by the 3CVM switching dynamics. This stems from the fact that Equation (A4) differ from those governing the mean-field dynamics in Refs. [66,67] for not having any non-noisy linear terms, and for the exogenous time dependence being stochastic (via the multiplicative dichotomous noise $\xi(t)$) rather than periodic.

3. Final State: Polarization and Consensus Probabilities

In its final state, the population is either in the polarized state pol-LR, or in one of its three absorbing/consensus states (all-L, all-R or all-C); see Figure 2. Here, P_{LR} denotes the probability to end up in the polarized final state pol-LR. The probabilities to reach the absorbing/consensus states all-C, all-L, and all-R are, respectively, denoted by P_C, P_L and P_R. The density of voters of each type is $\ell \equiv N_L/N$, $r \equiv N_R/N$ and $c \equiv N_C/N = 1 - \ell - r$, and initially the population consists of densities $x,\ y,\ z = 1 - x - y$ of L, R and C voters, respectively.

In the absence of environmental switching, the probabilities of ending in any of the absorbing/consensus or polarization states were found to depend non-trivially on the parameter $s \equiv Nb$ and (x, y, z) [53,54]; see Appendix B. Here, we are interested in finding how P_{LR}, P_C, P_L and P_R, as well as the final densities (ℓ, r, c) depend on ν under various scenarios. We consider the same initial density of L and R voters, i.e., $0 < x = y = (1-z)/2 < 1/2$, which suffices for the purposes of this study and simplifies the analysis. (Only in Appendix C, an example with $x \neq y$ is briefly considered.) By symmetry, $x = y$ implies $P_L(\nu) = P_R(\nu)$, i.e., the probability of L and R consensus is the same. When it occurs, polarization consists of a fraction $1/2$ of L and R voters; see Figure 2. We thus have: $P_{LR} + P_C + P_L + P_R = P_{LR} + P_C + 2P_L = 1$, and therefore focus on studying P_{LR} and P_C as functions of ν for different values of δ and z, treated as parameters, from which we obtain also the final densities: $(\ell, r, c) = (\ell,\ \ell,\ 1 - 2\ell) = ((1 - P_C)/2,\ (1 - P_C)/2,\ P_C)$; see Appendix A.2.

3.1. *Final State in the Regimes $\nu \to 0$ and $\nu \to \infty$*

The polarization and consensus probabilities can be computed analytically in the regimes $\nu \to 0$ and $\nu \to \infty$. For this, we take advantage of the results obtained in the *absence of external influences* for the polarization probability $\mathcal{P}_{LR}$, and the probabilities of C, L and R consensus, respectively, denoted here by $\mathcal{P}_C$, $\mathcal{P}_L$ and $\mathcal{P}_R$. In the absence of external influences, these quantities have been obtained in Refs. [53,54] and are summarized in Appendix B.1.

3.1.1. Polarization and Consensus Probabilities in the Regime $\nu \to 0$

When $\nu \to 0$, we can assume that there are no switches before attaining polarization or consensus. In this case, ξ is a quenched random variable, and the kinetics is the superposition, with probability $(1 \pm \delta)/2$, of the 3CVM dynamics in the stationary environment $\xi = \pm 1$. As a result, the polarization probability when $\nu \to 0$, P^0_{LR}, is the superposition of $\mathcal{P}_{LR}(\pm s, z)$, obtained in a static external state $\xi = \pm 1$, with probability $(1 \pm \delta)/2$:

$$P^0_{LR} = \left(\frac{1+\delta}{2}\right)\mathcal{P}_{LR}(s,z) + \left(\frac{1-\delta}{2}\right)\mathcal{P}_{LR}(-s,z), \tag{3}$$

which is readily obtained from Equation (A5) or (A7). Similarly, the consensus probabilities when $\nu \to 0$, are obtained from (A5), with Equations (A6) and (A8), and $\mathcal{P}_L(s,z) = (1 - \mathcal{P}_{LR}(s,z) - \mathcal{P}_C(s,z))/2$:

$$P^0_{C,L} = \left(\frac{1+\delta}{2}\right)\mathcal{P}_{C,L}(s,z) + \left(\frac{1-\delta}{2}\right)\mathcal{P}_{C,L}(-s,z). \tag{4}$$

With Equations (A9) and (4), we obtain the final density $\ell_0 = r_0$ of L and R voters when $\nu \to 0$:

$$\ell_0 = r_0 = \frac{1 - P_C^0}{2}. \tag{5}$$

3.1.2. Polarization and Consensus Probabilities in the Regime $\nu \to \infty$

When $\nu \to \infty$, so many switches occur before polarization or consensus that ξ self-averages [71,72,75–78]. In this case, ξ is an annealed random variable that can be replaced by its average: $\xi \to \langle\xi\rangle = \delta$. The switching dynamics of the 3CVM with $\nu \to \infty$ is thus the same as in Ref. [54], with $s \to s\langle\xi\rangle = s\delta$. In the limit $\nu \to \infty$, the polarization probability, P_{LR}^{∞}, is therefore obtained from Equation (A5) or (A7) according to

$$P_{LR}^{\infty} = \mathcal{P}_{LR}(s\delta, z). \tag{6}$$

Similarly, the consensus probabilities under high ν, $P_{C,L}^{\infty}$, are obtained from Equations (A6) and (A7):

$$P_{C,L}^{\infty} = \mathcal{P}_{C,L}(s\delta, z). \tag{7}$$

With Equations (A9) and (7), the final density $\ell_\infty = r_\infty$ of L and R voters in the regime $\nu \to \infty$:

$$\ell_\infty = r_\infty = \frac{1 - P_C^\infty}{2}. \tag{8}$$

Since $\mathcal{P}_{LR} \approx 1$ when $s \equiv Nb \gg 1$ and $\mathcal{P}_C \approx 1$ when $s < 0$ and $|s| \gg 1$ [54] (see Appendix B.1), the focus here is on the regime where the small influences bias affects a large but finite number of voters, i.e., $b \ll 1$ and $s \gg 1$, with $s\delta = \mathcal{O}(1)$. This allows us to highlight the effect of the external influences.

3.2. Polarization and Consensus Probabilities When $\delta > 0$

When $\delta > 0$, most of the time is spent in the external state $\xi = 1$, where influences favour polarization (pol-LR); see Figures 1 and 2.

In fact, as shown in Figure 3a, when z is not too close to 1, $P_{LR} > P_C$ for all values of ν. In this case, P_{LR} increases with ν over a large range of values (for $\nu \gtrsim b$), with $P_{LR} \approx P_{LR}^0 \approx (1+\delta)/2$ when $\nu \ll b$ and $P_{LR} \approx P_{LR}^\infty$ when $\nu \gg b$, whereas P_C is a decreasing function of ν, with $P_C \approx P_C^0 \approx (1-\delta)/2$ when $\nu \ll b$ and $P_C \approx P_C^\infty \ll 1$ when $\nu \gg b$; see Figure 3a. The fact that P_{LR} and P_C vary little with ν, and are quite close to $P_{LR}^{0,\infty}$ and $P_C^{0,\infty}$, indicates that the analytical predictions for $P_{LR}^{0,\infty}$ and $P_C^{0,\infty}$ not only apply to the limits $\nu \to 0, \infty$, but are also valid approximations of $P_{LR}(\nu)$ and $P_C(\nu)$ in the regimes of low and high switching rate (see also Section 4 below). We can also notice in Figure 3a, that P_{LR} exhibits a weak non-monotonic behaviour, with a "dip" for $\nu \sim b$. In this setting, the final fraction of L and R voters, given by $\ell = r = (1 - P_C)/2$ (see Appendix A.2), is an increasing function of ν, while the final density of centrists, $c = 1 - 2\ell = P_C$, decreases with ν; see inset in Figure 3a.

When there is a small initial fraction of L and R voters, with z close to 1, centrism can prevail over a large range of values of ν: $P_C > P_{LR}$ for $\nu \gtrsim b$, see Figure 3b. The probability P_{LR} thus decreases with ν, while P_C can have a non-monotonic behaviour as in Figure 3b where it exhibits a "bump" for $\nu \sim b$. In this case, centrists hold the majority opinion over an intermediate range of ν: $P_C > P_{LR} + P_L + P_R = 1 - P_C$, i.e., $P_C > 1/2$ when $\nu = \mathcal{O}(b)$; see inset in Figure 3b.

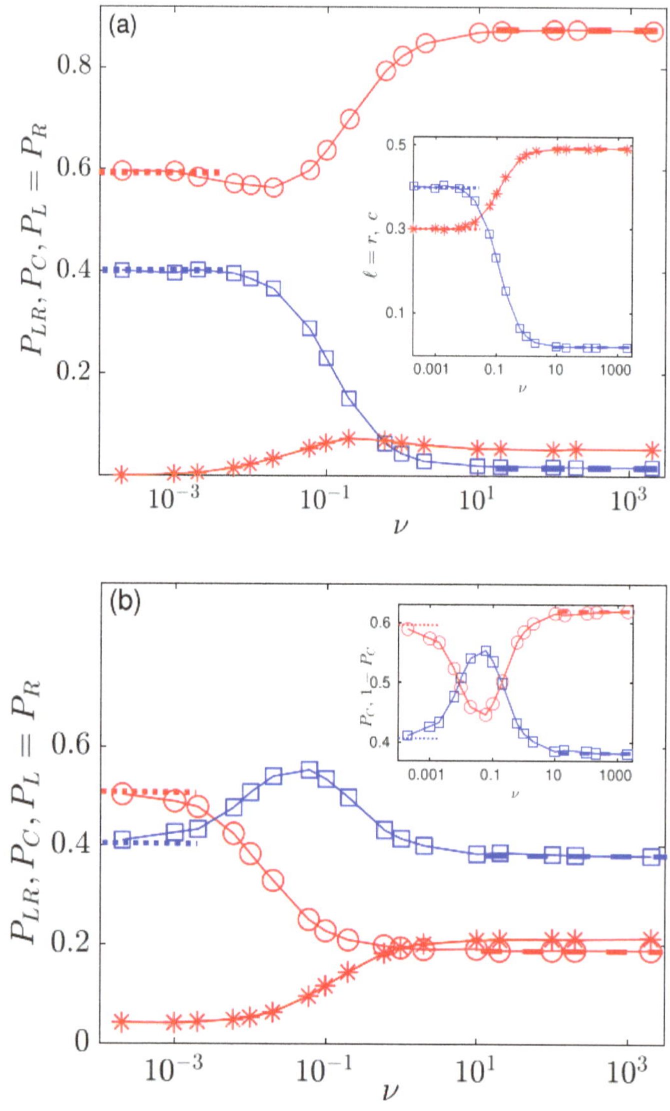

Figure 3. P_{LR} (circles), P_C (squares), and $P_L = P_R$ (stars), versus ν for $\delta = 0.2$ and different initial conditions. Here and in the other figures, symbols are from simulations (averaged over 10^5 samples). Thick dashed lines are eyeguides showing P_{LR}^{∞} (red) and P_C^{∞} (blue), and thick dotted lines are eyeguides showing P_{LR}^{0} (red) and P_C^{0} (blue), given by Equations (4), (3), (6) and (7). (**a**) Initial condition: $x = y = 0.25$ and $z = 0.5$. Inset: final densities $\ell = r$ (L/R-voters, stars) and c (centrists, squares) versus ν. The dashed lines show $\ell_\infty = r_\infty$ (red) and $c_\infty = 1 - 2\ell_\infty$ (blue) from Equation (8), and the dotted lines show ℓ_0 (red) and $c_0 = 1 - 2\ell_0$ (blue) from Equation (5). (**b**) Initial condition: $x = y = 0.06$ and $z = 0.88$. Inset: P_C (squares) and $1 - P_C$ (diamonds) versus ν: centrism is the majority opinion in the range of intermediate switching rate, $\nu \sim b = 0.1$, where $P_C(\nu) > 1/2$. The dashed lines show $1 - P_C^{\infty}$ (red) and P_C^{∞} (blue), and the dotted lines show $1 - P_C^{0}$ (red) and P_C^{0} (blue). In all panels and insets: $(N, b, s) = (200, 0.1, 20)$. See text for more details.

In Figure 3, the predictions (3), (4), (6) and (7) for $P_{LR/C}^{0}$ and $P_{LR/C}^{\infty}$ are in good agreement with simulation data when $\nu \ll b$ and $\nu \gg b$, respectively. The results reported in Figure 3 show that when $\delta > 0$ the main effects of the switching external influences is to tame the bias favouring polarization. P_{LR} and P_C can thus increase or decrease with ν, and even have a local extremum at a nontrivial switching rate. We explain this by solving, at fixed δ, $P_{LR}^{\infty} = P_{LR}^{0}$ and $P_{C}^{\infty} = P_{C}^{0}$ for z_{LR} and z_C with Equations (3), (4), (6) and (7):

$$\mathcal{P}_{LR}(s\delta, z_{LR}) = \left(\frac{1+\delta}{2}\right)\mathcal{P}_{LR}(s, z_{LR}) + \left(\frac{1-\delta}{2}\right)\mathcal{P}_{LR}(-s, z_{LR}),$$
$$\mathcal{P}_{C}(s\delta, z_{C}) = \left(\frac{1+\delta}{2}\right)\mathcal{P}_{C}(s, z_{C}) + \left(\frac{1-\delta}{2}\right)\mathcal{P}_{C}(-s, z_{C}). \tag{9}$$

Using (A5) and (A6), these equations are solved for z_{LR} and z_C. When $z < z_{LR}$, $P_{LR}^{\infty} > P_{LR}^{0}$ while $P_{LR}^{\infty} < P_{LR}^{0}$ when $z > z_{LR}$. Similarly, $P_{C}^{\infty} > P_{C}^{0}$ when $z > z_C$ and $P_{C}^{\infty} < P_{C}^{0}$ for $z < z_C$. In the examples of Figure 3, $z_{LR} \approx 0.708$ and $z_C \approx 0.887$, and therefore $z < z_{LR,C}$ in Figure 3a and $z_{LR} < z < z_C$ in Figure 3b, which agrees with $P_{LR}^{\infty} > P_{LR}^{0}$ and $P_{C}^{\infty} < P_{C}^{0}$ reported in Figure 3a, and with $P_{LR,C}^{\infty} < P_{LR,C}^{0}$ in Figure 3b.

The extrema of P_{LR} and P_C in Figure 3 at $\nu \sim b$ can be explained heuristically (and similarly for those in Figure 4, while there are no extrema in Figure 5; see below). In fact, when $\nu \to 0$, the population remains in the initial external state $\xi(0)$ until polarization or consensus occurs. When $s \gg 1$ and $\nu \ll 1$, polarization occurs with a probability close to 1 if $\xi(0) = 1$ and close to 0 otherwise, and we have $P_{LR}(\nu \ll b) \approx (1+\delta)/2$. We argue that $P_{LR}(\nu)$ decreases with ν when this rate is raised from $\nu \to 0$ to $\nu \lesssim b \ll 1$. For the sake of argument, we focus on the switching rate ν_* for which there is one external switch before reaching the final state. As discussed in Section 4.2 below, we have $\nu_* \lesssim b$ (see the inset in Figure 6b), with $\xi(t > 1/\nu^*) = -\xi(0)$, and the population settles in its final state after a time $T \sim (\ln N)/b$. Hence, if $\xi(0) = 1$, the final state is polarization if pol-LR is reached in a time $t \lesssim 1/\nu_*$, when still $\xi(t) = 1$. This occurs approximately with a probability $1/(T\nu_*)$. On the other hand, if $\xi(0) = -1$, the final state is polarization if pol-LR is reached after the external switch, when $\xi(t) = 1$, i.e., for $t \gtrsim 1/\nu_*$, and this occurs with an approximate probability $1 - 1/(T\nu_*)$. Hence, we estimate

$$P_{LR}(\nu_*) \approx \left(\frac{1+\delta}{2}\right)\frac{1}{T\nu_*} + \left(\frac{1-\delta}{2}\right)\left(1 - \frac{1}{T\nu_*}\right) = \frac{1}{2} + \delta\left(\frac{1}{T\nu_*} - \frac{1}{2}\right) < P_{LR}^{0},$$

which explains that $P_{LR}(\nu)$ decreases with ν when $\nu \lesssim b$. When ν is increased further above b, $P_{LR}(\nu)$ increases with ν and approaches $P_{LR}^{\infty} > P_{LR}^{0}$. This explains qualitatively the non-monotonic behaviour of $P_{LR}(\nu)$ in Figure 3a, with a dip at $\nu_* \approx 0.02$, and $P_{LR}(\nu_*) \approx 0.565 < P_{LR}^{0} \approx 0.6 < P_{L}^{\infty} \approx 0.876$, while the rough estimate, with $T \approx 63$ (see Figure 6a), gives $P_{LR}(\nu_*) \approx 0.56$. Similarly, in Figure 3b, $P_{C}^{\infty} \approx P_{C}^{0} < P_C(\nu_*)$, which results in a "bump" in $P_C(\nu)$ at some $\nu_* \lesssim b$.

3.3. Polarization and Consensus Probabilities When $\delta < 0$

When $\delta < 0$, most time is spent in the external state $\xi = -1$, where influences favour centrist consensus. Hence, when $s = Nb \gg 1$ and z is not too close to 0, centrism prevails over the other opinions: $P_C > P_{LR} \gg P_{L,R}$ for all values of ν; see Figure 4a. In this case, we find: $P_C \approx P_C^0 \approx (1+|\delta|)/2 > P_{LR} \approx P_{LR}^0 \approx (1-|\delta|)/2$ when $\nu \ll 1$, while $P_C \approx P_C^{\infty} \approx 1$ and $P_{LR} \approx P_{L,R} \approx P_{LR}^{\infty} \approx 0$ when $\nu \gg 1$. In Figure 4a, P_C increases with ν from $(1+|\delta|)/2$ to 1, while, P_{LR} decreases from $(1-|\delta|)/2$ to 0 as ν is raised. This results in a final state consisting of a majority of C voters, whose final density c increases from $(1+|\delta|)/2$ to 1, and a minority of L and R voters, whose final density $\ell = r$ decreases with ν from $(1-|\delta|)/4$ to 0; see the inset in Figure 4a. However, while all-C is the most likely final state, there is a finite probability of polarization at low and intermediate switching rate.

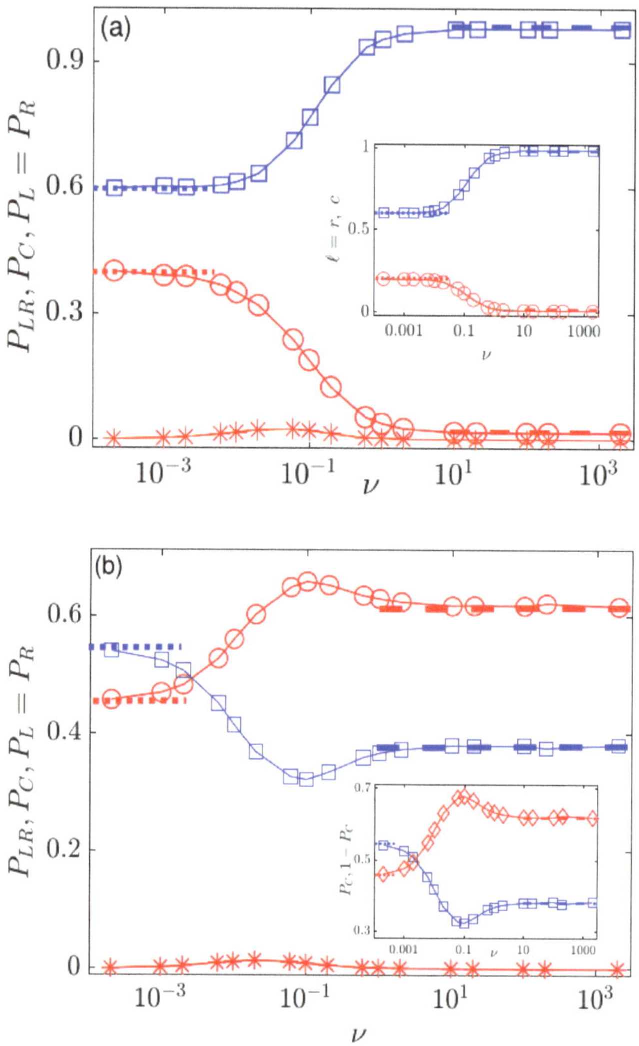

Figure 4. P_{LR} (circles), P_C (squares), and $P_L = P_R$ (stars) versus ν for $\delta = -0.2$, and different initial conditions. Symbols are results from simulations (averaged over 10^5 samples). Thick dashed and dotted lines are as in Figure 3. (**a**) Initial condition: $x = y = 0.25$ and $z = 0.5$. Inset: final densities $\ell = r$ (L/R-voters, stars) and c (centrists, squares) versus ν. (**b**) Initial condition: $x = y = 0.47$ and $z = 0.06$. Inset: P_C (squares) and $1 - P_C$ (diamonds) versus ν: centrists hold the majority opinion in the range of low switching rate ($\nu \ll b$) where $P_C(\nu) > 1/2$. In all panels and insets: $(N, b, s, \delta) = (200, 0.1, 20, -0.2)$. See text for more details.

When $s \gg 1$ and the initial population consists mainly of L and R voters ($z \ll 1$), polarization is the most likely final state, with $P_{LR} > P_C > P_{L,R}$ for $\nu \gg b$; see Figure 4b. Centrists thus generally hold the minority opinion when $\nu \gg b$, while C is the majority opinion ($P_C > 1/2$) only under low switching rate; see inset in Figure 4b. In the limiting regimes $\nu \ll b$ and $\nu \gg b$, P_{LR} and P_C, respectively, approach the values of $P_{LR}^{0,\infty}, P_C^{0,\infty}$. Here, again Equation (9) can be used to determine the initial density z_{LR}, such that $P_{LR}^{\infty} > P_{LR}^{0}$ if $z < z_{LR}$ and $P_{LR}^{\infty} \leq P_{LR}^{0}$ otherwise, and z_C such that $P_C^{\infty} > P_C^{0}$ if $z > z_C$ and $P_C^{\infty} \leq P_C^{0}$

otherwise. In the example of Figure 4, $z_{LR} \approx z_C \approx 0.112$. As for $\delta > 0$, in the regime of intermediate switching, as for $\delta > 0$, P_{LR} and $P_{L,R}$ can exhibit bumps and P_C a dip, see Figure 4b where P_{LR}, $P_{L,R}$ and P_C have modest extrema around $\nu \sim b$.

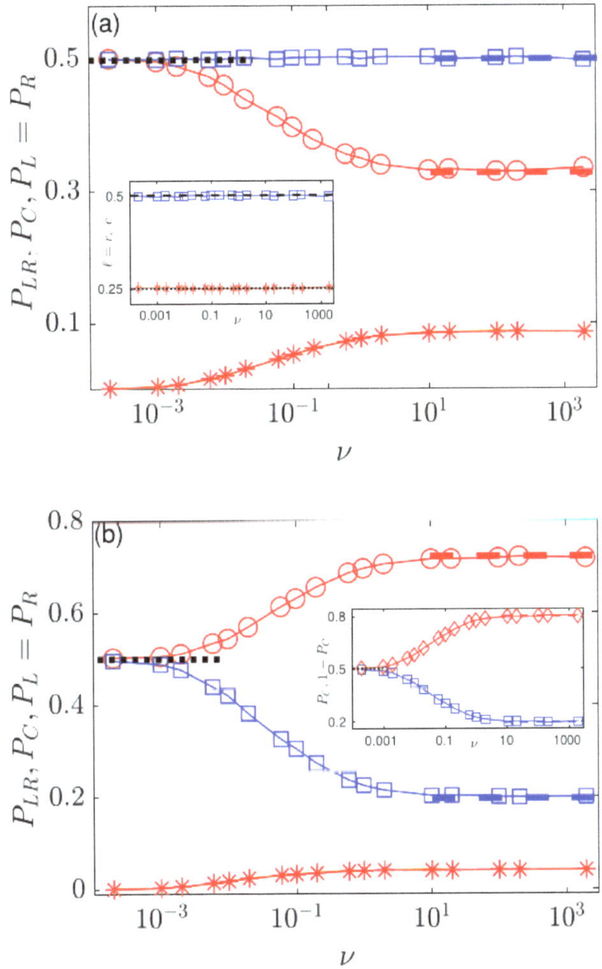

Figure 5. P_{LR} (circles) P_C (squares), $P_L = P_R$ (stars) versus ν for $\delta = 0$, and different initial conditions under symmetric switching. Symbols are results from simulations (averaged over 10^5 samples). Thick dashed and dotted lines are as in Figure 3. (**a**) Initial condition: $x = y = 0.25$ and $z = 0.5$. Inset: final densities $\ell = r$ (L/R-voters, stars) and c (centrists, squares) versus ν. In the special case $s \gg 1$, $\delta = 0$, $z = 2x = 2y$, $P_C(\nu) = c \approx 0.5$. (**b**) Initial condition: $x = y = 0.4$ and $z = 0.2$. Inset: P_C (squares) and $1 - P_C$ (diamonds) versus ν: under $\delta = 0$, centrists hold the minority opinion when $z < 1/2$. In all panels and insets: $(N, b, s, \delta) = (200, 0.1, 20, 0)$. See text for more details.

3.4. Polarization and Consensus Probabilities under Symmetric Switching ($\delta = 0$)

When $\delta = 0$, social switching is symmetric, and the same average amount of time is spent in $\xi = \pm 1$. External influences thus favour centrism ($\xi = -1$) and polarization ($\xi = 1$) in turn (see Figure 1c) and all realizations start in either of the external states $\xi = \pm 1$ with the same probability $1/2$.

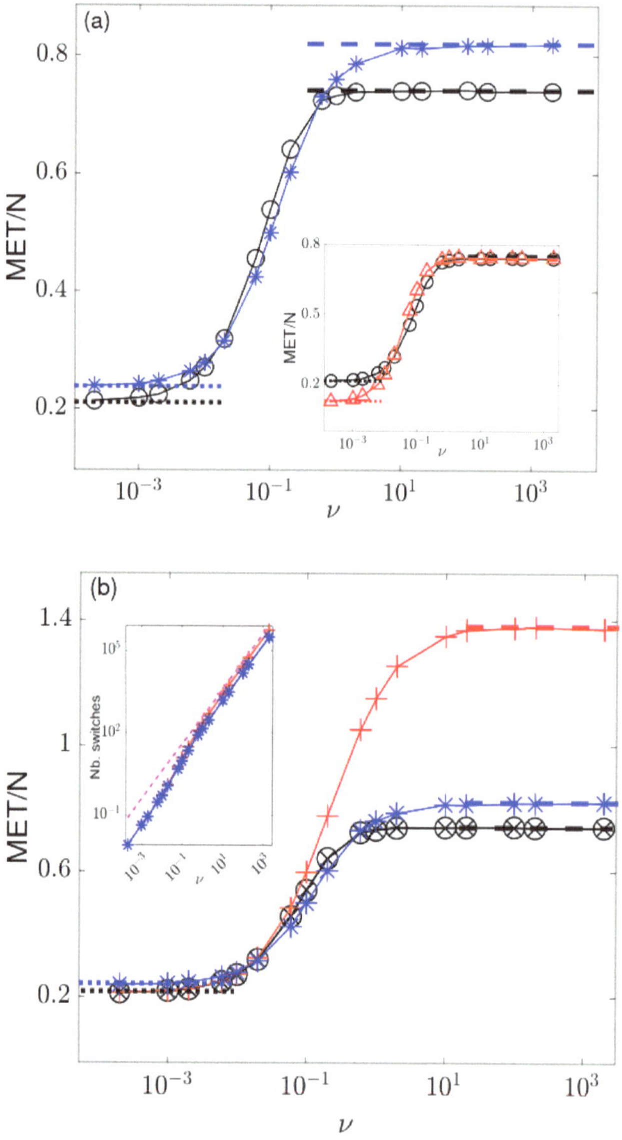

Figure 6. Scaled mean exit time (MET) versus ν: $T(\nu)/N$ for different N, δ and z. Symbols are from simulations (averaged over 10^5 samples). Dotted and dashed lines show T^0/N and T^∞/N from Equations (10) and (11), respectively. (**a**) $(N, b, s, \delta) = (200, 0.1, 20, 0.2)$, with $z = 0.5$ (black circles) and $z = 0.88$ (blue stars). Inset: $T(\nu)/N$ for $(N, b, s, \delta, z) = (500, 0.08, 40, 0.1, 0.5)$ (red triangles) and $(N, b, s, \delta, z) = (200, 0.1, 20, 0.2, 0.5)$ (black circles). MET scales as $T(\nu)/N \sim (\ln N)/(Nb)$ when $\nu \ll b$ and $T(\nu)/N = \mathcal{O}(1)$ when $\nu \gg b$. (**b**) $N = 200$, $b = 0.1$, $s = 20$, with $(\delta, z) = (0.2, 0.5)$ (black circles), $(\delta, z) = (-0.2, 0.5)$ (black crosses), $(\delta, z) = (0.2, 0.88)$ (blue stars), and $(\delta, z) = (0, 0.5)$ (red pluses). Inset: average number of switches before reaching the final state versus ν for the same parameters. Dashed line is an eyeguide of the slope $2N\nu$. See text for details.

In the regime where $\nu \ll b$, the final state is reached with a probability close to $1/2$ from either external state $\xi = \pm 1$. When $\xi = 1$, polarization is almost certain, whereas there is centrist consensus with probability 1 when $\xi = -1$. Hence, $P_{LR} \approx P_C \approx P^0_{LR} = P^0_C = 1/2$ and $P_L = P_R \approx 0$ when $\nu \ll b$. Under low switching rate, $\nu < b$, a small number switches can occur prior reaching the final state, but switching being symmetric, the net effect mostly cancels out and $P_{LR} = P_C \approx P^0_{LR} = P^0_C = 1/2$ when $s \gg 1$, as in Figure 5a,b.

When $\nu \gg b$, there are many switches before reaching the final state (see Section 4.2 below). This results in the self-averaging of ξ, yielding $\xi \to \langle \xi \rangle = 0$. Hence, when $\nu \gg b$ polarization and centrist consensus occur with probabilities:

$$P_{LR} \approx P^{\infty}_{LR} = 1 - \frac{1-(1-z)^2}{\sqrt{1+(1-z)^2}} \quad \text{and} \quad P_C \approx P^{\infty}_C = z,$$

where we have used Equations (6) and (7) with (A7).

We can determine when these probabilities increase with ν by solving Equation (9). When $s \gg 1$, we find $z_{LR} = (4 - \sqrt{18 - 2\sqrt{22}})/4 \approx 0.3621$ and $z_C = 1/2$. When $s \gg 1$ P_{LR} and P_C hence increase with ν when $z < z_{LR}$ and $z > 1/2$, respectively; see Figure 5. In the case $z < 1/2$, P_C is a decreasing function of ν and C is generally the minority opinion ($P_C < 1/2$); see the inset in Figure 5b. As shown in Figure 5a, when $z = 1/2$ then $P^{\infty}_C = P^0_C = 1/2$, and we find that the probability of centrist consensus remains essentially constant: $P_C(\nu) \approx 1/2$. When $s \gg 1$, $\delta = 0$ and $z = 2x = 2y = 1/2$, the final densities of voters are $c = P_C \approx 1/2$ and $\ell = r = (1 - P_C)/2 \approx 1/4$; see the inset in Figure 5a.

4. Mean Exit Time

The mean exit time (MET), here denoted by $T(\nu)$, is the average time to reach one of three absorbing/consensus states (all-L, all-R, all-C) or the polarization line (pol-LR). The MET hence complements the information provided by the polarization and consensus probabilities, and is therefore of great interest. Here, we study how the MET varies with ν for different given values of δ and z.

4.1. Mean Exit Time in the Regimes $\nu \to 0$ and $\nu \to \infty$

The MET is obtained analytically in the regimes $\nu \to 0$ and $\nu \to \infty$ in terms of $\mathcal{T}(s,z)$, its counterpart in the absence of external influences, as discussed in Appendix B.2.

4.1.1. MET in the Regime $\nu \to 0$

When $\nu \to 0$, there are no switches before reaching the final state. The MET, T^0, can thus be obtained by averaging $\mathcal{T}(s,z)$ over the stationary distribution of ξ. Since $\xi = \pm 1$ with probability $(1 \pm \delta)/2$, in this regime the MET reads:

$$T^0(s,\delta,z) = \left(\frac{1+\delta}{2}\right)\mathcal{T}(s,z) + \left(\frac{1-\delta}{2}\right)\mathcal{T}(-s,z). \tag{10}$$

$T^0(s,\delta,z)$ and its scaling are thus readily obtained from Equations (A10) and (A11). From the symmetry $\mathcal{T}(-s,z) = \mathcal{T}(s,1-z)$, we have $T^0(s,\delta,1/2) = T^0(s,-\delta,1/2) = \mathcal{T}(s,1/2)$ for $z = 1/2$ (the unwieldy expression of $\mathcal{T}(s,1/2)$ is given in Ref. [54]). From Figure 6, we find that the predictions of Equation (10) are in good agreement with simulation results when $\nu \ll 1$ in all scenarios ($\delta > 0$, $\delta < 0$ and $\delta = 0$).

4.1.2. MET in the Regime $\nu \to \infty$

When $\nu \to \infty$, many external influences switches occur before the population reaches its final state. This leads to the self-averaging of ξ, which can therefore be replaced by its average: $\xi \to \langle \xi \rangle = \delta$. Hence, under high switching rate, the influences bias is rescaled according to $b \to b\delta$, yielding $s \to s\delta$, at fixed N. When $\nu \to \infty$ the MET, T^{∞}, therefore satisfies Equation (A10) with s substituted by $s\delta$, and thus

$$T^{\infty}(s,\delta,z) = \mathcal{T}(s\delta,z). \tag{11}$$

In this regime, the MET follows readily from the solution of (A10), and its scaling is obtained from Equation (A11). Using the symmetry of $\mathcal{T}$, we have $T^{\infty}(s,\delta,z) = T^{\infty}(s,-\delta,1-z)$, which boils down to $T^{\infty}(s,\delta,1/2) = T^{\infty}(s,-\delta,1/2)$ when $z = 1/2$. Furthermore, when $\delta \to 0$, we obtain $T^{\infty}(s,0,z) = \mathcal{T}(0,z) = -2N[z\ \ln z + (1-z)\ \ln(1-z)]$, which is independent of the influences bias b. In Figure 6, the predictions of Equation (11) are in close to perfect agreement with the simulation results for $\nu \gg 1$ in all scenarios ($\delta > 0$, $\delta < 0$ and $\delta = 0$).

4.2. Mean Exit Time as Function of the Switching Rate

We now consider the dependence of the MET under arbitrary ν.

When the switching rate is low, $\nu \ll b$, it is unlikely that there are any switches before reaching the final states, and thus $T(\nu) \approx T^{0}(s,\delta,z)$; see Equation (10). When $b \ll 1$, $1 \ll s \ll N$, and z is not too close to 0 and 1, then $T(\nu) \sim (\ln N)/b$ (see Equation (A11)), and the average number of switches in this regime prior to reaching the final state is $\nu T(\nu) \sim (\ln N)\nu/b$; see Figure 6a,b. Hence, when $\nu \sim b$ there are on average $\mathcal{O}(\ln N)$ switches before reaching the final state.

Under high ν, the system experiences a large number of switches causing the self-average of ξ (with $\xi(t) \to \delta$). In this regime, $T(\nu) \approx T^{\infty}(s,\delta,z)$; see Equation (11), with $T(\nu) \sim N$. In this case, the average number of switches before reaching the final state is $\nu T(\nu) \sim \nu N$.

We can therefore revisit and characterise the three switching regimes:
(i) when $\nu \ll b$, the system is in the regime of "low switching rate" and the final state can possibly be reached without experiencing any switches;
(ii) when $\nu \sim b$, the population is in a regime of "intermediate switching rate", where there are typically $\mathcal{O}(\ln N)$ switches before reaching the final state;
(iii) when $\nu \gg b$, the system is in the regime of "high switching rate", where the external noise self-averages, and the number of switches before reaching the final state is $\mathcal{O}(\nu N)$ and hence grows linearly with νN, where $\nu N \gg Nb = s \gg 1$; see inset in Figure 6b.

When the switching rate is raised across the above regimes, from $\nu \approx 0$ to $\nu \gg 1$, the MET changes its scaling behaviour. In regime (i), the MET scales as $T \sim (\ln N)/b$ and therefore $T/N \ll 1$ when $s \gg 1$, while in regime (iii) the MET scales as $T \sim N$, and therefore $T/N = f(z)$ is a scaling function of the initial fraction z. In the intermediate regime (ii), when $\nu \sim b$, the MET increases steeply with ν and interpolates between $T(\nu)/N \ll 1$ and $T(\nu)/N \approx f(z)$ as ν sweeps from regimes (ii) into (iii). This picture is confirmed by the results shown in Figure 6a,b.

Figure 6a,b, illustrates that the initial fraction z has only a marginal effect on the MET in the limiting regimes $\nu \ll b$ and $\nu \gg b$, which is well captured by Equations (10) and (11). Figure 6b shows that, as predicted by Equation (11), the MET at fixed N, b, z is maximum when $\delta = 0$, with $T/N \approx T^{\infty}/N = -2[z\ \ln z + (1-z)\ \ln(1-z)]$; see Appendix B.2. The inset in Figure 6a illustrates that the MET scales linearly with N in the regime (iii), while N has a marginal effect on $T(\nu)$ in the regimes (i) and (ii): When $\nu \gg b$, we find $T(\nu)/N = \mathcal{O}(1)$ as in Ref. [54]. When $\nu \ll b$, we find that $T(\nu)/N$ decreases with N in agreement with $T(\nu)/N \sim (\ln N)/(Nb)$. The inset of Figure 6b confirms that when $\nu \gg b$, the average number of switches exhibits the same linear scaling with νN for different parameter sets.

Here, time-varying external influences are therefore responsible for a drastic change in the MET scaling: the MET scales as $(\ln N)/b$ under low switching rate, while it grows and scales linearly with N under high switching rate. When $s = Nb \gg 1$, reaching the final state thus takes much longer under $\nu \gg 1$ than $\nu \ll 1$.

5. Conclusions

Motivated by the evolution of opinions in a volatile social environment shaped by time-fluctuating external influences, arising, e.g., from news or social media, we have introduced a constrained three-state voter model subject to randomly binary time-fluctuating (switching) external influences. The voters of this population can hold either incompatible leftist or rightist opinions, or behave as centrists. The changing external influences (social environment) is modelled by a dichotomous random process that favours in turn the rise of polarization or the spread of centrism. The fate of this population is either to reach a consensus with leftists, rightists, or centrists, or to achieve polarization, which consists of a frozen state comprised of only non-interacting leftists and rightists. By combining analytical and computational means, we have investigated the effect of the time-fluctuating external influences on the population's final state under various scenarios. In particular, we have studied how the rate of switching, as well as the switching asymmetry and initial population composition, affect the fate of this population.

Focusing on the interesting case of a small influences bias affecting a finite, yet large, number of voters, we have shown that the consensus and polarization probabilities can vary greatly with the rate of change of the external influences: these probabilities can either increase or decrease with the rate of external variations, and can also exhibit extrema. Remarkably, when there is a large initial majority of voters holding the opinions opposed by the external influences, the majority can resist the influences: the opinions supported by the majority of agents and opposed by the influences are the most likely to prevail over a range of parameters characterising the external variations. When this occurs, the population settles in a final state in which a majority of voters holds the opinions opposed by the external influences, see Figures 3b and 4b. The study has also shown that time-switching influences are responsible for a drastic change in the scaling of the mean time to reach final state: the mean exit time is generally much bigger under high switching rate, when it scales linearly with population size, than under low switching rate.

The goal of this paper is to study how time-varying external influences may affect the social dynamics of an idealized population. It can be anticipated that the model analysed here is too simple to realistically capture the various complex effects of social and news media, and other time-varying external stimuli, on opinion dynamics. It is however natural to ask how this model can be generalized to become more realistic, and which kind of practical information it could then possibly provide. These points are addressed in Appendix C, where a potential application is briefly discussed. In fact, while the direct applications of the current model are admittedly limited, it is expected to still be useful since it sheds light on nontrivial effects that exogenous time-fluctuating influences can have on opinion dynamics. Hence, this work can be envisaged as a step towards more realistic modelling approaches to this challenging interdisciplinary problem.

Funding: This research received no external funding.

Data Availability Statement: Data used to generate Figures 3–6 are electronically available at the Research Data Leeds Repository [84].

Conflicts of Interest: The author declares no conflict of interest.

Appendix A. Master Equation and Mean-Field Limit When $N \to \infty$

In this Appendix, we discuss the master equation (ME) governing the model's dynamics, and then its description in the mean-field limit when $N \to \infty$.

Appendix A.1. Master Equation

The 3CVM switching dynamics in a finite population is a Markov chain defined by the transition rates:

$$W_L^{\pm}(N_L, N_R, \xi) = \frac{(1 \pm b\xi)}{2} \frac{N_L(N - N_R - N_L)}{N(N-1)}, \; W_R^{\mp}(N_L, N_R, \xi) = \frac{(1 \pm b\xi)}{2} \frac{N_R(N - N_R - N_L)}{N(N-1)}, \tag{A1}$$

such that $N_L \xrightarrow{W_L^{\pm}} N_L \pm 1$ and $N_R \xrightarrow{W_R^{\pm}} N_R \pm 1$. The associated ME gives the probability $P(N_L, N_R, \xi, t)$ to have N_L, N_R and $N_C = N - N_L - N_R$ voters in the population in the external state $\xi = \pm 1$ at time t [82], and here reads:

$$\begin{aligned}
\frac{\partial P(N_L, N_R, \xi = 1, t)}{\partial t} &= (\mathbb{E}_L^- - 1)\left[W_L^+ P(N_L, N_R, 1, t)\right] + (\mathbb{E}_R^- - 1)\left[W_R^+ P(N_L, N_R, 1, t)\right] \\
&+ (\mathbb{E}_L^+ - 1)\left[W_L^- P(N_L, N_R, 1, t)\right] + (\mathbb{E}_R^+ - 1)\left[W_R^- P(N_L, N_R, 1, t)\right] \\
&+ \nu[(1+\delta)P(N_L, N_R, -1, t) - (1-\delta)P(N_L, N_R, 1, t)], \qquad \text{(A2)} \\
\frac{\partial P(N_L, N_R, \xi = -1, t)}{\partial t} &= (\mathbb{E}_L^- - 1)\left[W_L^+ P(N_L, N_R, -1, t)\right] + (\mathbb{E}_R^- - 1)\left[W_R^+ P(N_L, N_R, -1, t)\right] \\
&+ (\mathbb{E}_L^+ - 1)\left[W_L^- P(N_L, N_R, -1, t)\right] + (\mathbb{E}_R^+ - 1)\left[W_R^- P(N_L, N_R, -1, t)\right] \\
&+ \nu[(1-\delta)P(N_L, N_R, 1, t) - (1+\delta)P(N_L, N_R, -1, t)], \qquad \text{(A3)}
\end{aligned}$$

where $\mathbb{E}_L^\pm$ and $\mathbb{E}_R^\pm$ are shift operators such that $\mathbb{E}_{L/R}^\pm f(N_{L/R}, N_{R/L}, t) = f(N_{L/R} \pm 1, N_{R/L}, t)$, and $P(N_L, N_R, \xi, t) = 0$ whenever N_L or N_R are outside $[0, N]$. $P(N_L, N_R, 1, t)$ and $P(N_L, N_R, -1, t)$ are coupled, and the last lines on the right-hand-side of (A2) and (A3) account for the random external switching. We notice that $W_{L/R}^\pm = 0$ when $N_L = 0, N_R = 0$ and $N_L + N_R = N$, corresponding to the three absorbing states all-R ($N_L = N_C = 0$), all-L ($N_R = N_C = 0$), all-C ($N_L = N_R = 0$, $N_C > 0$), and to the polarization state pol-LR ($N_L + N_R = N$, $N_C = 0$); see Figure 2. The multivariate ME (A2) and (A3) can be simulated exactly by standard methods, such as the Gillespie algorithm [85].

Appendix A.2. Mean-Field Limit When $N \to \infty$

In the mean-field limit where $N \to \infty$ and demographic can be ignored, the equation of motion of the densities are [54,82]:

$$\frac{d}{dt}\ell = W_L^+ - W_L^- = b\xi\, \ell(1 - \ell - r), \quad \frac{d}{dt}r = W_R^+ - W_R^- = b\xi\, r(1 - \ell - r), \qquad \text{(A4)}$$

where $c \equiv N_C/N = 1 - \ell - r$ is used, and the time dependence is omitted. Here, the mean-field limit means that the dynamics is aptly described by Equation (A4) where $\xi(t) \in \{-1, 1\}$ is a randomly switching multiplicative noise, see Equation (1). Equation (A4) are thus two coupled *stochastic* differential equations that have three absorbing steady states: $(\ell, r, c) = \{(1,0,0),\ (0,1,0),\ (0,0,1)\}$ associated respectively with the consensus states all-L, all-R and all-C. Equations (A4) also admit the line of steady states $(\ell, 1-\ell, 0)$, with $0 < \ell < 1$, associated with state of polarization pol-LR; see Figure 2. Equations (A4) conserve the ratio $\ell/r = x/y$, which implies that the final densities satisfy $\ell(\nu) = r(\nu)$ and $c(\nu) = 1 - 2\ell(\nu)$ when $x = y$.

It is useful to relate this result with the final densities in a *finite* population, in the case where $x = y$. When $N < \infty$, the final densities of R and L voters are $\ell(\nu) = r(\nu) = P_L(\nu) + P_{LR}(\nu)/2 = P_R(\nu) + P_{LR}(\nu)/2$ (as in the absence of external influences [53,54]). The first term comes from the probability of ending in L or R consensus (with $P_L = P_R$). The second term arises from the fact that, when $x = y$, the polarization state consists of half L and R voters. Since $P_L + P_R + P_{LR} = 1 - P_C$ and $P_L = P_R$, we find $\ell(\nu) = r(\nu) = [1 - P_C(\nu)]/2$, and the final density of C voters is: $c(\nu) = 1 - 2\ell(\nu) = P_C(\nu)$.

When $N \to \infty$, Equations (A4) hold and predict that the densities approach polarization when $\xi = 1$ and centrist consensus when $\xi = -1$ on a timescale $t \sim 1/b$. Hence, when $N \to \infty$, the probability of L and R consensus vanish, $P_L = P_R \to 0$, and the final densities thus satisfy $\ell(\nu) = r(\nu) \to P_{LR}(\nu)/2$ and $c(\nu) = 1 - 2\ell(\nu) \to 1 - P_{LR}(\nu)$.

Appendix B. Polarization, Consensus Probabilities, and Mean Exit Time in the Absence of Time-Varying Influences

In this Appendix, we reproduce the results obtained in Refs. [53,54] for the polarization and consensus probabilities in the absence of external influences, and used in Sections 3.1 and 4.1.

Appendix B.1. Polarization and Consensus Probabilities in the Absence of Time-Varying Influences

In the realm of the diffusion theory [68,69,82], when $N \gg 1$, $s \equiv Nb \neq 0$ and $x = y = (1-z)/2$, the polarization probability, here, denoted by $\mathcal{P}_{LR}$ is [54]:

$$\mathcal{P}_{LR}(s,z) = \frac{e^{sz}\sqrt{1-z}}{2}\sum_{k=0}^{\infty}\frac{(-1)^k(4k+3)}{(2k+1)(k+1)}\frac{(2k+1)!!}{(2k)!!}\frac{I_{2k+3/2}(s(1-z))}{I_{k+3/2}(s)}, \tag{A5}$$

where $I_k(\cdot)$ denotes the modified Bessel function of first kind and order k, and we used the property $P^1_{2k+1}(0) = (-1)^k(2k+1)!!/(2k)!!$ of the associated Legendre polynomials, $P^m_l(x)$ (with the convention $0!! = 1$). When $s < 0$ and $|s|(1-z) \gg 1$, Equation (A5) can be approximated by $\mathcal{P}_{LR} \approx (e^{2|s|(1-z)} - 1)/(e^{2|s|} - 1)$; see Ref. [54]. This simplified expression is particularly useful to approximate P^{∞}_{LR} when $\delta < 0$; see Equation (6).

In the realm of the diffusion theory, when $N \gg 1$, $s \equiv Nb \neq 0$, the C-consensus probability, here denoted by $\mathcal{P}_C$, is [54]:

$$\mathcal{P}_C(s,z) = \frac{e^{-2s(1-z)} - e^{-2s}}{1-e^{-2s}}. \tag{A6}$$

When $z \neq 0,1$, one has: $\lim_{s\to\infty}\mathcal{P}_{LR}(s,z) = 1$ and $\lim_{s\to-\infty}\mathcal{P}_C(s,z) = 1$. In the considered examples, when $s \gg 1$ and z is not too close to 0 or 1, polarization and centrist consensus are almost certain, i.e., $\mathcal{P}_{LR}(s,z) \approx 1$ if $s > 0$ and $\mathcal{P}_C(s,z) \approx 1$ when $s < 0$ [54].

When $b = s = 0$, and $x = y = (1-z)/2$, the probabilities $\mathcal{P}_{LR}$ and $\mathcal{P}_C$ become [53]:

$$\mathcal{P}_{LR}(0,z) = 1 - \frac{1-(1-z)^2}{\sqrt{1+(1-z)^2}}, \quad \mathcal{P}_C(0,z) = z. \tag{A7}$$

When $x = y = (1-z)/2$, the probability $\mathcal{P}_R = \mathcal{P}_L$ to end up in an L or R consensus is thus

$$\mathcal{P}_R(s,z) = \mathcal{P}_L(s,z) = \frac{1-\mathcal{P}_{LR}(s,z) - \mathcal{P}_C(s,z)}{2}, \tag{A8}$$

while there is the same average final fraction ℓ and r of voters of type L and R, given by

$$r = \ell = \frac{\mathcal{P}_{LR}(s,z)}{2} + \mathcal{P}_R(s,z) = \frac{1-\mathcal{P}_C(s,z)}{2}. \tag{A9}$$

Expressions (A8) and (A9) hold both when $s \neq 0$ with Equations (A5) and (A6), and when $s = 0$ with Equations (A7).

Appendix B.2. Mean Exit Time in the Absence of Time-Varying Influences

In Refs. [53,54], the (unconditional) mean exit time of the 3CVM in the absence of time-varying influences, here denoted by $\mathcal{T}$, was shown to satisfy

$$\frac{z(1-z)}{2N}\left[-2s\frac{d\mathcal{T}(s,z)}{dz} + \frac{d^2\mathcal{T}(s,z)}{dz^2}\right] = -1, \tag{A10}$$

with $\mathcal{T}(s,0) = \mathcal{T}(s,1) = 0$. (Note a typo in Equation (11) of Ref. [54] where a factor $1/2$ is missing; compare to Equation (A10).)

The symmetry of Equation (A10) under $(s,z) \to (-s,1-z)$ implies $\mathcal{T}(-s,z) = \mathcal{T}(s,1-z)$ [54]. When $s = Nb \neq 0$ and $|b| \ll 1$, with z not too close to 0 and 1, the MET scaling is [54,69,86]:

$$\mathcal{T}(s,z) \quad \sim \quad \begin{cases} N\ln N/|s| & \text{when } 1 \ll |s| \ll N \\ N & \text{when } |b| \ll 1 \text{ and } |s| = \mathcal{O}(1), \end{cases} \tag{A11}$$

yielding $\mathcal{T}(s,z) = \mathcal{O}(\ln N/b)$ when $|b| \ll 1$ and $1 \ll |s| \ll N$. When $b = s = 0$, the MET $\mathcal{T}(0,z)$ takes the closed form $\mathcal{T}(0,z) = -2N[z \ \ln z + (1-z) \ \ln(1-z)]$ [53], and in this case, it scales linearly with the population size: $\mathcal{T}(0,z) \sim N$.

Appendix C. Possible Generalizations and Applications of the Model

This Appendix is dedicated to a brief discussion of some possible generalizations and applications of the 3CVM with switching dynamics.

It is often hard to map complex real social systems onto idealized theoretical models: these commonly suffer from a number of limitations that make their use for the practical characterization of social behaviour difficult. The 3CVM with switching dynamics is no exception, and among its limitations, we can list the following: (i) it is doomed to end up in either consensus or L/R-polarization, but never admits the long-lived coexistence of the three opinions; (ii) the present model formulation assumes that all agents interact with all others on a complete graph (rather than on a complex dynamic network); (iii) as most classical voter models, all agents are identical in the 3CVM (there are no "zealots"); (iv) interactions are pairwise (no group pressure); (v) in many applications, there are more than three possible opinions/parties.

Since the 3CVM can be generalized to overcome the limitations (i)–(v) at the expense of its mathematical tractability, it is useful to discuss a possible application. The main challenge for this is to find data against which to calibrate the parameters ν and δ characterising the external time-varying influences. In this context, the formulation of the 3CVM suggests to test its use to describe the distribution of opinions in the readership of a newspaper such as *The Guardian* in the UK whose political backing of the Labour, Liberal (Lib Dem) and Conservative parties has changed on various occasions in the last 78 years.

We have used the dataset [87] referring to the 18 general elections held in the UK between 1945 and 2010 to try and estimate the parameters ν and δ for a population corresponding to a random sample of the readership of *The Guardian* (whose circulation between 1945 and 2021 has varied between 10^5 and 3×10^5). If we use the average time between each general election as unit of time, and notice that *The Guardian*'s political orientation changed 8 times between 1945 and 2010, in 1950, 1951, 1955, 1959, 1974, 1979, 2005, 2010. The average switching rate can thus be estimated as $\nu \approx 8/18$. By treating the backing of the Labour or Conservative party as being in the influences state $\xi = 1$, and representing the backing of the Lib Dem party by $\xi = -1$, we can estimate that on average $\langle \xi \rangle = \delta \approx 6/18$. Note that here the change from backing jointly two parties (e.g., Labour and Lib Dem) to supporting only one of those parties (e.g., Liberal Party) is considered as a "switch" (and the joint Labour/Liberal and Conservative/Liberal support is treated as an influence state $\xi = (1-1)/2 = 0$). The parameter b could in principle be estimated from the approach to the "final state" occurring on a timescale $\sim 1/b$, see Equation (A4). In this context, it is a difficult task to assess on what timescale consensus or polarization may occur, if ever. Here, for the sake of argument, we set $b = 0.01$. With $(b, \nu, \delta) = (0.01, 0.44, 0.33)$ and assuming an initial population consisting of 60% and 10% of labour and conservative supporters, respectively, and 30% of Lib Dem backers, the model predicts a final state consisting of 70% and 12% of Labour and Conservative supporters, respectively, and 18% of Lib Dem backers in a random sample of size $N = 200$. For a larger sample of size $N = 1000$ and the same above parameters, the model predicts a final state comprised of 85% and 14% of Labour and Conservative supporters, and only a small fraction of 1% backing the Lib Dem party. These figures, suggesting the slow evolution towards quasi polarization of *The Guardian*'s readership as N increases, with a raise of support for the Labour party, do not seem absurd but are not realistic as they underestimate the Lib Dem vote. Moreover, the 3CVM ignores entirely the existence of a small but non-negligible fraction of voters backing other parties (such as the Green party). A more realistic model would take into account more than three parties, and mechanisms ensuring the maintenance of long-lived coexistence of all opinions.

References

1. Granovetter, M. Threshold models of collective behavior. *Am. J. Sociol.* **1978**, *83*, 1420–1443. Available online: https://www.jstor.org/stable/2778111 (accessed on 9 April 2023).
2. Schelling, T.C. *Micromotives and Macrobehaviour*; W. W. Norton & Company, Inc.: New York, NY, USA, 1978.
3. Galam, S.; Gefen, Y.; Shapir, Y. Sociophysics: A new approach of sociological collective behaviour: I. Mean-behaviour description of a strike. *Math. J. Sociol.* **1982**, *9*, 1–13. [CrossRef]
4. Galam, S. Majority rule, hierarchical structures, and democratic totalitarianism: A statistical approach. *J. Math. Psychol.* **1986**, *30*, 426–434. [CrossRef]
5. Galam, S. Social paradoxes of majority rule voting and renormalization group. *J. Stat. Phys.* **1990**, *61*, 943–951. [CrossRef]
6. Galam, S.; Moscovici, S. Towards a theory of collective phenomena: Consensus and attitude changes in groups. *Eur. J. Soc. Psychol.* **1991**, *21*, 49–74. [CrossRef]
7. Castellano, C.; Fortunato, S.; Loreto, V. Statistical physics of social dynamics. *Rev. Mod. Phys.* **2009**, *81*, 591–646. [CrossRef]
8. Galam, S. *Sociophysics. A Physicist's Modeling of Psycho-Political Phenomena*; Springer Science+Business Media, LLC: New York, NY, USA, 2012. [CrossRef]
9. Sen, P.; Chakrabarti, B.K. *Sociophysics: An Introduction*; Oxford University Press: New York, NY, USA, 2014.
10. Perc, M.; Jordan, J.J.; Rand, D.G.; Wang, Z.; Boccaletti, S.; Szolnoki, A. Statistical physics of human cooperation. *Phys. Rep.* **2017**, *687*, 1–51. [CrossRef]
11. Schweitzer, F. Sociophysics. *Phys. Today* **2018**, *71*, 40–46. [CrossRef]
12. Jedrzejewski, A.; Sznajd-Weron, K. Statistical physics of opinion formation: Is it a SPOOF? *C. R. Phys.* **2019**, *20*, 244–261. [CrossRef]
13. Redner, S. Reality-inspired voter models: A mini-review. *C. R. Phys.* **2019**, *20*, 275–292. [CrossRef]
14. Li, Z.; Chen, X.; Yang, H.-X.; Szolnoki, A. Game-theoretical approach for opinion dynamics on social networks. *Chaos* **2022**, *32*, 73117. [CrossRef]
15. Liggett, T.M. *Interacting Particle Systems*; Springer: Berlin/Heidelberg, Germany, 2005. [CrossRef]
16. Glauber, R.J. Time-dependent statistics of the Ising model. *J. Math. Phys.* **1963**, *4*, 294–307. [CrossRef]
17. Asch, S.E. Opinions and social pressure. *Sci. Am.* **1955**, *193*, 31–35. Available online: https://www.jstor.org/stable/24943779 (accessed on 9 April 2023)
18. Milgram, S.; Bickman, L.; Berkowitz, L. Note on the drawing power of crowds of different size. *J. Personal. Soc. Psychol.* **1969**, *13*, 79–82. [CrossRef]
19. Lattané, B. The psychology of social impact. *Am. Psychol.* **1981**, *36*, 343–356. [CrossRef]
20. Axelrod, R. The dissemination of culture: A model with local convergence and global olarization. *J. Confl. Resolut.* **1997**, *41*, 203–226. [CrossRef]
21. Axelrod, R. *The Complexity of Cooperation: Agent-Based Models of Competition and Collaboration*; Princeton University Press: Princeton, NJ, USA, 1997. [CrossRef]
22. Castellano, C.; Marsili, M.; Vespignani, A. Nonequilibrium phase transition in a model for social influence. *Phys. Rev. Lett.* **2000**, *85*, 3536–3539. [CrossRef]
23. Klemm, K.; Eguiluz, V.M.; Toral, R.; San Miguel, M. Global culture: A noise-induced transition in finite systems. *Phys. Rev. E* **2003**, *67*, 045101. [CrossRef]
24. McPherson, J.M.; Smith-Lovin, L. Homophily in voluntary organizations: Status distance and the composition of face-to-face groups. *Am. Sociol. Rev.* **1987**, *52*, 370–379. [CrossRef]
25. Mobilia, M. Does a single zealot affect an infinite group of voters? *Phys. Rev. Lett.* **2003**, *91*, 028701. PhysRevLett.91.028701. [CrossRef]
26. Mobilia, M.; Georgiev, I.T. Voting and catalytic processes with inhomogeneities. *Phys. Rev. E* **2005**, *71*, 046102. 10.1103/PhysRevE.71.046102. [CrossRef] [PubMed]
27. Mobilia, M.; Petersen, A.; Redner, S. On the role of zealotry in the voter model. *J. Stat. Mech.* **2007**, *2007*, P08029. [CrossRef]
28. Mobilia, M. Commitment versus persuasion in the three-party constrained voter model. *J. Stat. Phys.* **2013**, *151*, 69–91. [CrossRef]
29. Mobilia, M. Nonlinear q-voter model with inflexible zealots. *Phys. Rev. E* **2015**, *92*, 012803. [CrossRef] [PubMed]
30. Galam, S.; Jacobs, F. The role of inflexible minorities in the breaking of democratic opinion dynamics. *Phys. A* **2007**, *381*, 366–376. [CrossRef]
31. Sznajd-Weron, K.; Tabiszewski, M.; Timpanaro, A.M. Phase transition in the Sznajd model with independence. *EPL (Europhys. Lett.)* **2011**, *96*, 48002. [CrossRef]
32. Acemoglu, D.; Como, G.; Fagnani, F.; Ozdaglar, A. Opinion fluctuations and disagreement in social networks. *Math. Op. Res.* **2013**, *38*, 1–27. [CrossRef]
33. Nyczka, P.; Sznajd-Weron, K. Anticonformity or independence?—Insights from statistical physics. *J. Stat Phys.* **2013**, *151*, 174–202. [CrossRef]
34. Masuda, N. Opinion control in complex networks. *New J. Phys.* **2015**, *17*, 033031. [CrossRef]
35. Li, X.; Mobilia, M.; Rucklidge, A.M.; Zia, R.K.P. How does homophily shape the topology of a dynamic network? *Phys. Rev. E* **2021**, *104*, 044311. [CrossRef]
36. Li, X.; Mobilia, M.; Rucklidge, A.M.; Zia, R.K.P. Effects of homophily and heterophily on preferred-degree networks: Mean-field analysis and overwhelming transition. *J. Stat. Mech.* **2022**, 013402. [CrossRef]

37. Castellano, C.; Muñoz, M.A.; Pastor-Satorras, R. Nonlinear q-voter model. *Phys. Rev. E* **2009**, *80*, 041129. [CrossRef]
38. Sznajd-Weron, K.; Sznajd, J. Opinion evolution in closed community. *Int. J. Mod. Phys. C* **2000**, *11*, 1157–1165. [CrossRef]
39. Slanina, F.; Lavicka, H. Analytical results for the Sznajd model of opinion formation. *Eur. Phys. J. B* **2003**, *35*, 279–288. [CrossRef]
40. Nyczka, P.; Sznajd-Weron, K.; Cislo, J. Opinion dynamics as a movement in a bistable potential. *Phys. A* **2012**, *391*, 317–327. [CrossRef]
41. Lambiotte, R.; Redner, S. Dynamics of non-conservative voters. *EPL (Europhys. Lett.)* **2008**, *82*, 18007. [CrossRef]
42. Slanina, F.; Sznajd-Weron, K.; Przybyla, P. Some new results on one-dimensional outflow dynamics. *EPL (Europhys. Lett.)* **2008**, *82*, 18006. [CrossRef]
43. Galam, S.; Martins, A.C.R. Pitfalls driven by the sole use of local updates in dynamical systems. *EPL (Europhys. Lett.)* **2011**, *95*, 48005. [CrossRef]
44. Timpanaro, A.M.; Prado, C.P.C. Exit probability of the one-dimensional q-voter model: Analytical results and simulations for large networks. *Phys. Rev. E* **2014**, *89*, 052808. [CrossRef]
45. Mellor, A.; Mobilia, M.; Zia, R.K.P. Characterization of the nonequilibrium steady state of a heterogeneous nonlinear q-voter model with zealotry. *EPL (Europhys. Lett.)* **2016**, *113*, 48001. [CrossRef]
46. Mellor, A.; Mobilia, M.; Zia, R.K.P. Heterogeneous out-of-equilibrium nonlinear q-voter model with zealotry. *Phys. Rev. E* **2017**, *95*, 012104. [CrossRef]
47. Deffuant, G.; Neau, D.; Amblard, F.; Weisbuch, G. Mixing Beliefs among Interacting Agents. *Adv. Complex Syst.* **2000**, *3*, 87–98. [CrossRef]
48. Hegselmann, R.; Krause, U. Opinion dynamics and bounded confidence models, analysis, and simulation. *J. Artif. Soc. Soc. Simul.* **2002**, *5*, 2. Available online: https://www.jasss.org/5/3/2.html (accessed on 9 April 2023).
49. Weisbuch, G.; Deffuant, G.; Amblard, F.; Nadal, J.P. Meet, discuss, and segregate! *Complexity* **2002**, *7*, 55–63. [CrossRef]
50. Ben-Naim, E.; Krapivsky, P.L.; Redner, S. Bifurcations and patterns in compromise processes. *Phys. D* **2003**, *183*, 190. [CrossRef]
51. Slanina, F. Dynamical phase transitions in Hegselmann-Krause model of opinion dynamics and consensus. *Eur. Phys. J. B* **2011**, *79*, 99–106. [CrossRef]
52. Vazquez, F.; Krapivsky, P.L.; Redner, S. Freezing and slow evolution in a constrained opinion dynamics model. *J. Phys. A Math. Gen.* **2003**, *36*, L61–L68. [CrossRef]
53. Vazquez, F.; Redner, S. Ultimate fate of constrained voters. *J. Phys. A Math. Gen.* **2004**, *37*, 8479–8494. [CrossRef]
54. Mobilia, M. Fixation and polarization in a three-species opinion dynamics model. *EPL (Europhys. Lett.)* **2011**, *95*, 50002. [CrossRef]
55. DellaVigna, S.; Kaplan, E. The Fox News effect: Media bias and voting. *Quart. J. Econ.* **2007**, *122*, 1187–1234. [CrossRef]
56. Gerber, A.S.; Karlan, D.; Bergan, D. Does the media matter? A field experiment measuring the effect of newspapers on voting behavior and political opinions. *Am. Econ. J. Appl. Econ.* **2009**, *1*, 35–52. [CrossRef]
57. Yang, J.; Leskovec, J. Patterns of temporal variation in online media. In Proceedings of the Fourth ACM International Conference on Web Search and Data Mining, Hong Kong, China, 9–12 February 2011; King, I., Nejdi, W., Li, H., Eds.; Association for Computing Machinery: New York, NY, USA, 2011; pp. 177–186. . [CrossRef]
58. Wettstein, M.; Wirth, W. Media effects: How media influence voters. *Swiss Polit. Sci. Rev.* **2017**, *23*, 262–269. [CrossRef]
59. Holbrook, T.M. *Altered States: Changing Populations, Changing Parties, and the Transformation of the American Political Landscape*; Oxford University Press: New York, NY, USA, 2016. [CrossRef]
60. Dewenter, R.; Linder, M.; Thomas, T. Can media drive the electorate? The impact of media coverage on voting intentions. *Eur. J. Polit. Econ.* **2019**, *58*, 245–261. [CrossRef]
61. Kaviani, M.S.; Li, L.; Maleki, H. Media, partisan ideology, and corporate social responsibility. *SSRN* **2021**, 3658502. [CrossRef]
62. Bhat, D.; Redner, S. Nonuniversal opinion dynamics driven by opposing external influences. *Phys. Rev. E* **2019**, *100*, 050301. [CrossRef]
63. Bhat, D.; Redner, S. Polarization and consensus by opposing external sources. *J. Stat. Mech.* **2020**, *2020*, 013402. [CrossRef]
64. Lorenz-Spreen, P.; Mørch Mønsted, B.; Hövel, P.; Lehmann, S. Accelerating dynamics of collective attention. *Nat. Commun.* **2019**, *10*, 1759. [CrossRef]
65. Helfmann, L.; Djurdjevac Conrad, N.; Lorenz-Spreen, P.; Schütte, C. Modelling opinion dynamics under the impact of influencer and media strategies. *arXiv* **2023**, arXiv:2301.13661. [CrossRef]
66. Hosseiny, A.; Bahrami, M.; Palestrini, A.; Gallegati, M. Metastable features of economic networks and responses to exogenous shocks. *PLoS ONE* **2016**, *11*, 1–22. [CrossRef]
67. Hosseiny, A.; Absalan, M.; Sherafati, M.; Gallegati, M. Hysteresis of economic networks in an XY model. *Phys. A* **2019**, *513*, 644–652. [CrossRef]
68. Crow, J.F.; Kimura, M. *An Introduction to Population Genetics Theory*; The Blackburn Press: Caldwell, NJ, USA, 1970. Available online: https://archive.org/details/an-introduction-to-population-genetics-theory-pdfdrive (accessed on 9 April 2023)
69. Ewens, W.J. *Mathematical Population Genetics. Volume 1: Theoretical Introduction*; Springer Science+Business Media New York: New York, NY, USA, 2004. [CrossRef]
70. Assaf, M.; Mobilia, M.; Roberts, E. Cooperation dilemma in finite populations under fluctuating environments. *Phys. Rev. Lett.* **2013**, *111*, 238101. [CrossRef] [PubMed]
71. West, R.; Mobilia, M.; Rucklidge, A.M. Survival behavior in the cyclic Lotka-Volterra model with a randomly switching reaction rate. *Phys. Rev. E.* **2018**, *97*, 022406. https://journals.aps.org/pre/abstract/10.1103/PhysRevE.97.022406. [CrossRef] [PubMed]

72. Horsthemke, W.; Lefever, R. *Noise-Induced Transitions: Theory and Applications in Physics, Chemistry, and Biology*; Springer: Berlin/Heidelberg, Germany, 2006. [CrossRef]
73. Bena, I. Dichotomous noise: Exact results for out-of-equilibrium systems. *Int. J. Mod. Phys. B* **2006**, *20*, 2825–2888. [CrossRef]
74. Ridolfi, L.; D'Odorico, P.; Laio, F. *Noise-Induced Phenomena in the Environmental Sciences*; Cambridge University Press: Cambridge, UK, 2011. [CrossRef]
75. Wienand, K.; Frey, E.; Mobilia, M. Evolution of a fluctuating population in a randomly switching environment. *Phys. Rev. Lett.* **2017**, *119*, 158301. [CrossRef] [PubMed]
76. Wienand, K.; Frey, E.; Mobilia. Eco-evolutionary dynamics of a population with randomly switching carrying capacity. *J. R. Soc. Interface* **2018**, *15*, 20180343. [CrossRef]
77. Taitelbaum, A.; West, R.; Assaf, M.; Mobilia, M. Population dynamics in a changing environment: Random versus periodic switching. *Phys. Rev. Lett.* **2020**, *125*, 048105. [CrossRef]
78. Taitelbaum, A.; West, R.; Mobilia, M.; Assaf, M. Evolutionary dynamics in a varying environment: Continuous versus discrete noise. *Phys. Rev. Res.* **2023**, *5*, L022004. [CrossRef]
79. Kalyuzhny, M.; Kadmon, R.; Shnerb, N.M. A neutral theory with environmental stochasticity explains static and dynamic properties of ecological communities. *Ecol. Lett.* **2015**, *18*, 572–580. [CrossRef]
80. Hufton, P.-G.; Lin, Y.-T.; Galla, T.; McKane, A.J. Intrinsic noise in systems with switching environments. *Phys. Rev. E* **2016**, *93*, 052119. [CrossRef]
81. Hidalgo, J.; Suweis, S.; Maritan, A. Species coexistence in a neutral dynamics with environmental noise. *J. Theor. Biol.* **2017**, *413*, 1–10. [CrossRef]
82. Gardiner, C. *Stochastic Methods: A Handbook for the Natural and Social Sciences*; Springer: Berlin/Heidelberg, Germany, 2009.
83. Chakrabarti, B.K.; Acharyya, M. Dynamic transitions and hysteresis. *Rev. Mod. Phys.* **1999**, *71*, 847–859. RevModPhys.71.847. [CrossRef]
84. Mobilia, M. *SimData_Figs3to6*; University of Leeds: Leeds, UK, 2023. [CrossRef]
85. Gillespie, D.T. A general method for numerically simulating the stochastic time evolution of coupled chemical reactions. *J. Comput. Phys.* **1976**, *22*, 403–434. [CrossRef]
86. Blythe, R.A.; McKane, A.J. Stochastic models of evolution in genetics, ecology and linguistics. *J. Stat. Mech.* **2007**, *2007*, P07018. [CrossRef]
87. *The Guardian*. Datablog. Newspaper Support in UK General Elections [Dataset]. 2010. Available online: https://www.theguardian.com/news/datablog/2010/may/04/general-election-newspaper-support (accessed on 9 April 2023).

MDPI AG
Grosspeteranlage 5
4052 Basel
Switzerland
Tel.: +41 61 683 77 34

Physics Editorial Office
E-mail: physics@mdpi.com
www.mdpi.com/journal/physics